AF530363

Seabuckthorn

Seabuckthorn

(*Hippophae* spp.): The Golden Bush

Editors:
Sanjai K. Dwivedi
T. Parimelazhagan
Shashi Bala Singh
Z. Ahmed

2009

SATISH SERIAL PUBLISHING HOUSE
403, Express Tower, Commercial Complex, Azadpur, Delhi-110033 (India)
Phone: 011-27672469, 27672852, Fax: 91-11-27672046
E-mail: info@satishserial.com, hkjain1975@yahoo.com
Website: www.satishserial.com

Published by :

SATISH SERIAL PUBLISHING HOUSE
403, Express Tower, Commercial Complex, Azadoyr, Delhi-110033 (INDIA)
Phone : 011-27672852 Fax : 91-11-27672046
E-mail : info@satishserial.com, hkjain1975@yahoo.com

ISBN 81-89304-60-7

Typeset at: Laxmi Art Creations, Delhi
Printed at: Salasar Imaging Systems, Delhi-35

डॉ० डब्ल्यू. सेल्वामूर्ति
विशिष्ट वैज्ञानिक एवं
मुख्य नियंत्रक अनुसंधान तथा विकास

Dr. W. Selvamurthy
Distinguished Scientist &
Chief Controller Research & Development

भारत सरकार, रक्षा मंत्रालय
रक्षा अनुसंधान तथा विकास संगठन
201, डी.आर.डी.ओ. भवन, नई दिल्ली–110011
Government of India, Ministry of Defence
Defence Research & Development Organisation
201, DRDO Bhawan, New Delhi-110011

FOREWORD

Seabuckthorn (*Hippophae rhamnoides L.*) is a unique and valuable crop for cold arid region. Interest on seabuckthorn in India is recent, although the plant has been used traditionally in Ladakh since long. Growing of seabuckthorn in marginal soil of the cold desert has immense environmental benefits in the fragile ecosystem. Root nodule of seabuckthorn has symbiotic association with bacterium belonging to the genus *Frankia* having the ability to fix 180 kg of nitrogen per ha per year. Berries of seabuckthorn are among the most nutritious of all berries and have immense medicinal properties. Vitamin C represents a major nutrient in the berries because of its presence in large quantities ranging from 200-2, 500 mg/100g. Seabuckthorn oil has unique properties to absorb ultraviolet light and promotes healthy skin, which is being recognized and sought by the cosmetic industry. Seabuckthorn leaves contain many nutrients and bioactive susbstances and used for preparing herbal tea and animal feed. The plant has been extensively exploited for treatment various ailments.

Defence Institute of High Altitude Research, Defence Research and Development Organisation (DRDO) has done pioneering research on seabuckthorn since 1992. The institute has developed, patented and commercialized seabuckthorn beverage, which is popular in Indian market under the brand name ' Leh Berry', 'Ladakh Berry', 'Power Berry' etc.

Similarly, herbal tea developed by the institute with seabuckthorn leaves as major constituent is gaining popularity among the consumers. Several other products developed by DRDO are at different popularity among the consumers. Several other products developed by DRDO are at different stages for commercialization. DRDO through its R & D in various laboratories has made a modest beginning for utilizing the shrub for increasing endurance and operational efficiency of the Indian Defence Forces. The spin-off benefits have immensely helped in improving socio-economic condition of the people in cold deserts of Ladakh.

I am happy that an exclusive book, which embodies the wealth of knowledge on various aspects of seabuckthorn, through manuscripts contributed by eminent researchers has been compiled. I compliment the editors for their tremendous efforts in bringing out this comprehensive publication on seabuckthorn. It is hoped that the publication will provide scientific guidelines for the researchers, academicians, policymakers, environmentalists and industrialists in exploiting the potential of the wonder plant of Himalayas.

W. Selvamurthy

Place : New Delhi
Date : 30, July 2008

PREFACE

Seabuckthorn (*Hippophae* species) is a miraculous plant of the cold arid Himalayan regions. It is also grown in various countries of Asia, Europe and America and has recently attracted the attention of scientific community and policy-makers on account of its innumerable useful properties. It is a rich reservoir of vitamins like C, A, B, K & E and possess numerous bio-active compound which is a rare combination to be found in the plant kingdom. In addition, a plethora of organic acids, carotenoids, antioxidants, flavonoides and sterols etc., impart immense nutraceutical, pharmaceutical and cosmoceutical values to the plant enhancing its potential for industrial exploitation for human welfare. The plant has aptly been called 'Golden Bush' and is also termed as " store house of vitamins". Medicinal properties of seabuckthorn are documented by the name "*Amlavetas*" in the ancient Indian Medical Classic "*Ayurveda*" as well as "*tSer-Mang*" in Amchi system of Medicine, prevalent in Indian Himalayas.

Seabuckthorn blossomed in isolation in the high reaches of Himalayas. Earlier it was considered a thorny menace in India. It was even burnt to control its profuse suckering. Its potential was realized when it was used by Defence Institute of High Altitude Research, DIHAR (Formerly, Field Research Laboratory) DRDO, Leh in the year 1992 for greening of Ladakh under the cold desert afforestation programme. Since then, the institute has been working on various Research and Development projects related to commercial cultivation/ scientific propagation, evaluation and conservation of genetic diversity and post harvest management/ value addition in seabuckthorn. The biggest achievement of the institute by far has been the development, patenting and subsequent commercialization of a herbal beverage from seabuckthorn fruits in August, 2001 which is readily available in the Indian market under various trade names viz, *Leh Berry, Ladakh Berry, and Power Berry* etc. Today, seabuckthorn beverage is perhaps the most sought after health drink all over India. Besides this, DIHAR also has to its credit the development of scientific propagation techniques, identification and conservation of promising genotypes, development of mechanical harvester for seabuckthorn and preparation of a herbal tea with seabuckthorn leaves.

The principal aim of this book has been to collate the state-of-the art review, latest information and viewpoints of eminent investigators, who excell in their own areas of research. An attempt has been made to encapsulate scientific information pertinent to cultivation, utilisation and processing of seabuckthorn viz., its distribution, botany, genetic diversity, properties/ phytochemistry, propagation, cultivation, harvesting, conservation, characterisation through biotechnological tools, utilization and processing technologies of seabuckthorn. In the book, 19 chapters from 23 different institutions involving 47 researchers eminent in their field have been compiled. The chapters have been critically reviewed by the experts to improve the quality. We hope that the present compilation will be extremely useful for the scientists, academicians, industrialists, environmentalists and people associated with ethnobotany, traditional medical system and natural products. It is sincerely hoped that the book will be widely circulated and used for the development of seabuckthorn resources, carrying out scientific investigations, development of various health products and environmental conservation around the globe and will go a long way in strengthening scientific cultivation and utilization of seabuckthorn. Since this is an edited book, the views and opinions expressed by the authors, who belong to different organizations, are entirely their own and neither DRDO nor the editors/publisher takes any responsibility for the same.

We are grateful to all the eminent contributors, who cooperated all along and enthusiastically contributed chapters, often bearing our constant reminders gracefully. The editors would like to convey their sincere thanks to Defence Research and Development Organization (DRDO) for supporting the project. The editors are also grateful to Shri M. Natarajan, SA to RM, Dr. W. Selvamurthy, CC R&D (LS&HR), Dr. Narendra Kumar, Director, DOP, Dr R.C. Sawhney, Director, DLS and Dr. A.K. Singh, Director, CEPTAM (DRDO), Delhi for their guidance, blessings and encouragement. The support rendered by Dr. Sandhya Gupta, Senior Scientist, National Bureau of Plant Genetic Resources, New Delhi and Dr. Deepa Dwivedi, BBA University, Lucknow in critically reviewing the chapters is duly appreciated. Constant support provided by our colleagues of DIHAR, Leh and CEPTAM, Delhi are also gratefully acknowledged. Our publisher and his team need special appreciation for their sincere efforts in bringing out the book on time. We also solicit the co-operation of erudite readers to derive full benefit from the views of the learned contributors and also welcome suggestions from them.

Delhi **Editors**
15 August, 2008

LIST OF CONTRIBUTORS

A Banerji
Phytochemical Research Unit
Regional Research Laboratory
Thiruvananthapuram-695019 (India)
*Presently at Centre for Biotechnology
Amrita Vishwa Vidyapeetham
Kollam-690525 (Kerala)

A K Shukla
Department of Forestry
Guru Ghasidas University
Bilaspur-495009 (CG)

A Ranjith
Agro-processing & Natural
Products Division
Regional Research Laboratory (CSIR)
Thiruvananthapuram-695019, Kerala

A S Bawa
Defence Food Research
Laboratory (DRDO)
Siddarthnagar, Mysore-570011

Abhilasha Sharma
Dept. of Basic Principles
Factory of Ayurveda
Intitute of Medical Sciences
BHU, Varanasi (UP)

Aruna Agrawal
Dept. of Basic Principles
Factory of Ayurveda
Intitute of Medical Sciences
BHU, Varanasi (UP)

B C Basistha
State Council of Science & Technology
for Sikkim, Development Area
Gangtok-737101, Sikkim

Basant Ballabh
Defence Institute of High
Altitude Research (DRDO)
(Formerly, Field Research Laboratory, FRL)
Leh-194101(J&K)

Bipin Kumar Sharma
Molecular Microbial Ecology Laboratory
Department of Forestry
North Eastern Regional Institute of
Science and Technology
Nirjuli-791109, Itanagar
(Arunachal Pradesh)

C. Arumughan
Agro-processing and Natural products
Division, Regional Research
Laboratory (CSIR)
Thiruvananthapuram-695019
Kerala

Deepa H Dwivedi
Department of Applied Plant Sciences
BBA University, Vidya Vihar
Lucknow-226025(UP)

Desy Rani
Department of Animal Nutrition
College of Vety. & Animal Science
CSK HPKV, Palampur – 176 062 (HP)

Farhath Khanum
Defence Food Research Laboratory (DRDO)
Siddarthnagar, Mysore-570011

G.Victor Rajamanickam
Dean & coordinator, CARISM
SASTRARA, Thanjavur (TN)

Gautam Biswas
Osaw -Agro Industries Pvt. Ltd.
Post Bag No.-5, Jagadhari Road
Ambala Cantt-133001 (Haryana)

GP Dubey
C/o Dept. of Basic Principles
Fac. of Ayurveda, Intitute of Medical
Sciences, BHU, Varanasi (UP)

HB Naithani
C/o Systemic Botany Division
Forest Research Institute
New Forests, Dehradun, (Uttarakhand)

Hridip Kumar Sharma
Molecular Microbial Ecology Laboratory
Department of Forestry
North Eastern Regional
Institute of Science
and Technology, Nirjuli-791109
Itanagar (Arunachal Pradesh)

J S Chauhan
Department of Animal Nutrition
College of Vety. & Animal Science
CSK HPKV, Palampur – 176 062 (HP)

K P Suja
Phytochemical Research Unit
Regional Research Laboratory
Thiruvananthapuram-695019 (Kerala)

K S Pradeep Kumar
Phytochemical Research Unit
Regional Research Laboratory
Thiruvananthapuram-695019 (Kerala)

K. Sarinkumar
Agro-processing and Natural Products
Division, Regional Research
Laboratory (CSIR)
Thiruvananthapuram-19 (Kerala)

N S Bist
Systemic Botany Division
Forest Research Institute, New Forests
Dehradun (Uttarakhand)

O P Chaurasia
Defence Institute of High Altitude
Research (DRDO), (Formerly, Field
Research Laboratory, FRL)
Leh-194101(J&K)

P C Sharma
University School of Biotechnology
Guru Gobind Singh Indraprastha
University, Delhi-110 006

Padma Gurmet
Sowa-Rigpa (Amchi) Research Centre
CCRAS, Ministry of Health and FW
Govt. of India, Leh-194 101 (Ladakh)

R. Ghangal
University School of Biotechnology
Guru Gobind Singh Indraprastha
University, Delhi-110006

Ranjit Singh
Defence Institute of High Altitude Research
(DRDO), (Formerly, Field Research
Laboratory, FRL) Leh-194101(J&K)
Presently at Defence Agricultural
Research Laboratory (DRDO)
Pithoragarh (Uttarakhand)

Rohit Bisht
College of Forestry & Hill Agriculture
G.B. Pant University of Agriculture
& Technology, Hill Campus
Ranichauri-249 199
Tehri Garhwal (Uttarakhand)

S C Tiwari
Molecular Microbial Ecology Laboratory
Department of Forestry, North Eastern
Regional Institute of Science
and Technology, Nirjuli-791109
Itanagar (Arunachal Pradesh)

S K Sharma
College of Forestry & Hill Agriculture
G.B. Pant University of Agriculture &
Technology, Hill Campus
Ranichauri-249 199
Tehri Garhwal (Uttarakhand)

S N Mishra
Department of Biosciences
Maharashi Dayanand University
Rohtak-124001(Haryana)

S. Raghuvanshi
Plant Molecular Biology
University of Delhi, South Campus
Delhi-110024

Sandhya Gupta
Tissue Culture & Cryopreservation Unit
National Bureau of Plant Genetic
Resources, Pusa, New Delhi-11012

Sanjai K Dwivedi
Defence Institute of High Altitude Research
(DRDO), (Formerly, Field Research
Laboratory, FRL) Leh-194101(J&K)
Presently at-Centre for Personnel Talent
Management (DRDO), Metcalfe House
Delhi-110054

Sanjeev Sagar
Osaw -Agro Industries Pvt. Ltd.
Post Bag No.-5, Jagadhari Road
Ambala Cantt-133001 (Haryana)

Shashi Bala Sigh
Defence Institute of High Altitude Research
(DRDO), (Formerly, Field Research
Laboratory, FRL) Leh-194101(J&K)

ShushmaTiwari
Dept. of Basic Principles
Fac. of Ayurveda
Intitute of Medical Sciences
BHU, Varanasi (UP)

T Parimelazhagan
Defence Institute of High
Altitude Research
(DRDO), (Formerly, Field Research
Laboratory, FRL) Leh-194 101
Presently at Deptt of Botany
Bharatiyar University
Coimbatore (Tamil Nadu)

T Stobdan
Defence Institute of High
Altitude Research
(DRDO), (Formerly, Field Research
Laboratory, FRL) Leh-194101(J&K)

V K Rao
College of Forestry & Hill Agriculture
G.B. Pant University of Agriculture &
Technology, Hill Campus
Ranichauri-249 199
Tehri Garhwal (Uttarakhand)

V K Sah
College of Forestry & Hill Agriculture
G.B. Pant University of Agriculture &
Technology, Hill Campus
Ranichauri-249 199
Tehri Garhwal (Uttarakhand)

V K Sharma
Department of Animal Nutrition
College of Vety. & Animal Science
CSK HPKV, Palampur - 176 062 (HP)

V K Yadav
College of Forestry & Hill Agriculture
G.B. Pant University of Agriculture &
Technology, Hill Campus
Ranichauri-249 199
Tehri Garhwal (Uttarakhand)

V V Venugopalan
Agro-Processing & Natural Products Division, Regional Research Laboratory (CSIR) Thiruvananthapuram-695019 (Kerala)

Virendra Singh
CSK Himachal Pradesh Agricultural University, Palampur – 176 062 (HP)

Zakwan Ahmed
Defence Institute of High Altitude Research (DRDO), (Formerly, Field Research Laboratory, FRL) Leh-194101(J&K) Presently at Defence Agricultural Research Laboratory (DRDO) Pithoragarh, Uttarakhand

CONTENTS

1

Botany of Seabuckthorn

O.P. Chaurasia, Basant Ballabh and Zakwan Ahmed

CONTENTS

1.1 INTRODUCTION

Seabuckthorn (*Hippophae sp.*) is widely distributed shrub in trans-Himalayan zones of Ladakh under Jammu & Kashmir (J&K) and Lahaul-Spiti under Himachal Pradesh (H.P.). Seabuckthorn the wonder plant of cold arid Himalaya, grows in wastelands, dry rocks and along river belts in Indus, Nubra, Changthang, Suru, Zanskar and Lahaul-Spiti valleys in India (Singh *et al.*, 2000).

The genus *Hippophae* L. belongs to the family Elaeagnaceae, and it is the most useful genus in the family. As the name indicated Seabuckthorn, the plant has been found growing from sea lavel to high altitude mountain regions but mainly distributed in higher elevations of Indian Himalayas, from Ladakh (J&K) to Lahaul-Spiti (H. P.), Kumaon and Garhwal in Uttranchal, Sikkim and Arunachal Pradesh (Chauhan, 1995). Under natural conditions, seabuckthorn grows on slopes, well drained soil with silt, riverbanks, lakes, seashores and wastelands. It is well adapted to dry conditions and it has root nodules due to presence of nitrogen fixing bacteria, *Frankia*. Seabuckthorn has been reported from about 30 countries of Asia and Europe and it has been introduced in North and South America. About 80-90 percent availability of seabuckthorn (*i.e.* about 1.2 million ha) is mainly in China and the rest resources are found in India, Russia and Mangolia.

1.2 TAXONOMY OF *HIPPOPHAE*

Taxonomically the genus *Hippophae* L. was founded by C. Linnaeus with *H. rhamnoides*, as the typical species, including only two species *H. rhamnoides* L. and *H. canadensis* L. which were recorded in *Species Plantarum* in 1753. Due to difference from genus *Hippophae* L., *H. canadensis* was transferred to genus *Shepherdia* Nutt. by Nuttall because of the presence of 4-fid calyx and eight stamens. In 1825, D. Don. described a Himalayan species *H. salicifolia* D. Don. from Nepal. In 1863, Von Schlechtendal, D.F.L. described another species *H. tibetana* Schlecht from Qing-Zang Plateau. In 1908 and 1909, C. Servettlaz merged *H. salicifolia* and *H. tibetana* with *H. rhamnoides* and treated it as subspecies. However, most taxonomists did not agree with this view. A. Rousi (1971) considered the type specimen of var. *procera*, a mixture and listed var. *procera* as a basionym of *H. rhamnoides* subsp. *sinensis*. Based on his studies with 2200 specimens, collected in 33 herbaria, 34 populations growing in gardens and 8 mass collection in nature, published "The genus *Hippophae* L. A taxonomy study". In his classic monograph, he recognized 3 species, *H. rhamnoides* L., *H. salicifolia* D. Don., and *H. tibetana* Schlecht. The first species was divided into 9 subspecies, 7 of which, subsp. *carpatica, cauasina, turkestanica, mongolica, sinensis, gyantsensis and yunnanensis* were newly described subspecies. In 1978, the Chinese taxonomists, S.W. Liuand and T.N. studied the plants of this genus, collected from Qing-Zang Plateau and published "The genus *Hippophae* from Quing-Zang Plateau". They recognized 3 species *H. rhamnoides* L., *H. tibetana* Schlecht., and *H. neurocarpa* S.W. Liu et T.N. He.

There are also some confusing names presented in the earlier literature, such as *Osyris rhamnoides* Scopoli (1772) *Rhamnoides hippophae* Moench (1794), *H. littoralis* Salisbury (1796), *H. rhamnoides* B.H. *sibirica* (1821) *H. sibirica* (1841), *Elaeagnus rhamnoides* (L.) A. Nelson (1935), *E. salicifolia* (D. Don.) A. Nelson (1935). Besides this two Chinese researchers found a new species and two new subspecies, *H. goniocarpa* S.W. Liu et. K. Sun, *H. goniocarpa* subsp. *litangensis* Lian et X.L. Chen and *H. neurocarpa* S. W. Liu et T.N. He subsp. *stellatopilosa* Lian et X.L. Chen. (Lian *et al.,* 2003).

1.3 BOTANICAL CHARACTERISTICS

Botanically *Hippophae* is a very characteristic genus and is a deciduous shrub or alpine scrub, small tree or tree, often with branched thorns. All organs are usually covered with scales-hairs of stellate hairs. The chromosome number is 2n = 24. Seabuckthorn shows a great range of morphological variations in growth pattern, bark, leaves, flower buds, flowers, pollen morphology, fruits in shape, size, colours and seeds. The details are as under :

1.3.1 Growth Pattern

The tendancy of growth shows both the form of bush, alpine scrub and tree in most of the species, however, *H. neurocarpa* and *H. tibetana* grow consistently as bushes may be due to habitat adaptation. Some of the taxa illustrate better growth potential of a woody plant with maximum height of 5-7 meter. *H. rhamnoides* and *H. tibetana* have distinct branching at knobs and twisted portions. The young branches have hairiness which also separates these taxa from others. The typical thorns on branches is special character in *H. rhamnoides* ssp. *turkestanica*. The sex can be identified only after bud formation of 3 to 4 years old plants.

1.3.2 Bark

The *H. rhamnoides* ssp. *turkestanica* Rousi, has whitish bark because of its thick cork layer as illustrated by Rousi (1971). However, there is a variation in colours (black or brown to gray bark) in other taxa.

1.3.3 Leaf

Generally leaves are alternate. Leaves are opposite in *H. rhamnoides* ssp. *sinensis*, although in the latter not strictly regularly. Leaves in whorls are an autapomorphy for *H. tibetana*. The simple, alternate, opposite or three leaves in whorled with short petioles are 13 mm long, ventrally green and silvery at dorsal side. The width and length of leaves is also a distinct character for each taxon. The main lateral veins are clearly visible in most of the taxon. In majority of mature leaves stellate hairs with long arms are clearly distinct. The abundant brown stellate hairs are found in young leaves of European subspecies of *H. rhamnoides*.

1.3.4 Clustered Bud

Buds in clusters are typical for Elaeagnaceae, but are lacking in some of the subspecies of *H. rhamnoides*. As described by Servettaz (1909) the family is not homologous with true accessory buds. The clustered buds correspond with the mode and site of the flower development. In *H. rhamnoides*, flowers are born in basal parts of the shoots of the previous year. *Hippophae* species show variation in scales of winter buds and distinguished by two groups of species one with few scales (two to three on female and less than 10 in male plants) and other with numerous scales (more than six on female and 15-35 on the male plants).

1.3.5 Flower

Flower bud is conspicuous and stable. The male flower buds are 4-angled, tower-shaped, spiral tower-shaped, cruciform or bifid ovate, whereas the female flower buds are cruciform, low spiral tower-shaped, ovate or bifid-ovate. Flowers are unisexual and dioecious, flowering before the leaf bud burst.

1.3.5.1 Male Flower

The male flowers appear from the axils of deciduous bracts, sessile, calyces almost 2-parted, 4 stamens, 2 of which alternate and the other 2 opposite with the lobes of calyx, short filaments and anthers oblong. The male flower buds are 4-angled tower-shaped.

1.3.5.2 Female Flower

The female flowers are solitary, axillary or clustered into inflorescence like, short pedicles, calyces pocket like, 2 teeth lobed at apex, superior ovary with single carpel and locules and with one ovule, styles lightly exerted. Female flower buds cruciform, near cruciform or bifid ovate.

1.3.6 Pollen

Pollen grains are 3 or 4 colporate, with a psilate or granulate or verrucate ornamental surface. Pollen morphology of *Hippophae* was studied by Sorsa (1971) and chemical variation and chromosomes by Rousi *et al.* (1978, 1980, 1982).

1.3.7 Fruit

Fruits included within the succulent calyx-tube, drupe-like carpodermis membraneous or thin leathery, separating from combing with the seed coat, single seed with bony seed coat. The average weight of fruits varied from 5.6-11.0 g to 8.2-17.4 g. The fruits are oval, extended oval, round oval, rounded in shape. Sometimes other forms such as obtuse-conical, round-cylindrical and five ribbed, elliptical, egg-shaped, narrow-oval, extended egg-shaped are also found. The dominating fruit colour is orange and its various shades that reach to yellow and red stages. The yellow, light-yellow and lemon yellow shades are found in some places. The fruits and leaves are rich in many bioactive substances. With xeromorphic structure of above ground parts of plants, seabuckthorn has high tolerance to atmospheric drought.

1.3.8 Seed

The seeds are usually glossy, oval shaped, elliptical, rounded oval, narrow oval and egg shaped in shape and brown, light-brown to dark brown in colour. Seeds show great variations in length, thickness, seed width, seed form and mean seed weight.

1.4 CLASSIFICATION

The genus *Hippophae* is classified according to Benthum & Hooker's System of classification as follows :

Kingdom : Plantae

Sub-kingdom : Phanerogamia (includes all seed producing plants)

Division : Angiospermae (ovules enclosed in definite ovaries so that the seeds are covered with a fruits wall and are not naked)

Class : Dycotyledons (all angiosperms in which the embryo possesses two cotyledons, leaves with reticulate venation and vascular bundles are arranged in a ring and are open)

Subclass : Monochlamydeae (Petals lacking)

Series : Dephnals (Gynaecium monocarpellary, with one or paired ovules, flowers perfect, mostly woody plants)

Family : Elaeagnaceae (Stems and leaves covered with silvery hairs, flowers bisexual or unisexual, perianth fused in a single whorl, fruit one seeded, often enclosed by the thickned lower part of persistent perianth)

Genus : *Hippophae* (Flowers one sexed, occurring on different plants, borne on previous year's shoots. Flowers with perianth 2-lobed)

Species : *rhamnoides* Linn.

subsp. *sinensis* Rousi.
subsp. *yunnanensis* Rousi.
subsp. *turkestanica* Rousi.
subsp. *mongolica* Rousi.
subsp. *caucasina* Rousi.
subsp. *carpatia* Rousi.
subsp. *fluviatilis* Rousi.
salicifolia D. Don.
goniocarpa Lian, X.L. Chen. *et* K. Sun.
gyantsensis (Rousi) Lian.
neurocarpa S.W. Liu. *et* T.N. He.
tibetana Schlecht.

1.5 BOTANICAL ENUMERATION TO THE SPECIES

(i) *Hippophae rhamnoides* Linn. ssp. *turkestanica* Rousi.

Name : Seabuckthorn , Tsermang or tastalulu

Distribution : From Pakistan to H.P., C. Asia, Europe & N. W. Himalayan regions

Habitat : River belts & wasteland slopes, Leh, Nubra, Zanskar, Suru, Changthang and Lahaul-Spiti valleys between 2,600-4,000 m

Plant Characters : A dwarf, very thorny, perennial shrub or small tree. Stem woody, erect, long upto 10-15 ft with silvery-waxy covering on the old shoots. Shoots often modified into spines. Leaves variable, oblong blunt to green above and silvery scaly on both sides. Flowers very small greenish or yellowish. Male flowers in clusters at the shoots. Tepals free and suborbicular. Female flowers dull yellow, short stalked. Fruits urticles, subglobose, succulent, orange or red. Seeds solitary, unequally bilobed, light black, stony.

Flowering & Fruiting : May & September

Uses : The whole plant has got enormous medicinal properties and is the most commercially exploited species. The fruits, leaves and seeds are considered as an anti-aging, anti-cold, restore memory and energetic. The fruits are antitusassive, blood purifier, heals peptic ulcer, chilblains, oral mucocitosis, rectal mucocitosis, lungs disorders, cuts and wounds.

(ii) *Hippophae rhamnoides* Linn. ssp. *sinensis* Rousi.

Name : Seabuckthorn

Distribution : Mountaineous regions of China and Mongolia

Habitat : Widely grown species along river belts & wasteland slopes, around Sichuan, Qinghai, Shaanxi, Beijing between 600-3,700 m

Plant Characters : A perennial shrub or small tree reaching 2-18 m high. The young branchlets are stiff. Leaves variable, opposite or suboppoite, the midrib on the above impressed into sulcus but upto top shallowed and vanished, the scale hairs on the lower surface, white or rarely rusty red. Male flowers buds 4-angled tower shaped, female flower buds cruciform. Seeds solitary with shining surface.

Flowering & Fruiting : April & September

Uses : In China, this species is the main resource for seabuckthorn development and for the industry. All the parts viz. fruits, leaves, seeds have got commercial importance.

(iii) *Hippophae rhamnoides* Linn. ssp. *yunnanensis* Rousi.

Name : Seabuckthorn

Distribution : Mountaineous regions of China and Mongolia

Habitat : Found along river belts & wasteland slopes, around Yunan, Qinghai, Xizang between 600-3,700 m

Plant Characters : A perennial shrub or small tree reaching over 20 m high. The young branchlets are soft. Most leaves variable, alternate, the midrib on the above impressed and upto top appearing as deeper sulcus, with rusty red scale hairs on the lower surface. Male flowers buds are inconspicuous, 4-angled tower.shaped, female flower buds near cruciform or ovate bifid. Fruits usually yellow. Seeds usually flat.

Flowering & Fruiting : April & September.

Uses : All the parts viz. fruits, leaves, seeds have got medicinal importance.

(iv) *Hippophae rhamnoides* Linn. ssp. *caucasia* Rousi.

Name : Seabuckthorn

Distribution : Russia, Turkey, Iran, and Armenia

Habitat : Found along river belts and wasteland slopes, between 1,000-2,500 m

Plant Characters : A perennial shrub or tree reaching over 10 m high. Branches are brown, with less or no branched thorns. Leaves variable, alternate, the widest commonly below the middle, abaxial usually green, adaxial surface silvery, acuminate at apices.

Flowering & Fruiting : June & September

Uses : The plant has got medicinal importance.

(v) *Hippophae rhamnoides* Linn. ssp. *carpatica* Rousi.

Name : Seabuckthorn

Distribution : Romania, Ukraine, Hungary, Austria, Germany

Habitat : Found along seashore upto 380 m

Plant Characters : A perennial shrub or small tree. Branches are straight. Leaves variable, alternate, adaxial surface is usually mixed with most rusty red scales. Seeds flat.

Flowering & Fruiting : April & September

Uses : All the parts viz. fruits, leaves, seeds have got medicinal importance.

(vi) *Hippophae rhamnoides* Linn. ssp. *fluviatilis* Rousi.

Name : Seabuckthorn

Distribution : France, Austria, Spain, Italy

Habitat : Found growing in open lands, river banks, hillsides, between 100-1,900 m

Plant Characters : A perennial shrub or small tree. Leaves variable, alternate. Fruits are 5-6 mm long, 4-6 mm wide. Seeds ovate.

Flowering & Fruiting : April & September

Uses : All the parts viz. fruits, leaves, seeds have got medicinal importance.

(vii) *Hippophae rhamnoides* Linn. ssp. *mongolica* Rousi.

Name : Seabuckthorn

Distribution : Mountaineous regions of China, Mongolia and Russia

Habitat : Found along river belts and wasteland slopes between 1,200-2,000 m

Plant Characters : A perennial shrub reaching over 10 m high. Branches are brown, with less and no branched thorns. Leaves variable, alternate, the widest usually above the middle, above green, adaxial surface silvery, acuminate at apices. Flower buds are bifid ovate or near cruciform.

Flowering & Fruiting : April & September

Uses : All the parts viz. fruits, leaves, seeds have got medicinal importance.

(viii) *Hippophae salicifolia* D. Don.

Name : Seabuckthorn, ChharamaChuk, Amli, Kalabis

Distribution : Pakistan, H.P., Uttarakhand (India) and other parts of Asia

Habitat : On slopes, Lahaul valley of H.P. and Badrinath of Uttarakhand (India), around 4,390 m

Plant Characters : A deciduous thorny willow like and erect shrub or small tree with rusty scaly branches. Leaves 5-10 cm long, linear-lanceolate, softly white-pubescent, above when young, densely white-tomentose or stellately-hairy beneath, margins recurved. Flowers in clusters. Male flowers yellowish brown, with 2 scaly lobes, stamens usually 4, female flowers short stalked, 2-lobed, stigma exserted. Fruits ovoid, yellow in axillary clusters. Seeds solitary, light black with shining surface.

Flowering & Fruiting : May & September.

Uses : All the parts have got medicinal value. Plant is also rich in various vitamins and carotenoids and have commercial importance like *H. rhamnoides.*

(ix) *H. goniocarpa* Lian, X.L. Chen *et* K. Sun.

Name	:	Seabucktrhorn
Distribution	:	Higher mountains of China
Habitat	:	It grows on mountain slopes, river banks, flood lands and valleys at an elevation of 2,635-3,745 m around Sichuan, Qinghai
Plant Characters	:	A small tree, 5-8 m high. Branches soft, branchelets of the current year red-brown or dark brown. Leaves adaxial, surfaces are densely covered with scale-hairs, leaf margin usually explanate, never manifestly revolute, the midrib on the above impressed into sulcus but up to top shallow or absent. Buds of male flower are cruciform, of female flower near cruciform (bifid ovate, the second pair of the bud is clearly visible). Fruits are terete, juicy, apricot-coloured or straw-yellow, (5.5) 6-10 mm long, (3.4) 4-5.8 mm wide, (1.25) 1.4-2.3 times longer than wide. Seeds are slightly flat, with 3-5 inconspicuous longitudinal angles.
Flowering & Fruiting	:	May & September
Uses	:	The whole plant is economical, but fruit, seeds and leaves are used in medicine.

(x) *H. gyantsensis* (Rousi) Lian.

Name	:	Seabuckthorn
Distribution	:	China to Mongolia
Habitat	:	It grows near river banks, flood lands and valley terraces at an elevation of 2,625-5,000 m
Plant Characters	:	A small tree, 5-8 m high. Branches are soft, branchlets of the current year brown-yellow. Leaves abaxial surface are scattered stellate hairs, adaxial surface is densely covered with scale-hairs. Flower buds are ovate bifid. Fruits

are yellow, longitudinal angles almost developed in towering-shaped. Seeds are plano-convex, with 6 longitudinal angles nearly dihedral.

Flowering & Fruiting : May & September

Uses : The whole plant is used for household uses and leaves, fruits and seeds are used in preparation of certain medicinal products.

(xi) *H. neurocarpa* S.W. Liu *et* T.N. He.

Name : Seabuckthorn

Distribution : China, Tibet and Mongolia

Habitat : It grows near river banks, flood lands and river terrace between 2,710-4,126 m

Plant Characters : A small size plant, 1-3.5 m high, the crown cover appearing is as platform in the adult plants. Branches are stiff, branchlets of the current year are gray-white. Leaves adaxial surface are densely covered with scale-hairs, commonly explenata never manifestly revolute at margins, the midrib on the above impressed into sulcus but up to top shallowed to vanished. Flower buds are ovate to ovate-bifid. Fruits are black-brown, curved prism, with little or very little juice, one end thinner with 5-7 longitudinal angles, 7.7-8.4 mm long, 2.9-3.4 mm wide, 2.5 times longer than wide.

Flowering & Fruiting : May & September

Uses : The fruits, seeds and leaves have got medicinal properties.

(xii) *Hippophae tibetana* Schlecht.

Name : Seabuckthorn

Distribution : Pakistan, Himachal Pradesh & Uttranchal (India) & other parts of Asia

Habitat	:	Along river belts and wastelands, Zanskar and Spiti valleys around 4,230 m
Plant Characters	:	A small, dwarf, much branched perennial and alpine scrub with long stout terminal spines formed from the tips of old branches. Leaves variable, oblong blunt to green. Flowers very small greenish or yellowish. Fruits subglobose, succulent, orange or yellow. Seeds light black to brown.
Flowering & Fruiting	:	June & September
Uses	:	Like other species this species is a source of vitamins and carotenoids.

1.6 CONCLUSIONS

Seabuckthorn (*Hippophae*) is commonly found in high altitude regions of Himalaya and polar sea belts of the world which is abundantly used genus of the family Elaeagnaceae. *H. rhamnoides,* the most common species, is commercially exploited plant from Asian, European and other western countries since a long back. Taxonomically, *Hippophae* is a very complex genus and comprises of 6 species and 7 subspecies, distributed throughout Asian mountains and Europe of the world. The Indian cold desert Himalaya is represented by three species namely *H. rhamnoides, H. salicifolia* and *H. tibetana* found growing in higher mountains of trans Himalaya.

Botanical characteristics are the most essential attributes basically for the correct identification. Morphological characters such as roots, stem, leaves, stipule, thorns, flowers, fruits, seeds, pollen, plant habit, habitat, and flowering & fruiting time play important role in correct identification. Therefore, botanical characteristics are the major plant identification norms which will provide a readymade data on identification of *Hippophae* species for future use.

References

Albrecth, H.J. 2003. Research on seabuckthorn in Germany - A multipurpose wonder plant (Ed. Singh *et al.*), Indus Pub. Company, New Delhi.

Annonymous. 1948. Wealth of India: Raw Materials (Council of Scientific and Industrial Research). New Delhi.

Basistha, B.C., Adhikari, I.M., Deokata, K.P. & Thapa, K.K. 2001. A case study of *Hippophae* L. with special reference to agro-technique in the Sikkim Himalayas. In. Souvenir of International workshop on seabuckthorn, February 18-21, 2001, New Delhi.

Benthum, G. & Hooker J.D., 1862-1883. *Genera Plantarum*, (1-3), London.

Chauhan, N. S. 1999. Medicinal & Aromatic plants of Himachal Pradesh, Indus Pub. Company, New Delhi.

Chaurasia, O. P. & Singh, B. 1996-2001. Cold Deserts Plants Vol.-I-V. Field Research Laboratory C/O 56 APO.

Korovina, M.A., Igoshina, V., Fefelov, V.A., Yeliseev, I.P. & Gogoleva, G.A. 1993. Geography of genetic resources for selection of seabuckthorn. In. Proc. International symposium on seabuckthorn, pp. 162-170, Russia.

Lian, Y.S. 1990. A Study on Classification Position of *Hippophae gyantsensis* (Rousi) Lian. *Hippophae* 1, 24-30.

Liu, S.W. & He, T.N. 1978. The Genus *Hippophae* from Qing-Zang Plateau. *Acta Phytotaxonomica Sinica* 16(2), 106-108.

Rousi, A. 1971. The Genus *Hippophae* L. A Taxonomic Study. *Ann. Bot. Fennici* 8, 177-277.

Rousi, A., Karunen, P., Eronen, P. & Hakala, I. 1978. Comparative chomatographic study of Eurasian *Hippophae* populations. *Ann. Bot. Fennici*. 15, 73-79.

Rousi, A. & Arohonka, T. 1980. C. bands and ploidy level of *Hippophae rhamnoides*. *Hereditus* 92, 327-330.

Rousi, A., Karunen, P. & Hakala, I. 1982. Chemical variation in Eurasian population of *Hippophae rhamnoides* revealed by two-dimensional thin-layer chromatography. *Ann. Bot. Fennici* 19, 219-224.

Ruiling, P., Zhemin, Z., Yingcai, M. Zhenghua & Bingsheng, D. 1989. The distribution characteristics of seabuckthorn (*H. rhamnoides* L.) and its research progress in China. In : Proceedings of International Workshop on Seabuckthorn, pp. 1-16, China.

Singh, B. & Chaurasia, O.P. 2000. Medicinal flora of Indian cold desert : Proceedings of XXV International Horticultural Congress Part 13, *Acta Horticulture* No. 523, pp 65-72.

Singh, V., Singh B. & Awasthi, C.P. 1995. Distribution, Taxonomy and nutritional value of Seabuckthorn (*Hippophae* L.) and its Research Progress in China. In : Proceedings of International Workshop on Seabuckthorn, p 52-59, Beijing, China, p.206.

Sorsa, P. 1971. Pollen morphological study of the Genus *Hippophae* L., Including the new taxa recognized by A. Rousi. *Ann. Bot. Fennici* 8, 228-236.

Zhuode, Y., Fu, A. & Yongshan, L. 1989. Discussion on the problems of origin, classification, community and resource of seabuckthorn in China. In : Proceedings of International Workshop on Seabuckthorn, p. 21-30, China.

□□□

2

Morphology and Anatomy of Seabuckthorn

Ranjit Singh, S.N. Mishra, Sanjai K. Dwivedi and Zakwan Ahmed

CONTENTS

2.1 INTRODUCTION

Seabuckthorn (*Hippophae rhamnoides* L.) is a member of family Elaeagnaceae of order Daphnales. Today six species of *Hippophae* are known in the world, namely (i) *H. salicifolia,* (ii) *H. rhamnoides* with eight subspecies namely, ssp. *sinensis* Rousi, ssp. *yunnanensis* Rousi, ssp. *turkestanica* Rousi, ssp. *mongolica* Rousi, ssp. *caucasia*.Rousi, ssp. *carpatica* Rousi, ssp. *rhamnoides Rousi*, ssp. *fluviatilis* van Soest, (iii) *H. gonicarpa.* (Lian & .Chen., 1995), with two sub species ssp. *litagenesis* Lian *et* X.L.Chen. and ssp. *goniocarpa,*(iv) *H. gyantsensis* (Rousi) Lian, (v) *H. neurocarpa* S.W.Liu *et* T.N. He with two subspecies, ssp. *stellatopilosa* Lian *et* X.L.Chen and ssp. *neurocarpa* and (vi) *H. tibetana* Schlecht (Lian *et al.*, 2003). Three species of seabuckthorn are reported from India, *Hippophae salicifolia* in Uttranchal and North East region, *H. rhamnoides* in Ladakh and Himachal Pradesh and *H. tibetana* in small areas of Ladakh (Chaurasia & Singh, 1996). In Ladakh *H. rhamnoides* is the most commonly found plant along the river banks, tributaries of rivers, along Nalas etc. *H. tibetana* is found in Zanskar valley of Ladakh.

Seabuckthorn is a dwarf, sometimes tall, much branched, nitrogen fixing, temperate shrub found throughout Himalayan region. It is distributed widely in the cold arid regions of Himachal Pradesh, Uttaranchal, and Ladakh region of Jammu & Kashmir and in some parts of Sikkim. On the basis of fruit colour, shape size, leaf colour, thorn density etc. seventeen morphotypes have been identified from naturally growing seabuckthorn in Nubra valley (Singh *et al.*, 2005 a). Seed anatomy of seabuckthorn indicates some difference in seabuckthorn morphotypes (Singh *et al.*, 2005 b).

2.2 DISTRIBUTION PATTERN

H. rhamniodes L. is a xerophytic plant and can grow in all kinds of soil in Ladakh Indis. It is found growing in fertile as well as rocky, sandy, marshy areas but mostly along the river sides. It was also found growing in both acidic and saline soils. It is an effective soil binder and help in sand dune fixation in Nubra valley, especially in between Hunder and Diskit villages. It is also found growing in crevices on barren mountain slopes near Warsi village. It is found growing near Panamik hot spring where highly acidic and hot water flows. Moreover, it is also found growing in plenty near hill locked lake, where soil is sandy and highly basic, a thick white crust is deposited over soil in that area. It was able to thrive well in sandy, even stony and poor soils.

All the villages of Nubra valley (Ladakh) were extensively surveyed and the frequency, density, abundance distribution pattern etc were analyzed. The population survey on those locations in various villages revealed that seedlings and juvenile/vegetative plants of *H. rhamnoides* L. has status of mostly present in most of villages except Changmar and Bogdang, while Changmar is completely devoid of any seabuckthorn and Bogdang had seldom present status of seabuckthorn. The mature plants have also constantly present status in all the villages of Nubra valley except Changmar and Bogdang.

Distribution pattern of seedlings and juvenile/vegetative plants on the basis of Abundance/Frequency ratio, generally showed contagious distribution pattern except in Tershey, Warsi, Pinchimic and Youlkam where random distribution pattern was observed. In case of mature plants distribution pattern was essentially contagious in all villages except Pinchimic, where random distribution pattern was observed. Seedlings frequency of vegetation or nature varied from 60-100 % in most of the villages except Chatti and in Bogdang it was 20-40 % only (Table 2).

Density of seedlings of *H. rhamnoides* L. varied from 2.2 -5.6 plants/m^2, except in Chatti (1.4) and in Bogdang (1.3). Vegetative/Juvenile plants density varied from 1.0- 6.6 plants/m^2. Mature plants showed 6.1-12 plants/m^2, except 0.5 in Chatti and 1.1 in Bogdang. Average male/ female ratio of the plants in the natural population was found to be 41:59. In addition to these villages, *H. rhamnoides* L. was also found growing in villages like, Kowait (11,200 ft), Kinchu (11,600 ft) , Eye (11,400 ft), Aranuk Uokma (10,400 ft) and Aranuk gogma (10,500 ft), these villages are on the other side of the Nubra river.

2.3 MORPHOLOGY

Nubra valley of Ladakh was surveyed and it was observed that there is a marked difference in morphology of seabuckthorn growing at various places and even in plants at the same site. Plants were either dwarf or tall (4-24 ft in height) much branched, thorny shrub but sometimes attain the height of a tree. At some places, plants were bushy and thorny with profuse suckering, with scaly young twigs. White hair and scales were prominent over young shoots and leaves. Brown scales were common over mature stem, leaves and flowers. All plants were having a silver waxy or thick brown covering on the older shoots. Thorns were many and every branch was terminating into a thorn only.

Table 2.1: Distribution Pattern of Flowering/Fruiting Plants of Seabuckthorn in various Villages of Nubra Valley (Ladakh), India.

Village	Frequency*	Density* (plants/m²)	Abundance	A/F Ratio	Frequency Class	Distribution Pattern
Tyakshi	100±3.4	9.6±1.2	9.6±1.2	0.096	CP	CN
Turtuk (Y)	90±2.1	6.6±1.4	7.3±0.9	0.08	CP	CN
Turtuk (F)	80±1.8	9.8±1.1	12.25±1.3	0.153	MP	CN
Chalunka	90±1.5	6.5±0.6	7.22±1.1	0.080	CP	CN
Bogdang	40±1.1	1.1±0.01	2.75±0.3	0.068	CP	CN
Changmar	0	0	0	0	0	0
Udmaru	80±1.8	4.8±0.2	6.0±0.2	0.075	MP	CN
Tershey	90±1.4	4.8±0.4	5.33±0.4	0.059	CP	CN
Skuru	90±2.7	6.1±1.1	6.77±0.5	0.075	CP	CN
Thoise	90±2.2	7.2±1.3	6.87±1.1	0.058	CP	CN
Skampuk	90±2.1	9.9±2.1	11.0±2.1	0.122	CP	CN
Partapur	70±2.5	2.0±0.1	2.85±0.4	0.040	MP	RD
Hunder	100±3.4	12±3.1	11.2±3.2	0.120	CP	CN
Diskit	100±3.4	10.3±2.5	10.3±1.2	0.103	CP	CN
Khalsar	100±3.2	11.9±3.1	11.9±2.4	0.119	CP	CN
Chatti	20±1.1	0.5±0.05	2.5±1.2	0.12	RR	CN
Tirit	80±.14	7.0±1.2	8.75±2.4	0.109	CP	CN
Lakjung	90±2.4	6.8±1.4	7.55±1.3	0.083	CP	CN
Tegar	80±2.1	8.8±2.1	11.0±2.1	0.137	OP	CN
Sumur	90±3.2	7.7±1.3	8.55±1.6	0.092	CP	CN
Pinchimic	80±2.1	3.0±0.3	3.75±1.3	0.046	MP	RD
Sasoma	90±3.2	7.5±1.1	8.33±2.4	0.092	CP	CN
Terisha	80±2.1	6.5±1.0	8.12±1.5	0.101	MP	CN
Youlkam	90±2.4	6.1±0.8	6.77±1.6	0.075	CP	CN
Chamlang	90±2.4	7.1±0.4	7.88±1.02	0.087	CP	CN
Tongseth	80±2.1	6.9±0.9	8.62±2.12	0.107	MP	CN
Panamik	90±3.5	7.3±1.1	8.11±2.06	0.090	CP	CN
Sataksha	100±3.4	7.5±1.0	7.5±1.76	0.075	CP	CN
Warsi	80±2.1	6.7±1.3	8.37±1.56	0.104	OP	CN

Values ± SD Where CP - Constantly present, MP Mostly present, RR - Rarely present CN - Contagious distribution, RD - Random distribution.

2.3.1 Root

Roots were tap, much branched, having primary, secondary, tertiary root with hairs. It generally grew up to 4-5 meters below the ground bearing prominent root nodules, due to *Frankia actinirhizal* symbiotic association, which is an adaptation to survive in poor soils in cold deserts like Ladakh and Lahaul-Spiti. The root nodules were more prominent in sandy areas. Root can produce new generation plants at length through root turions. A single plant root had potential to give rise to hundreds of plants of different generations. Root nodules were observed in few plants only at particular site.

2.3.2 Stem and Thorn

Stem is hard, woody, much branched with varied number of thorns (12-28 thorn/m^2) and covered with thick brown bark. Branching was alternate and opposite but sometimes irregular. Young stem bear silvery white coating and scales over it. Each branch of the stem always terminates into a thorn. Many white hairs and scales were found on the young stem.

Thorns of *H. rhamnoides* are appendages from stem only and are of complex type i.e. thorn were further divided into thorn. It had thick cuticular deposition and were alternate in arrangement on the stem. Some of plants were having soft and less thorn on stem.

2.3.3 Leaf

H. rhamnoides leaf is dorsiventral, simple (undivided lamina), waxy, glandular and herbaceous, pinnate (with one midrib), petiolate. Leaf laminas were acicular, reticulate venation (reticulate unicostate), acute apex, sometimes obtuse. Phyllotaxy was spiral or alternate. Leaves were generally 1.5 to 5.6 cm long, small, elliptic, glabrous, hairy and xerophytic in nature. Leaves were covered with silvery white scales and hairs on the both sides present but more at the lower side. Leaf was light green on the dorsal side but silvery white on the ventral side. However, the plants found in Nubra valley of Ladakh showed variation in leaf structure and colour.

2.3.4 Trichomes (Scale and Hair)

Scales are epidermal appendages and they were peltate type, present on all parts of *H. rhamnoides*, but prominently on stem, leaf and fruits. A scale consisted of a discoid plate of about 50-55 cells which are pointed at the tip, borne on a stalk or some times directly to the foot. Trichome was initiated as protrubrence from epidermal cells; the protrubence elongates and developed into a multicellular structure. Prominent function of scales was protection from cold and dry climate and to control the rate of water loss during transpiration. Hair are multicellular long epidermal appendages present on the leaf and young shoots and tender roots.

2.3.5 Fruit

Fruits are small drupe (Utricle) in all plants. Fruit size on an average was 6.5 mm (ranging from 4.2 mm to 10.45 mm) and having 5.2 mm width (ranging from 3.4 mm to 7.1 mm). Fruit shape varied from round to oval, tubular to capsular in plants examined at different habitat. According to Royal Horticulture Colour Chart London (2001) the fruit colour was in three categories i.e. yellow, orange and orange red and variation within these categories. Different morphotypes exhibit significant difference in bunching patterns of fruits among morphotypes, generally fruits were found in the bunch of 3, 4, 5 & 6 fruits. Raw fruits are sour to highly acidic in juice but on ripening fruits become slightly sweeter. 100 g contains about 661 fruits.

2.3.6 Seed

Seeds are dark brown or light brown, ovoid to elliptical in shape and on an average with crustaceous shiny testa. Seeds are scantily endospermic with two cotyledons, expanded, exalbuminous, and embryo achlorophyllous. Seeds are very hard and containing oil (8-18 %), depending upon the plants type. 100 seeds weight is generally 1.24 ± 0.13 g. Length of a seed is generally 4.46 ± 0.22 mm and width is generally 2.51 ± 0.17 mm. Seabuckthorn seeds are having a single rib or furrow in almost the centre of the vertical length. All the seeds of seabuckthorn are having a sheath over them; due to the presence of this sheath the fruits of seabuckthorn are called utricle.

2.4 ANATOMY

The internal structure of plant parts viz. stem, root, leaf, thorn, male flower, female flower, scales were studied in detail in common seabuckthorn and same was compared in all the morphotypes of seabuckthorn to establish variation also at histological level (Essau, 1939) in morphotypes.

2.4.1 Stem

Transverse section of stem of *H. rhamnoides* L. showed following internal structures:

Epidermis: Epidermis is made up of single layer of cells, with very thick cuticular thickenings over it. It consisted of tangentially flattened cells, which are closely united with each other on their radial walls. Epidermal hair are present in young stems but lacking in older ones. Chloroplast and stomata are always absent.

Cortex: The tissue beneath the epidermis is called cortex and is differentiated into three regions:

Hypodermis: was outermost part of the cortex and it had 3-4 layers of collenchyma cells. The intercellular spaces are absent due to the thickenings of the cells in the corners.

General Cortex: it had 4-5 layers of parenchymatous cells, with some intercellular spaces.

Endodermis: formed the innermost layer of the cortex, it consisted of barrel shaped, closely fitted cells without any intercellular spaces.

Pericycle: tissue between the vascular bundles and the endodermis consisted of sclerenchyma alternating with parenchyma, it indicated as heterogenous pericycle. The sclerenchyma associated with phloem designated as hard bast of *H. rhamnoides.*

Vascular bundles: The vascular bundles of *H. rhamnoides* are conjoint, bicollatral (a bundle having the phloem on both sides of xylem) and open (vascular bundle having vascular cambium). Xylem is of endarch type.

Pith: Pith is of parenchymatous cells and having less intercellular spaces than other dicot stems.

2.4.2 Root

Root of *H. rhamnoides* is well developed. The young root also show deposition of cuticulare thickening considered to be an adaptation to the rockey nature of soil. Internal structure of the root comprised following distinct feature such as:

Epidermis: was single layered and having cuticle over them, sometimes epidermis in the root is called epiblema. The initials of root hairs were also present.

Cortex: is made up of parenchyma cells with intercellular spaces in between them. The cells of cortex contained leucoplast and starch. Sometimes when epidermis degenerates due to some reason, the outer layer of cortex becomes cuticularized and becomes exodermis, the condition common in hilly plants.

Endodermis: is the innermost layer of the cortex and form a ring of compact barrel shaped cells around the stele. Both radial and inner walls are thickened. The radial thickenings are in the form of casparian strips. Intercellular spaces are absent. The otermost layer of stele lying just below the endodermis is called as pericycle, consisted of thin walled cells with full of cytoplasm.

Vascular bundles: Vascular bundles are having xylem alternate with phloem. The protoxylem is exarch type and consisted of annular and spiral vessels. The metaxylem has reticulate and pitted vessels. The phloem consist of sieve tubes, companion cells and phloem parenchyma.

Pith: The pith is made up of parenchyma cells but it changes into thick walled tissue after juvenile phase.

2.4.3 Root Nodules

H. rhamnoides is a non-leguminous nitrogen fixing plant, four morphotypes have very prominent root nodules, sometimes more prominent than the leguminous plants. *H. rhamnoides* has *Frankia,* an actinomycete in symbiotic association. The anatomy of the root nodule showed that the actinomycete (endophyte) was having hypha of different shapes. *Frankia* enters the host root through infection thread, which advances in the root cortex, then ultimately the sheath bursts and the bacteroids like cells of *Frankia* are liberated inside the host cell. Then inside the host cell they get covered by a membrane, multiply, enlarge and swell and further infect the new cells.

The mature nodules have a vascular strand and an apical meristem. It has a central infected zone with a surrounding cortex. The cells of the infected zone are highly enlarged and vacuolated. They contain *Frankia's* cells. The *Frankia* draw their nutrition from the host. They also require host's organic acids for assimilation of ammonia (NH^+_4), produced during nitrogen fixation. The actinomycete fixes elemental nitrogen into amino acids. Most of it is passed out of *Frankial* cells into host cells which are connected with vascular strand with quick translocation to other parts of plant. The nodules may excrete some of the nitrogenous compounds into the soil. Excretion is considerable under the low photosynthetic activity of the host. Internal anatomy of nodule is similar to that of lateral roots with a cortical cylinder of vascular tissue, a cortical region in which the infected cells are found and a typical outer periderm layer. Most nodule lobes have a determinate meristem at the distil tip. *Frankia* exists in the vegetative mycelial state in the nodules.

2.4.4 Leaf

Leaf of *H. rhamnoides* is dorsiventral or bifacial and can be differentiated into a dorsal and ventral side. A typical *H. rhamnoides* leaf showed following internal structures in the transverse section.

Epidermis: formed of a single layer of cells, are closely fitted and has outer thick wall. The outer wall was cutinized. The chloroplast and stomata are not present in the epidermal cells. Large numbers of white hair are present above the epidermis.

Mesophyll: is the main part of leaf tissue and is differentiated into two regions.

Palisade tissue: composed of two layers of closely arranged columnar cells. A large no of chloroplasts are present in each cell. The palisade parenchyma was present just below epidermis appeared to be meant for photosynthesis. Although palisade tissues are closely fitted but they are still exposed to intercellular spaces hence suitable for receiving the proper gas supply.

Spongy tissue: lying just below the palisade parenchyma and was made of parenchyma cells but they had irregular arrangement. The spongy cells form a network of interconnected cells. Because of prominent intercellular spaces, these cells are spongy parenchymatous. The spongy parenchyma cells contained lesser chloroplasts than palisade cells.

Vascular bundle: are arranged in the spongy parenchyma. A midrib is also present. The structure of vascular bundles is essentially similar to midrib region. The vascular bundles are interconnected to each other in the median plain of the leaf blade forming a network. Each vascular bundle is conjoint, collateral and closed. Xylem is present towards upper epidermis and phloem towards lower epidermis. Xylem is consisted of annular and spiral vessels, tracheids, wood fibres and xylem parenchyma. The vascular bundle is enclosed in a parenchymatous bundle sheath, made up of compactly arranged lignified cells.

Midrib region: Palisade and spongy parenchyma cells are absent. It is made up of collenchyma cells followed by parenchyma cells above vascular bundle.

Lower epidermis: was similar to upper epidermis except that traversed by stomata.

Stomata: in *H. rhamnoides* leaf are of shrunken type present on lower surface i.e. hypostomatic. Each stomata is surrounded by two guard cells, which are placed in a pit followed by substomatal cavity, a lesser prominent in *H. rhamnoides* than other dicots. Stomata are surrounded by numerous trichome hairs or scales. Though numbers of hair were present on the lower epidermis.

2.4.5 Thorn

Anatomically thorns are essentially similar to the stem, except they have thick deposition of cutin over epidermis.

The cortex was more sclerenchymatous than pith. A typical thorn showed following internal structure in a transverse section:

Epidermis: is made up of single layer of cells, with a very thick cuticle on the outer surface. It consisted of compact tangentially flattened cells, which are closely united with each other on their radial walls. Epidermal hair are absent in mature thorn but scales are present over thorn.

Cortex: tissue beneath the epidermis is differentiated into following regions:

Hypodermis: is the outermost part of the cortex and it has 3-4 layers of collenchyma cells. The intercellular spaces are absent due to the thickenings of the cells in the corners.

Endodermis: forms the innermost layer of the cortex, it consisted of barrel shaped, closely fitted cells without any intercellular spaces.

Pericycle: the tissue between the vascular bundles and the endodermis is known as pericycle. It consisted of sclerenchyma alternating with parenchyma, exhibiting heterogenous pericycle. The sclerenchyma associated with phloem can be called as hard bast of *H. rhamnoides.*

Vascular bundles: The vascular bundles in thorn are conjoint, bicollatral (a bundle having the phloem on both sides of xylem) and open (vascular bundle having vascular cambium). Xylem is of endarch type

Pith: Pith is of sclerenchymatous cells and having lesser intercellular spaces than stem.

2.4.6 Trichome

Trichome is initiated as protuberance from epidermal cells, the protrubence elongates and it develops into a multicellular structure. Cell walls of trichomes are cellulosic and covered with cuticle. Their main function is protection from cold and dry climate. *H. rhamnoides,* have two types of trichomes, which were present in all the morphotypes.

Scales and hair are epidermal appendages, which are of peltate type. They occur on all parts of *H. rhamnoides*, but more prominent on stem, leaf and fruit. A scale consisted of a discoid plate of cells (50-55 cells) which are pointed at the tip, borne on a stalk or some times directly to the foot. Hairs are multicellular long epidermal appendages present on the leaf and young shoots and tender roots. Such hair had a foot embedded in the epidermis and body projecting out.

2.4.7 Seed

Seeds of *H. rhamnoides*, are very hard and have to be soaked in water before section cutting for anatomical studies. A longitudinal section showed that seed had three main regions:

Seed wall: Seed wall is very complicated and highly cuticularised and brown in colour. Seed wall is made up of three regions comprised of, outer wall, made up of sclerenchymatous cells with heavy deposition with cutin, middle wall, made up of parenchymatous cells and thicker than outer and inner wall and an inner wall, made up of sclerenchymatous cells with cutin deposition.

Connective tissue or Intermediate wall: made up of collenchyma cells and these cells are with a sticky proteinaceous substance. Vacuoles were also present in between connective tissue.

Seed embryo: The L.S of seed embryo exhibited three regions: (i) Pre radicle made up of meristamatic cells which give rise root on germination, (ii) Apical meristamatic region made up of meristamatic cells and which was considered as future stem, and (iii) cotyledons two in number which become the first leaves of the plant.

2.5 FLORAL ANATOMY

2.5.1 Female Flower

Female flowers are minute having 3.37-3.40 mm length and 0.82-0.89 mm width and made up of stigma, style and ovary. Stigma was slightly curved towards, either side rough and sticky. Ovary is having a single basal ovule, which is of anatropous type, having two integuments and it was crassinucellate type having short and broad funicle. The body of the ovule consists of a central solid ovoid mass of soft, juicy thin walled parenchyma, known as nucellus, which is the actual megasporangium. The integuments aroused from the base of the nucellus, called chalaza, however the integuments did not cover the nucellar completely, rather left with a small pore, the micropyle at one end. This end is called the micropylar end of the ovule. The nucellus was free near the micropylar end. The entire surface of all the parts of the ovule was covered with cuticle.

In *H. rhamnoides* anatropus ovule bends over due to the curvature of the funicle at its apex. Consequently it lies against the funiculus with which it is fused throughout its length. The fused portion is called the raphe. In anatropous ovule the body is straight but not upright in posture. It is inverted. The micropyle and helium were near each other and were parallal. The chalaza was situated away from the helium at the opposite end.

2.5.2 Male Flower

Male flowers of *H. rhamnoides are* very minute (length 3.20-3.42 mm and width 2.70-2.75 mm). Microscopic examination indicated male flower having four stamens, arranged at equal distance from each other and are slightly curved inwards. Dehiscence through longitudinal slits. T. S. of anther showed that they were trilobed and having a central conducting strand. It showed three microsporangia, each having innumerable anthers.

2.6 SECONDARY GROWTH

The meristem producing the secondary xylem and phloem is the vascular cambium, occupied the lateral position in the axis hence, also called as lateral meristem. The cambial cells differ from normal meristams cells, in being highly vacuolated. These occur in two forms, the fusiform initials, which was much longer than wide and the ray initials, which was almost isodiametric. The fusiform initials constitute the axial system, which gave rise to axial system of secondry xylem, and the ray initials forms the radial system of zone of cambial initials.

At the time of secondary growth, the cells of medullary rays, in a line with the fascicular cambium (cambium of vascular bundle), become meristamatic and form new strips of secondary meristem, called the interfascicular cambium. These strips joined the strips of vascular cambium on both the sides and form a complete ring, called cambium ring. Though formation of secondary tissues started at the vascular bundle, but later on the cambium ring as a whole forms new cells on both the sides by tangential divisions, the activity was more on the inner side and these cells were gradually modified into secondary xylem cells, which consisted of wood vessels, wood fibers, tracheids, and wood parenchyma. The cells formed on the outer surface of cambium got modified into secondary phloem, having sieve tube, sieve plate, companion cells, bast fibres and phloem parenchyma.

As the activity of cambium is more on the inner side, more of the secondary xylem is formed. At maturity this secondary xylem formed the main bulk of the stem. By its pressure, the cambium and phloem were pushed outwards and hence the primary phloem had been crushed, however the primary xylem did not lose its identity, it was recognized as conical protrubances in the pith. The secondary phloem formed on the outside, also forced the primary phloem to outside which later on becomes non functional. The cortex cells divide anticlinally for some time so as to increase the circumference. To keep pace with the increasing width, the secondary tissue was crushed at randomly and the epidermis also increased in circumference by the anticlinal divisions.

At some places, adjacent to the ray initials, the cambium formed narrow band of radially elongated parenchymatous cells both on the inner and outer side instead of xylem and phloem. Hence, these formed the secondary medullary rays and were one to few cells in thickness and in height. As the secondary xylem cylinder expanded, cambial ring also expanded by occasional radial divisions in addition to the usual tangential divisions. However due to severe winters in Ladakh, no autumn or late wood was formed therefore only one ring was formed in a year, and is called as the annual ring of *H. rhamnoides*. Thus, it differed from other dicot plants in having only autumn wood in the annual ring. Because a ring is formed in a year the age of the plant can also be determined by a transverse section of the tree trunk.

2.6.1 Activity of Cambium

The secondary tissue formed by the activity of cambium (Banman, 1953) exerted a pressure on the outer tissues and the epidermis might be ruptured, hence the hypodermis and cortex were also adversely affected. To replace this peripheral protective tissue, new secondary tissues were developed and known as periderm, which had three parts:

Cork cambium or *Phellogen*: which originated in the outer layer of hypodermis cells, consisted of narrow row of thin walled and roughly rectangular cells which were living and active. The cork cambium became meristamatic and formed new cells on both the sides. *Secondary cortex* or *phelloderm:* newly formed cells of the cork cambium on inner side converted into parenchyma and form the secondary cortex.

Cork (Phellum): new cells formed on the outside of the cork cambium lost their contents, became filled with air and were arranged in rows at right angles to the surface, forming a dead layer. The cell wall became suberised and the intercellular spaces were lost. The dead tissue probably acts as a protective coating and is known as cork or phellum.

The cork cells being suberized, check the outwards passage for water thus forming bark. The outer tissues became dead and thus act as a bark. The bark included all the dead tissues lying outside the cork cambium. It continued epidermis, hypodermis, a part of cortex, secondary cortex, phloem and cork.

2.6.2 Secondary Growth in Root

Secondary growth in root of *H. rhamnoides* was observed due to the activity of the cambium and the cork cambium. Secondary growth starts at an early stage (after 2-3 months of age) that it is very difficult to get a root without secondary growth.

Some of the cells of conjunctive tissue became meristamatic and form strip of cambium, the number of strips formed depends upon the number of phloem bundles. In *H. rhamnoides* root was a polyarch i.e. having number of strips. These strips were extended both ways in between phloem and xylem and ultimately unite with the pericyclic cells lying just outside the protoxylem. The pericyclic cells divide tangentially and form two layers of which the cells of inner layer also become meristematic and unite with the strips of cambium and thus a continuous wavy band of cambium is formed extending down the phloem and over the xylem. It divides by periclinal divisions and then by anticlinal divisions for increase in circumference. Phloem and cambium strip are pushed outwards and the wavy band of cambium now became circular to form a cambium ring. The secondary xylem consisted of comparatively large vessels, tracheids, a little wood fibre and well developed xylem parenchyma. Thus after secondary growth, the xylem formed the main bulk of the root and present in the form of solid core. The primary xylem bundles and pith were crushed. The annual rings in the root are not visible like stem.

The secondary vascular tissues did not form a continuous ring and interrupted by the bands of radially elongated parenchymatous cells, called as primary medullary rays. These are formed above each primary xylem patch and extended up to the phloem. Sometimes smaller medullary rays also develop and called as secondary medullary rays. The number of rays goes on increasing with the increase in the size of the vascular cylinder. Secondary xylem had tracheids and vessels.

The cells of pericycle divided tangentially and formed layer of cork or phellum on the outside and secondary cortex or phelloderm on the inner side. The bark in root included cork, endodermis, cortex and epiblema.

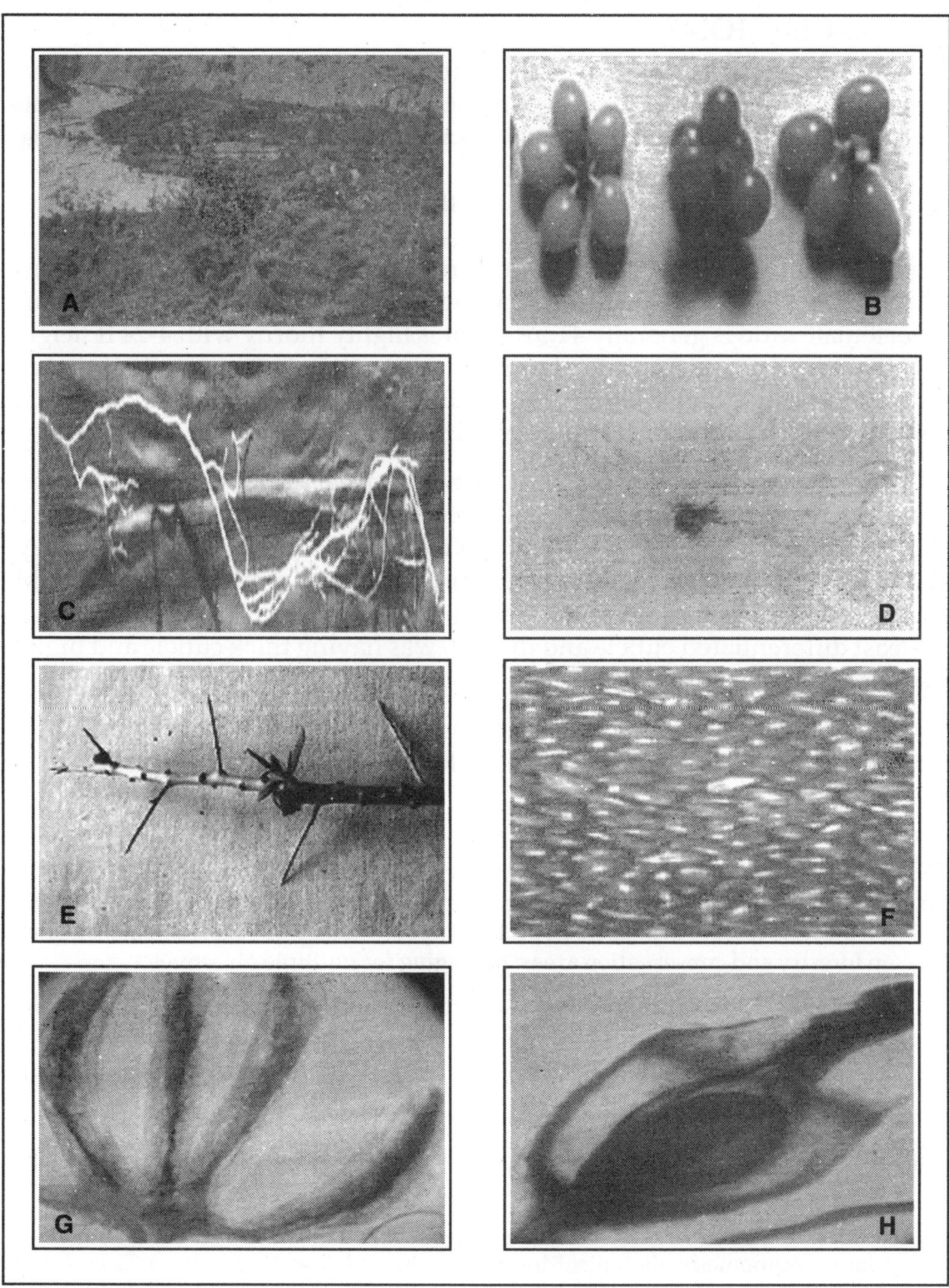

Fig. 2.1. Seabuckthorn morphological aspects (A-H): A–Habitat, B-Fruits, C-Root, D-Scales, E-Thorns, F-Seed, G-Male flower (W.M), H-Female flower (W.M)

2.7 CONCLUSIONS

On the basis of the study it is concluded that seabuckthorn has considerable variation in plant morphology of leaf, bark, thorn, stem, roots, fruits etc. Lots of variation in size, shape and colour of the fruits has also been recorded. On the basis of the findings of the morphological study, seventeen morphotypes has been designated for the convenience. The seabuckthorn population survey revealed that seedlings/juvenile plants/flowering fruiting plants have mostly present status except in Changmar, Bogdang and Pinchimic villages. Density of the mature plant varied from 6-12 plants/m^2. Average male: female ratio is generally 41:59. Plant is highly thorny with 4-24 ft height. Root is tap root and runs horizontally up to 4-5 meters producing new generation plants. Root is having root nodules, which due to *Frankia-ctinorhizal* symbiotic association fix the atmospheric nitrogen and enrich the soil.

Anatomy of the stem, thorn, root, flower and leaf also suggested difference in some morphotypes at histological level like cuticle was least prominent in M-4, 5 and 11. Sclerenchyma was present in patches in all morphotypes except M-2 where it formed a ring. Thorn of all morphotypes were similar except M-1, 14 & 16. M-1 has larger vascular tissue, double layered cuticle while M-14 has least differentiated cuticle and in M-16 was having thick cuticle and highly sclerenchymatous cortex.

The phenotypic variations are substantial which indicate presence of entirely new species or subspecies, which is needed to be confirmed at molecular level.

References

Baker, R.M. 1996. The future of the invasive shrub seabuckthorn (*Hippophae rhamnoides*) on the west coast of Britain. In: Aspects of applied biology 44 vegetation management in forestry and conservation areas: managing for multiple objectives.

Banman, M.W. 1953. Anticlinal divisions and the organisation of conifer, cambium. *Bot. Gaz.* 129, 137-143.

Chaurasia, O.P. & Singh, B. 1996. Cold desert plants, Vol. I. Field Research Laboratory (DRDO), C/O 56 APO.

Claphan, A.R., Tutin, T.G. & Warburg, E.F. 1957. Flora of the british Iseles, Cambridge, p.596.

Duhoon, S.S., Kopper, M.N. & Chandra, U. 1996. Seabuckthorn a less known wonder plant of ethnomedicobotanical importance in cold deserts of India. *J. Econ. Taxon. Bot.* 12, 43-45.

Esau, K. 1939. Development and structure of the phloem tissue. *Bot.Rev.* 5, 373-432.

Jasra, A W. & Khan S.A. 2001. Seabuckthorn a multidimensional development approach in Pakistan. In: *Proceedings of International Workshop on Seabukthorn*, p.25-31, New Delhi, p274.

Korovina, M.A. & Feflov, V.A. 2003. Morphological variations in the fruits of seabuckthorn (*H. rhamnoides*. L.) in different regions of CIS states. In: *Seabuckthorn a multipurpose plant Vol-I. Botany Harvesting and Processing Technologies*. pp. 72-83. (Eds.V. Singh *et al.*) Indus publishing company, New Delhi.

Li, T.S.C. & Schroeder, W.R. 1996. Seabuckthorn (*Hippophae rhamnoides*) A multipurpose plant. *Horticulture Technology* 6, 370-380.

Lian, Y.S. & Chen X.L. 1995. New discovery of genus *Hippohae* In: *Proceedings of International Workshop on Seabuckthorn* p 62-66. Beiging, China.

Lian, Y, Chen, X. & Lian, H. 2003. Taxonomy of seabuckthorn *Hippophae* L. In: Seabuckthorn (*Hippophae* L.) A multipurpose wonder plant, Vol.I Botany.

Rongsen, Lu. 1996. Feasibility study on seabuckthorn development in Pakistan. ICIMOD, Kathmandu, Nepal.

Rousi, A. 1971. The genus *Hippophae* L.- A taxonomic Stduy. *Ann Bot. Fenn*.8, 177-227.

Royal Horticultural Society . 2001. London, Practical Manual Colour Chart. RHS, London.

Singh.V. 2003. Geographical adaptation in distribution of seabuckthorn resources. In : Seabuckthorn (*Hippophae* L.) A multipurpose wonder plant , Vol.I Botany, Harvesting and Processing technologies. (V. Singh *et al.*, Eds.). pp.21-34. Indus publishing company, New Delhi, India.

Singh, R., Dwivedi, S.K. & Mishra, S.N. 2002. Seabuckthorn: A Cold Desert Gold, *The Botanica* 52, pp 71-74.

Singh, R., Dwivedi, S.K. & Mishra, S.N. 2003. Ethnobotany of Seabuckthorn '*Hippophae*' in Ladakh. *Ethnobotany* 15, 1-5.

Singh R, Dwivedi, S.K. & Mishra S.N. 2005a. Morphology and Distribution of seabuckthorn (*Hippophae rhamnoides* L.) in Nubra Valley of Ladakh. *Phytomorphology* 55 (1&2), 123-132.

Singh R, Dwivedi, S.K. & Mishra S.N., Ahmed Z. 2005b. Comparative study of seed morphology and Anatomy of three species of *Hippophae* in India. *Plant Archives* 5, 569-572.

Srodon, A. 1970. *Hippophae rhamnoides* L. in the quaternary of Poland. *Acta Geography* Lodz 24, 415-424.

Tiwari, S.C. & Singh, S.S. 2000. A report on seabuckthorn from Arunachal Pradesh, Northeast India. *Souvenir of International workshop on seabuckthorn* p 72-77, New Delhi.

❏❏❏

3

Seabuckthorn in Ladakh

Sanjai K. Dwivedi, T. Stobdan and Shashi Bala Singh

CONTENTS

Seabuckthorn (*Hippophae spp L.)* is an important plant of cold arid zones of Himalayas in India. It has been reported from about 30 countries of Asia, Europe, South and North America. It is found in European countries like Russia, Belarus, Ukrain, Germany, Britain, Sweden, Poland, Hungry, Norway, Czech, Finland, Netherlands and Romania. In these countries seabuckthorn has been distributed along the coastal areas of seas (Singh & Dogra, 1996). *Hippophae* has also been reported from Russia's Kalinigrad, Subcaucasus and southern part of Siberia. Though *Hippophae* is not growing naturally in Canada and America but was introduced in Canada in 1930 by L. Skinner for effective snow trapping in winter (Singh& Dogra, 1996).

In Asia, it grows mainly in countries like China, India, Mongolia, Russia, Nepal, Bhutan, Pakistan, Afghanistan etc. China possesses world's maximum seabuckthorn resource of about 1.2 million hectare followed by India (1.0 million ha). In Asian mountains, it has been found growing at an elevation range of 5000 ft to 16000 ft AMSL. Seabuckthorn (*H. rhamnoides* ssp *turkestanica, H. salicifolia* and *H. tibetana)* has been found growing in various parts of Pakistan, Nepal and Afganistan. The higher mountaneous regions of Mangolia, lying between China and Russia in central Asian part, are well convered with natural vegetation of seabuckthorn (Rousi, 1971). In Pakistan, it is distributed Baltistan and Gilgit (Jasra, 2001). In Afghanistan, *Hippophae rhamnoides*.L is found in Bamian, Kabul, Parwan, Gulbahar and Nuristan (Rongsen, 1992). The global distribution of seabuckthorn has been presented in Table-3.1.

Table 3.1: The global distribution of *Hippophae*

Taxons	*The Areas of Distribution*	*The Status of Utilization*
1. H. rhamnoides. subsp. rhamnoides	Scandinavian countries, Baltic Sea countries, Germany, Belgium, Netherlands, Ireland, Poland, U.K. France, Russia	Many varieties are cultivated in some European countries and Canada.
2. H. rham. subsp. sinensis	The North, Northwest, Southwest of China	Wild resources are used for ecological restoration and berries are processed for products. Some new varieties are in tests.
3. H. rham. subsp. yunnanensis	Sichuan, Yunnan, Tibet of China	Wild resources are used for ecological restoration only.
4. H. rham. subsp. mongolica	Siberia of Russia, Mongolia, Xinjiang of China	More than 60 varieties are cultivated in Russia, Mongolia, many East European counties. Many West European counties, Canada and China introduced the varieties for test

contd....

table contd....

5. *H. rham. subsp. turkestanica*	India, Pakistan, Afkhanistan,Turkmenistan, Kirghizstan, Uzbekistan, Kazakhstan, Iran, Turkey, Xinjiang , Tibet of China	Wild resources are used for ecological restoration and berries are processed for various products on commercial level in India for the production of food, medicine and cosmetics.
6. *H. rham. subsp. fluviatilis*	Around Alps Mountains: Germany, France, Switzerland, Austria, Czech, Slovakia, Italy,	Most of wild resources are protected as forest species. Some berries are collected for processing products
7. *H. rham. subsp. carpatica*	The Capathinan Mountains, Transsylvanian Alps,the valley and the mouths of the Donube and its tributary.	Most of wild resources are protected as forest species. Some varieties are cultivated for processing products
8. *H. rham. subsp. caucasica*	The Caucasus Mountains, Georgia, Azerbaijan, Armenia, Ukraine, Romania, Turkey, Bulgaria, Iran, Russia.	Most of wild resources are protected as forest species. Some selected varieties are cultivated for test.
9. *H. goniocarpa*	Sichuan, Qinghai of China	Most of wild resources are protected as forest species. Very few studies have been don on it.
10. *H. goniocarpa subsp. litangensis*	Sichuan, Qinghai of China	Most of wild resources are protected as forest species. Very few studies have been don on it.
11. *H. neurocarpa*	Sichuan, Qinghai,Gansu of China	Most of wild resources are protected as forest species. Very few studies have been don on it.
12. *H. neurocarpa subsp. stellatopilosa*	Sichuan, Qinghai, Tibet of China	Most of wild resources are protected as forest species. Very few studies have been don on it.
13. *H. tibetana*	Sichuan, Qinghai, Gansu, Tibet of China, Nepal, India	Most of wild resources are protected as grassland species. Very few studies have been conducted on it.
14. *H. gyantsensis*	Tibet of China	Most of wild resources are protected as forest species. Some berries are collected for producing Tibetan medicine.
15. *H. salicifolia*	The southern slope of Himalayan Mt. Tibet of China, Bhutan, Nepal, India	Most of wild resources are protected as forest species. Some berries are collected for processing/medicinal purposes.

In India, seabuckthorn grows mainly in high altitude, cold arid conditions of Ladakh (Leh and Kargil), Lahaul-Spiti, parts of Chamba (Pang) and upper Kinnaur districts of Himanchal Pradesh and Badrinath and Chamoli in Uttarakand (Dwivedi *et al*, 2006,a). Seabuckthorn has also been reported from Sikkim (Basistha *et al* 2001) and in Dibang valley in Arunachal Pradesh (Tiwari & Singh, 2001). General characteristics and distribution of seabuckthorn in India has been presented in Table-3.2 & 3.3.

Table 3.2: General characteristics of *Hippophae* species found in India

Species	Distribution	Growing Altitude(m)	Plant height	Flowering time	Fruit Ripening time
Hippophae rhamnoides	India, China, Tibet, Kyrghystan, Kazakistan, Uzbegistan	600-4200	5-6 m	May	September-October
Hippophae salicifolia	India, Tibet, Bhutan, Nepal	2700-3700	3-10 m	June	October
Hippophae tibetana	India, Tibet, China	3000-5200	0.8-1.2 m	May	August-September

Table 3.3: Distribution of Seabuckthorn species in India

Species/ Local Name	Ladakh (*Tsermang*)	Himanchal Pradesh (*Chharma*)	Uttaranchal (*Ames, Chuk or Chu*)	North East Region (*Tare, Taroobo*)
Hippophae rhamnoides subsp turkestanica	Indus valley Nubra valley Suru valley Changthang valley	Kukumsari Lahaul Kaza Tabo	?	
sub Sp gytansis	Turtuk (?) (Nubra valley)	--------	-------	Nathula (Sikkim)
Hippophae salicifolia	----------	Lahaul valley, Kaza	Yamunotri, Badrinath, Harhindun (Garhwal) and Gori valley, Kali valley, Dharma, Buddi, Bagnitiyar (Kumon)	Lachen, Lanchug, Domang (North Sikkim) Arunachal Pradesh
Hippophae tibetana	Zanskar	Sangrum (Pin Valley) Kibbar Takcha (Spiti)	Gomukh Ranimini Niti Nelong Bital Shinla Milan	North Sikkim

3.1 AREA AND DISTRIBUTION

In Ladakh in the state of Jammu and Kashmir(India),seabuckthorn is known as "*Tsermang*" and its fruits are called as "*Tsestalullu*". Plant is being used as fence and for fodder since ages (Dwivedi *et al.*,2002,c). Two species *H. rhamnoides ssp turkestanica* and *H. tibetana* have been reported *from* Ladakh. Among *H rhamnoides ssp turkestanica* is widely distributed species of Ladakh and whereas *H tibetana* has been distributed at higher elevations in Zanskar valley of Ladakh. *Hippophae rhamnoides* L. grows naturally in all five valleys of Ladakh viz Indus, Nubra, Suru, Zanskar, and Changthang having altitude ranging from 8500 to 12500 feet above mean sea level. Limited patches of the species have been noticed at Shani and Manda Villages (Zanskar) at 13600 feet above mean sea level and Tangtsey (Changthang) at an altitude of 15400 feet above mean sea level. The maximum biodiversity of seabuckthorn has been observed in Nubra, Leh and Suru valleys of the region. Seabuckthorn has been growing naturally mainly along the rivers, channels and around the agricultural fields. It has also been found growing in rockey, barren, wasteland and even in salt affected soils.(Singh *et al*,.2005,a). In Nubra valley, it grows along the banks of Shyok and Nubra rivers. The major growing areas have been observed in Warsi, Khalser, Panamik, Sumur, Tegar, Thoise, Diskit, Turtuk and Pachathang villages in Nubra valley. It is an effective soil binder and help in sand dune fixation in Nubra valley, especially in between Hunder and Diskit villages. It is also found growing in crevices on barren mountain slopes near Warsi village (Singh, 2004). It is growing near Panamik hot spring where highly acidic and hot water flows. Moreover, it is also found growing in plenty near hill locked lake, where soil is sandy and highly saline. In Indus valley, it grows mainly on the banks of Indus river. The major population of seabuckthorn has been recorded around the villages like Stakana, Chhuchot, Thiksey, Shey, Choglamsar, Spituk, Saboo, Leh, Phyang, Nimoo, Bazgo etc. It also grows on the banks of Suru and Drass rivers in Kargil district. Thus, seabuckthorn has ubiquitous distribution in Leh and Kargil districts of Ladakh with excellent ability to tolerate abiotic stresses like soil, moisture and nutrient besides tolerating -30°C temperature during winter months (Dwivedi *et al.*,2002,c). Most of the populations of seabuckthorn in Ladakh are wild and are of seedling origin. However, dense and inpnetrable population is due to profuse suckering ability of the plant. Seabuckthorn being a cross pollinated plant, considerable variations have been recorded in the region. There is no record about the genesis and introduction of the crop in Ladakh; however it has been proved that seabuckthorn has been growing in the area since time immemorial. However, it is believed by locals that migratory birds from Siberia might have introduced seabuckthorn in the region. Satellite imagery conducted by Field Research Laboratory (DRDO), Leh in collaboration with Defence Electronics Application Laboratory (DRDO), Dehradun has shown about 11,500 hectare areas purely and 30,000 ha under mixed forest platation in the Ladakh region (Dwivedi *et al.*,2003).

3.2 FODDER VALUES

In order to evaluate the fodder value of seabuckthorn leaves, samples of seabuckthorn leaves from male plants were collected from naturally seabuckthorn growing areas of Leh and Nubra velley. The collected samples were sun dried and analysed separately for their nutritive value. The nutritive analysis indicated that seabuckthorn leaves are potential nutritive fodder (Table 3.4) for sheep, goat and cattle etc (Dwivedi *et al.*, 2004,b) and is useful fodder of the region.

Table 3.4: Leaf Biochemical constituents in the leaf of seabuckthorn.

Parameters	Content
Moisture	52-69%
Total Ash	1.76%
Crude Protein	2.2-2.4
Crude Fibre	4.6-4.85 %
Ether Extract	6.6-6.94 %
Total Carbohydrate	32-37%
Calcium	69 mg/100g

3.3 GENETIC DIVERSITY

Extensive studies conducted by Field Research Laboratory (DRDO) have helped to identify 200 genotypes from the natural population of seabuckthorn in Ladakh which have shown wide variations in seabuckthorn (Dwivedi *et al.*, 2005,b & c). It has also been recorded that tree height varied from 6-24 ft and it is inversely proportional with altitude at which the plant is growing. Seabuckthorn is generally bush in habit however, plants having proper trunk and a tree structure have also been observed. Variations in the leaf intensity, leaf size, shape, etc have been recorded (Dwivedi *et al.*,2007,a). Variation in physico-chemical properties of the ripe fruits have also been studied and have been presented in Table-5. Based on the study 19 promising selections have been identified and conserved in field gene bank at Leh.

Table 3.5: Physico-Chemical parameters of fruits morphotypes of Seabuckthorn (*Hippophae rhamnoides* ssps *turkestanica*).

Parameters	Range	Mean Value	SEm
Fruit Length(mm)	5.42-9.27	7.34	0.0889
Fruit Width (mm)	4.67-7.16	5.91	0.0736
Weight of 100 fruits (g)	7.61-20.54	14.07	10.69
Volume of 100 fruits (ml)	7.9-24.1	16.0	0.314
TSS (°B)	10.2-18.4	14.3	0.189
pH	1.50-2.3	2.15	0.016
Acidity (%)	1.46-3.621	2.54	0.103
Sugar (%)	3.65-5.23	4.44	0.097
Vitamin 'C'(mg/100g)	138-1400	76.9	26.36
Juice (%)	68.18–80.85	74.5	0.655
Seed (%)	4.44-8.65	6.54	0.377
Pomace (%)	11.4-27.5	19.45	0.585
Seed weight(g/100 seeds)	1.13-1.35	1.24	0.021

3.4 SCIENTIFIC METHODS OF PROPAGATION

Seabuckthorn has been growing wild under natural conditions, has been planted by local farmers or by Forest department for protection of flood/ soil erosion on river banks, sand dunes fixation or for fencing the new plantation or agricultural fields where fruit yield and its quality has no importance. However, for commercial cultivation, it is imperative that the crop be multiplied from the best material which has been selected for its fruit yield and other desirable parameters. Seabuckthorn can be propagated by various methods. A brief description of various methods of propagation are given below:

3.4.1 Propagation Through Seeds

Seabuckthorn can be propagated successfully by seed which are viable for a period of two years, after which they lose viability rapidly. Seabuckthorn seeds are small (4.5 mm in length) with brown and hard seed coat.(Singh *et al*,.2005,b). One gram seed may contain about 100-105 seeds (Dwivedi *et al.*, 2005, a). The freshly harvested seeds have a short physiological dormancy and thus can not germinate immediately after harvesting. May is the best time for sowing of seeds in nursery beds. Before sowing the mature seeds should be stratified for about 25 days by keeping in a box in alternate layers of moist sand. These boxes are kept under cool and shady condition for about 25 days. This method gives over 85 % germination (Dwivedi et al.,2001,b).

Seed can also be sown directly in nursery beds in the month of November where cold treatment is given before germination under natural conditions. Both male and female plants are borne from seeds. The method is useful only in raising large-scale nurseries for afforestation for conservation of mountainous wastelands.

3.4.2 Propagation Through Suckers

Seabuckthorn has profuse suckering habit and has been reported to yield 13-40 suckers per plant depending upon the growing conditions and age of the plants. These suckers are true to the type which can be separated along with the roots from the mother plants and planted directly at the planting site. The best time for separation of suckers and planting is in March before bud break (Dwivedi *et al.*, 2001, a). Care should be taken to identify and tag the plants during the fruiting season in order to avoid mixing of male and female suckers. This being an asexual method of propagation produces true to type plants. However, the major disadvantage is difficulty in operation due to thorny plant structure.

3.4.3 Propagation Through Hard Wood Cuttings

Seabuckthorn can be propagated successfully by stem cuttings and this method has been recommended for commercial propagation. It helps in development of proper canopy of the plant and maintenance of male/ female ratio in the orchard. One to two year old stem cuttings having pencil thickness are ideal for its propagation. The stem cuttings are taken from identified male and female plant to get elite planting material. The best time for taking cutting is in the month of March. The side branches and thorns are removed and 30 cm long cuttings are made. The cuttings are then placed in running water for a period of 2-3 days. Seabuckthorn cuttings are not easy to root on their own, thus the rooting hormone indole butyric acid (IBA) is applied @ 500 ppm before planting it in the nursery beds. The best time for planting of cuttings has been found to be the month of April (Dwivedi *et al.*, 2002,a). Properly treated cuttings should be planted soon after treatment in well prepared nursery beds. Planting should be done in rows 30-45 cm apart and proper spacing between the cuttings (Dwivedi *et al.*,2004,b). Seabuckthorn normally takes 60-75 days in rooting depending upon the soil and climatic conditions. About 85 percent success has been recorded in rooting of stem cuttings under Ladakh conditions. These cuttings are ready for plantation in the main field in March of the following year. (Dwivedi *et al.*, 2006,a).

3.4.4 Propagation Through Soft Wood Cuttings

Field Research Laboratory (DRDO) has successfully standardized asexual method of propagation of seabuckthorn through soft wood cuttings using 200-300 ppm NAA. The technique has yielded over 90 percent rootability in cuttings with excellent root quality under mist condition having course, sand rooting media. The plantlets get ready for transplantation in poly bags within 90 -100 days after planting of cuttings (Dwivedi, *et al.*2007,b). The method is much faster than hard wood cuttings with superior root quality and better establishment percentage and also has no chance of transmission of diseases through bud wood. The method can be adopted for quick and fast multiplication of disease free quality planting material of seabuckthorn for its commercial cultivation.

3.4.5 Propagation Through Micropropagation

Micro propagation is the most modern method of propagation and fast multiplication of elite planting material. It helps in production of disease free planting materials throughout the year and in conservation of valuable, exotic and indigenous materials under laboratory conditions. Shoot proliferation in seabuckthorn has been achieved by using actively growing shoot tip in Murashige and Skoog's (MS) nutrient media having 30 g sucrose, 08 g agar and plant bio regulators viz, 0.1 ppm IBA and 2.0 ppm BA for shoot proliferation and 2.0 ppm NAA for rooting the plantlets. About 64% plantlets may be obtained after hardening in greenhouse, for commercial plantation (Anon, 2006).

3.5 NURSERY MANAGEMENT

Light and well drained soils are ideal for nursery area. Soil should be at least 1-2 m deep. Heavy soils having poor drainage facilities should not be selected for nursery purposes. The selected site should be in a sunny location with assured irrigation facilities. Salt effected soils are not suitable for nursery. Nursery bed should be rich in organic matter and other nutrients.Before sowing seeds in nursery the bed should be ploughed deep and all stones, roots, etc., should be removed. Well rotted FYM @ 4 kg/m^2 should be incorporated the soil 20-30 days before sowing/planting. Beds should be dug properly and should be free from stones, pebbles, roots weeds, etc. Soil should be light and organic manure should be mixed properly. The bed should be 2 meter wide since this helps in ease in cultural operations. Beds can be of any desirable length, as per the slope and method adopted for irrigation. Before planting the cuttings or sowing of seeds, the bed should be prepared of fine tilth and adequate arrangements should be made for assured irrigation and protection of the beds from animals. In heavy soils, additional FYM and sand should be added to make the soil light (Dwivedi, *et al*,2006,b).

3.6 ORCHARD ESTABLISHMENT

Seabuckthorn prefers sunny, well drained, deep soils with assured irrigation facilities.It has deep root system (3-5 m) and for its optimal growth soil must have 2-3m depth which should be free from stones/hard rocks etc. Seabuckthorn has been found growing in salt effected soils having pH value of 8.7. However, cultivation of seabuckthorn in salt affected soils should be avoided. Seabuckthorn has not been cultivated under planned orcharding. It has been growing wild in nature or as forest crop. However, for its scientific production, it should be planted scientifically by providing appropriate space between plants to develop proper shape and give maximum yield, quality and fruits per unit area and per unit time. It also helps in facilitating cultural operation and scientific harvesting by minimizing time loss. It is recommended that seabuckthorn generally be planted in hedge row system providing more space in between the rows and lesser space between plants. Single or double hedge row systems can be adopted which provide enough space to approach each plant easily for training, pruning, harvesting and other cultural operations. In single hedge row system the distance between row to row is maintained at 2-4 m for pure cultivation and 4-5 m for inter cropping. The seabuckthorn plants are normally planted in rows at a distance of 1.0 meter. However, method and distance of planting can be modified as per the climatic condition, fertility status of soil, varietal features, orchard management practices adopted and harvesting requirements, etc. The pits for plantation should be of 60 cm diameter and 60-80 cm deep to remove the stones, roots etc from root zone. The best time for digging of pits is end October or early November. If they are not dug during this period, it must be dug at least 45 days before planting (Dwivedi *et al.*, 2006, a&b). After digging of pits the soil should be cleaned properly and well rotted FYM @ 2 Kg per pit should be mixed properly before filling the pits. In heavy soils, sand should also be mixed to make soil light and porous. The pits should be opened for solarisation for about 10-15 days and thereafter, they should be filled properly with the prepared mixture. Seabuckthorn can be planted in autumn (Oct-Nov) and spring (March-April) after onset of dormancy and before sprouting of buds, respectively. Spring planting has been found to give better results with regards to plant establishment and its survival (Dwivedi *et al.*, 2006, b). At the time of planting, it must be ensured that root is not coiled since it may lead to the death of the plant.The most important point to be kept in mind is that seabuckthorn is a dioecious plant, where male and female plants are separate, and is generally pollinated by wind. Thus, at least 10 % male plants must be planted in the orchard, distributed uniformly all over the orchard to ensure pollination of the female plants.

3.7 PRUNING AND TRAINING

The basic aim of pruning is to strike a balance between the root and shoot (vegetative) growth of the plant by heading back or thinning out the excessive vegetative growth of the plant. It is also practiced to induce fresh growth, encourage regeneration to remove diseased or dried branches, to promote flowering/ fruiting, reduce crop load, to improve the quality and colour of the fruits, provide shape and strong structure to the plant and remove undesirable growth and interfering branches. Data on pruning in seabuckthorn is not available since no systematic study has been conducted. However, it is essentially required for canopy management, removal of profuse suckers and induction of new growth which is essential to induce flower buds in next season. Seabuckthorn should be pruned regularly for ease in harvesting and exposure of plant to sun for production of quality fruits. The best time for pruning of seabuckthorn in Ladakh is the month of March, before sprouting of the plant in spring. Pruning should be done with a sharp secature or saw (Dwivedi *et al.*,2006,b). Power pruners are required for pruning the crop under natural conditions. The cut end should be pasted with paste of fungicide to avoid infection and reduce moisture loss from the plants. Pruning should not be done once the plant/ buds start sprouting.. Intensity of pruning depends upon the age and growth of plant and soil conditions. However, 50 % growth of 3-4 year old shoots should be removed.

Training is an essential operation to provide a desirable shape to the plant to make cultural operations viz, irrigation, weeding, spraying, harvesting, etc., easy and also provide a strong and scientific structure to the plant for quality production of fruits. Training helps in exposure of the fruits to sun for better color development and enable the plant to bear maximum crop load. Evaluation of different systems of training need to be done for seabuckthorn since there is no standard system recommended for the crop. However, it has been observed that seabuckthorn, being a thorny bush, should be restricted to a plant height of less than 2 meter for ease in harvesting (Dwivedi *et al.*, 2006, a). Normally properly managed seabuckthorn plants develop a crown of 2-3 m diameter at the age of four years after planting.

3.8 FLOWERING

Seabuckthorn is a dioecious plant where male and female plants are separate. Flowering in seabuckthorn has been generally noticed on one to two year old shoots. The flower buds are developed in the axils of leaf of each node of shoot. Plant bear floral buds mixed with vegetative buds on shoot decreasing from base towards the tips. Seabuckthorn generally flowers in first fortnight of May under Ladakh conditions. However, time and duration of flowering largely depends on the cultivar, prevailing weather conditions and altitude of the region. Flowering coincides with leaf expension after breaking of dormancy. Under Ladakh conditions, male plant flower earlier than female plants. Male plants generally flowers in end of April while female flowers appear in first fortnight of May. Male flowers produce pollen grains in abundance and they are transported in groups. Seabuckthorn is a wind pollinated crop. The germination of pollen grains occurs in 3-4 hours of pollination and fertiiization take place 5-10 days after pollination (Singh *et al.*,2004,a&b).

3.9 YIELD

Vegetatively propagated seabuckthorn plants normally start bearing fruits at the age of 3-4 years after plantation, while seedlings take 5-6 years to come into fruiting. Seabuckthorn orchard starts commercial production only after 8 years of plantation. Yield of seabuckthorn plant varies according to the age, managemental practices adopted variety, etc. Cultivated varieties have been reported to yield about 10-15 ton/hectare fruit under scientifically managed orchard conditions. However, under Ladakh conditions, yield of ripe fruits ranges from 0.6-2.0 kg per plant, since plants are growing wild under natural conditions. About 200-300 qtl of the fresh leaves can be harvested per hectare per year after 8 years of plantation (Dwivedi *et al*,.2006a &b).

3.10 POST HARVEST MANAGEMENT

Seabuckthorn is a non-climacteric fruit which ripens on the plant and can not be ripen after harvesting. Maturity indices of the crop have not been studied in detail. Ripening of fruits generally sets in by mid to end August under Ladakh conditions. Ripening of fruits can be judged on the basis of surface colour, texture, juice content and total soluble solids. The green fruits turn pale-yellow, yellow or orange or deep orange/red on ripening and become very soft and juicy. A sharp increase in the TSS of the fruits has been recorded after ripening and it reaches 12^0B and even above. Fruit juice content reaches above 70 %. Ripe fruits remain laden on plants even through out winter up to April and even beyond. However, it is advisable to harvest the fruits before onset of winter (Sub zero temperatures) which otherwise damage the texture and quality of the fruits. Fruits of seabuckthorn are very soft, delicate, juicy and small. Fruits are highly perishable and cannot be transported over long distances. Thus, after harvesting fruit needs to be processed as soon as possible. Ripe fruits should be processed within 24 hours of harvesting. Fruits are collected in plastic baskets which should not have a capacity more than 10 kg in order to avoid the damage to fruits by pressure. Moreover, the ripe fruits of seabuckthorn have a muskey odour which must be reduced before processing. Washing with cold water helps in reducing the odour of fruits. Data on storage of seabuckthorn fruits is not available and needs to be studied in detail. However, for storage and distance transportation of ripe fruits the recommended temperature is 4-6°C to reduce post harvest losses. Quick freezing of fruits and storage at -20°C has also been found effective for long term storage of fruits. Other post harvest managemental practices viz. sorting, grading, packing etc., have not been developed so far.

3.11 PROCESSING OF SEABUCKTHORN

One kg of ripe seabuckthorn fruit may contain 7407-7913 numbers of fruit. A pulper/ juicer can extract 61.42-72.07 percent of juice from ripe fruits. The seed content in ripe fruit generally varied from 7.46-8.31 percent. The pH of fresh juice has been found ranging from 1.49-1.57, total soluble solids from 12.33-20.75 percent and acidity from 3.40-4.01 percent. (Dwivedi *et al*,.2002,b). Seabuckthorn fruits are rich in flavonoids, polyphenols, vitamin C, vitamin E, other antioxidants, unsaturated fatty acids, etc.The leaves are rich source of polyphenols, flavonoids, antioxidants and dietary fiber. The plant is reputed to have considerable medicinal value being useful for the treatment of various health problems. Owing to these attributes seabuckthorn could become a food ingredient of high value for the development of functional/health foods, nutraceuticals and other value added products. Because of the high content of vitamin C, seabuckthorn fruit is generally used to make soft drinks and other similar food products. The juice is a good source of several amino acids and minerals. The pulp and the seed of seabuckthorn contain high quality oil, which is considered to be a carrier of many concentrated bioactive substances.Seabuckthorn products prepared from fruit pulp or juice are rich in vitamin C, provitamin A and other vitamins, especially of the B group and phytochemicals and are therefore considered to have therapeutic and chemopreventive activities. Seabuckthorn juice is known to block the endogenous formation of nitroso compounds more effectively than ascorbic acid and thereby prevents tumor production. Seabuckthorn fruits are used for preparation of jelly are used against pulmonary complaints. About 33.5 percent rich yellow pigment frorm fruits of seabuckthorn may be obtained which is widely used as a food additive (Bawa, 2004).

3.12 VALUE ADDED PRODUCTS

Fruits of seabuckthorn are rich in various nutrients and also possess medicinal property. However, they are very perishable in nature and cannot be stored for more than 2-3 days. Process for preparation of seabuckthorn squash has been developed successfully without adding any colouring and flavouring agent. The stability of the beverage during storage was found upto 12 months at ambient temperature under Ladakh conditions (Dwivedi *et al*,.2004,a). The observations recorded at various posts in Siachen glacier (12,000 to 21,000 ft above MSL) highlight that seabuckthorn squash does not freeze even upto minus (-) 22°C and remained in liquid form, thus it could be consumed without thawing. Process for preparation of seabuckthorn nectar has been patented which is commercially available in the market in the name of Leh Berry, Ladakh Berry, Power Berry, Siachen Berry etc. (Dwivedi *et al*,.2006,c). Delicious jam can also be prepared using seabuckthorn pulp. The jam is a rich source of vitamin A, B_1, B_2, K, E and other mineral nutrients.For preparation of seabuckthorn sauce a suitable base material is essential so that it can contain a minimum of 30 per cent total soluble solids. The product is prepared by maceration of spices, herbs and fruit pulp by boiling them in vinegar and cooking for a long period to get a smooth consistency. Appropriate quantity of sugar and salt are also added to develop its taste (Dwivedi *et al*,.2004,c)Semi ripe seabuckthorn fruits are ideal for making pickle. The clean harvested fruits should be completely immersed in edible oil. The pickle is preserved in salt which improves the taste and flavour and also helps in preserving nutrients and preventing spoilage. The pickle should contain 15 per cent salt. Spices like red chili powder, cinnamon, cumin, cardamom (large), black pepper powder and headless clove are also added to improve the taste and flavour of the product (Dwivedi *et al*,.2004,d). Seabuckthorn vinegar, concentrated juice and seabuckthorn powder can also be prepared from seabuckthorn, which contain various kinds of vitamins, amino acids and trace elements.

Tender and healthy leaves of (male) seabuckthorn are used for production of refreshing tea. seabuckthorn tea is rich in flavonoides, vitamins etc., and has medicinal value viz; shoothing, energizer, anti stress, stimulant, restores memory useful in cold and cough and is a health tonic (Chaurasia *et al*,2004). Other high altitude herbs like salam panja (*Dactylorhiza hatigeria*), Local tea (*Bidens pilosa*), local caraway (*Carum carvi*), black caraway (*Bunium persicum*), origano (*Origanum vulgare*), local mint (*Mentha longifolia*), yarrow (*Achillea sp*), rose root (*Rhodiola sp*) etc have also been added in the tea (Anon, 2006).

3.13 CONCLUSION

Seabuckthorn, a neglected plant of the cold arid regions, is a goldmine with an untapped potential. Maiden efforts towards preparation of products from seabuckthorn have suitably highlighted its immense potential commercially. The need of the hour is to establish commercial orchards on scientific methods and harvest its economic potential to the full by further developing seabuckthorn based industry. Development of high yielding cultivars, standardization of agro techniques, standardization of harvesting, post harvesting and processing technologies,value addition to plant produce through standardization of seabuckthorn food products and cosmetics are some important areas where research studies need to initiated systematically.

References

Anonymous 2006. Annual Report-2005. Field Research Laboratory (DRDO) Leh pp-64-67.

Bawa, A.S.2004. Seabuckthorn - Potential functional ingredients for food industry. In: Souvenir and Book of Abstracts of National Seminar on Cultivation, Harvesting and Scientific Exploitation of Seabuckthorn, p.5-6, August 26-27, 2004, FRL (DRDO), Leh, India. 136p.

Basistha, B.C. Adhikari, I.M. Deokata, K.P. & Thapa, K.K. 2001. A case study of Hippophae L. with special reference to agro-technique in the Sikkim Himalayas. In : Seabuckthorn-A Resource for Health and Environment in Twenty First Century (V. Singh and P.K. Khosla, Eds.). Souvenir of International Workshop on Seabuckthorn, February 18-21, 2001, New Delhi.

Chaurasia O P, Basant Ballabh, Perimalazhagan T and Raut B 2004. Seabuckthorn (*Hippophae rhamnoides var turkestanica*) of Indian Cold Desert.Paper presented in national seminar on Seabuckthorn-2004 at Field Research Laboratory (DRDO) Leh pp-24-25.

Dwivedi SK, Singh R & Ahmed Z 2007 a. Identification and evaluation of biodiversity of Seabuckthorn in cold arid regions of Ladakh *J. of Agric Sc.* 77(7): 427-431.

Dwivedi SK, Hemraj & Ahmed Z 2007 b. Standardisation of a fast method of vegetative propagation of Sea buckthorn through soft wood cutting in India. Paper presented in the 3 rd International Seabuckthorn Association conference held at Quebec, Canada,August,12-16,2007.Paper No83.

Dwivedi, S.K.,Singh, Ranjit and Ahmed, Z. 2006 a. The Seabuckthorn. Field Research Laboratory, DRDO, Leh, Ladakh (J&K) India.

Dwivedi, S.K. and Ahemed Z 2006 b. Seabuckthorn. In: Advances in Arid Hort Vol.II. Eds Saroj PL & Awasthi, OP. IBDC, Lucknow. pp 465-482.

Dwivedi, SK and Ahmed, Z 2006 c. Seabuckthotn – Ideal for production of neutraceuiticals and functional foods. In: Workshop on Science and Technology of neutraceuiticals and functional foods. The Current Scenario. Organised by DIPAS (DRDO) Delhi on 07th March, 2006. Abstract pp 16-17.

Dwivedi S.K, Singh R and Ahmed Z 2005 a. Sea buckthorn- The golden Bush of India cold deserts. *SAIC News Letter* 15(2):5-6.

Dwivedi, S.K., Singh, R. and Ahmed, Z. 2005 b. Morpho-biochemical characteristics of seabuckthorn growing in Ladakh. In: Seabuckthorn vol II. Ed. V. Singh. Daya Publishers, New Delhi.

Dwivedi, S.K and Raut B 2005 c. Biodiversity in fruit corps in cold and regions of Indias. In: National Seminar on Commercialization of Horticulture in Non-Traditional Areas at CIAH – Bikaner from 5-6 Feb. 2006. Book of Abstract. pp 12.

Dwivedi, S.K., Attrey, D.P. and Chaurasia, O.P. 2004 a. Studies on preparation and preservation of seabuckthorn squash beverage. *Ind. J. Hort.* 61(1): 78-80.

Dwivedi, S.K., Singh, Ranjit and Raut, B. 2004 b. Present status and future thrust on seabuckthorn research in Ladakh. In: National Seminar on Seabuckthorn at Field Research Laboratory (DRDO), 26-27 Aug, 2004.Leh. pp. 38-44.

Dwivedi S.K, Chaurasia OP and Raut B 2004 c. Seabuckthorn the Golden bush In : Seminar on seabuckthorn, DLS, DRDO HQ, New Delhi, 12-13 Mar 2004.

Dwivedi S.K, Singh Ranjit and Ahmed Z 2004 d. Seabuckthorn *(Hippophae spp)* A potential fruit crop for the Industry. Paper presented in first Indian Horticulture Congress. 6-9 Nov 2004, Pusa, New Delhi. pp-38.

Dwivedi, S.K., Singh, Ranjit and Raut, B. 2003. Studies on distribution of seabuckthorn through remote sensing and its genetic diversity in Ladakh. In: International Congress on Seabuckthorn. Seabukthorn a reosource of Health, a challenge o modern technology. Berlin (Germany), 14-18 September, 2003. Abst p. 29.

Dwivedi, S.K., Paljor Eli, Attrey, D.P. and Singh, B. 2002 a. Propagation of common seabuckthorn through hard wood cuttings in Ladakh. *Ind. J. Agril. Sciences*. 72 (4): 228-229.

Dwivedi, S.K., Attrey, D.P. and Kareem Abdul 2002 b. Post harvest management of seabuckthorn (*Hippophae spp.*) in cold arid regions of Ladakh. In: National workshop on post-harvest management of horticultural produce organized by National Productivity Council, 7-8 Feb. 2002 at Hotel Shivalikview, Chandigarh. pp. 89-98.

Dwivedi, S.K., Singh, B. and Attrey, DP. 2002 c. Studies on distribution, propagation and ultilisation of seabuckthorn in Ladakh. Seabuckthorn-A resource for health and environment in 21st Century. Eds. V. Singh and P. K. Khosala. Proceedings of International Workshop on Seabuckthorn, Feb. 18-21, 2001. CSKHPKVV-Palampur, India. pp. 20-24.

Dwivedi, S.K., Singh, B. and Attrey, D.P. 2001 a. Studies on propagation and utilization of seabuckthorn in Ladakh. In: International workshop on seabuckthorn orgtanised by HPKVV-Palampur, 18-21 Feb. 2001 at India International Centre New Delhi. Soveniour, session III.

Dwivedi, S.K. and Attrey, DP and Chaurasia, OP 2001 b. Seabuckthron - A Hope for greening of Ladakh. Field Research Laboratory (DRDO) - Leh .

Jasra, A.W. 2001. Seabuckthorn-a multidimensional development in Pakistan. In: International Workshop on Seabuckthorn, 18-21 Feb 2001. Session I.

Rongsen, Lu 1992. Seabuckthorn - A multipurpose plant for fragilemountains.Occasional paper No 20. ICIMOD-Kathmandu.

Rousi, A 1971. The genus *Hippophae*. L - A taxonomic study. *Ann. Bot. Fennici*, 8:177-227.

Singh R 2004. Ethnobotany, Anatomy and Biochemical studies on Seabuckthorn in Nubra valley of Ladakh" PhD thesis "Submitted to Maharashi Dayanand University, Rohtak (Haryana).

Singh R, Dwivedi SK, Mishra SN and Raut B 2004 a. Floral characteristics of seabuckthorn (*Hippophae rhamnoides L)* in Ladakh. In souvenir of national seminar on Seabuckthorn-2004 at Field Research Laboratory (DRDO) Leh pp-64-67.

Singh R, Dwivedi S K, Mishra SN & Ahmed Z. 2004 b. Floral Biology of Seabuckthorn (*Hippophae rhamnoides* L). A multipurpose plant of cold arid Ladakh. *Journal of Eco and Taxo Bot.*

Singh, R., Dwivedi, S.K. and Mishra, S.N. 2005 a. Distribution and morphology of seabuckthorn in Nubra valley of Ladakh. *Phytomorphology*. 55(1&2): 123-132.

Singh, R., Dwivedi, S.K., Mishra, S.N. and Ahmed, Z. 2005 b. Comparative study of seed morphology and anatomy of three species of seabuckthorn in India. Plant Archives. 5(2). 569-572.

Singh, V. & Dogra, K.K. 1996. Characterstics, distribution, biomass, utilization, regeneration and nutritional values of seabuckthorn (Hippophae L.). *Indian Forester* 122 (6): 486-491.

Tiwari, S.C. & Singh, S.S. 2001. A report on seabuckthron from Arunanchal Pradesh, north-eastern India. In: Seabuckthorn-A Resource for Health and Environment in Twenty First Century (V. Singh and P.K. Khosla, Eds.) Souvenir of International workshop on February 18-21 New Delhi.

□□□

4

Seabuckthorn in Himachal Pradesh

Virendra Singh

CONTENTS

4.1 INTRODUCTION

The dry temperate and cold deserts of Himachal Pradesh (H.P.), comprising Lahaul-Spiti, upper parts of Chamba and Kinnaur districts, are characterized by extreme climatic conditions, high rates of soil erosion and land slides, shortage of fuel wood, timber and fodder, low productivity of agricultural lands and sparse vegetation. The past efforts of afforestation by poplar and willow trees in the region have met only mixed successes for the want of a multipurpose plant species that can simultaneously satisfy the long-term conservation as well as the short-term economic needs of the people. Seabuckthorn is a multipurpose species, which is indigenous to the region and meets above requirements. In view of high economic and ecological potential and increasing trade in seabuckthorn, in 1993, it was decided to start the studies with conducting surveys of resources, followed by standardization of propagation and cultivation technologies, selections of promising forms and their evaluation and introduction of high yielding form from other countries, biochemical characterization and development of high yielding forms with desirable characteristics.

4.1.1 Distribution

Surveys were conducted in seabuckthorn growing areas of H.P. during 1995. In Himachal Pradesh, seabuckthorn grows mainly in the districts of Lahaul-Spiti, parts of Kinnaur and Chamba, which are surrounded by Tibet state of China and Ladakh region of Jammu and Kashmir state in the North on the southern side. Rainfall fluctuates from 50 to 700 mm per year depending on place to place. The mean temperature in Kelong, district headquarter and center place of Lahaul, shows considerable variation in temperature throughout the year (a maximum of 27 °C in July and minimum of -16 °C in February). During the winter this region is affected by a series of western disturbances, which cause heavy snowfall (200-400 cm/yr). Seabuckthorn grows at 2,500-3,300 m amsl in Lahaul. In Lahaul, it grows throughout the length of the Chandra valley in the villages, i.e. Sissu, Khorpani, Raling, Khongsar, Gondhla and Dalang. Several good stands of *Hippophae rhamnoides* sub species *turkestanica,* grows on both sides of Bhaga river (Tinu, Gemur, Jispa and Darcha). In Chandra-Bhaga (Chenab) valley, seabuckthorn is found as component of agroforestry systems in the villages Kirting, Sansa, Jhalma, Jhunda, Jasratha and Madhgaon. In Mayar valley, seabuckthorn is spread in Urgos and Chaling villages (Singh *et al.*, 1995). Prach (1994) studied the vegetation succession in river gravel bars in Lahaul. It was found that early successional stages, dominated by the typical river gravel species *Myricaria germanica* and *H. rhamnoides,* were highly uniform among the study sites throughout the mountains.

Spiti, a cold desert valley of district Lahaul-Spiti, which adjoins Tibet on its eastern border and Ladakh on its North. At base, the altitude varies from 3,120 m at Sumdoh to 4,500 m amsl near Kunjam pass. Spiti is a broad and flat valley, which is intersected by high vertical cliffs of black, grey and brown sand. Rainfall rarely crosses to 100 mm per year. *H. rhamnoides* grows widely from Sumdoh (3,100 m) to upper area (3,900 m amsl). There are excellent stands of this species at Shego-Lara (3,760 m), Lingthi (3,889 m) and Shichling (3,560 m). It grows mainly on riversides and beds; some farmers has also raised windbreaks around their fields. *H. rhamnoides* have also been reported growing in upper areas of Spiti in villages like Kiato, Kiamal, Morang, Sumling and Rangrik. *H. tibetana* has been reported growing in higher altitude areas at Losar (4,100 m) to Takcha (4,500 m amsl), Kibbar (4,100 m) and Chicham (4,200 m). *H. tibetana* has shown excellent growth at the undisturbed flat river bed at Takcha, whereas it has been destroyed in the villages of Kibbar and Chicham. Both *H. rhamnoides* and *H. salicifolia* have been found in upper Kinnaur. Seabuckthorn has been also found in upper areas of Shimla and Kullu districts. There are about 1,200 ha area under seabuckthorn in Lahaul-Spiti and upper Kinnaur, however, most of them are scattered and thin stands, except some stands growing in Bhaga valley in Lahaul and Lingthi, Shchling and Shego-Lara in Spiti, where quite good stands (about 300 ha) are growing (Singh *et al.,* 1995).

4.2 MORPHO-BIOCHEMICAL CHARACTERIZATION

Among the three species, *H. rhamnoides* was a shrub or small tree form (2-5 m), *H. salicifolia* was a small or large tree (4-9 m), whereas *H. tibetana* was a under shrub (0.2-0.6 m). Fruit size varied from 8 to 27 g/100 fruits in *H. rhamnoides,* 30-35 g in *H. salicifolia* and 15-21 g in *H. tibetana.* Colour of the fruit varied from red to reddish orange in *H. rhamnoides,* yellow in *H. salicifolia* and orange with green strips in *H. tibetana.* Seed size also varied from 0.718 to 1.103 g/100 seeds in *H. rhamnoides,* from 0.943 to 1.142 g in *H. salicifolia* and from 1.050 to 1.331 g in *H. tibetana.* The seed colour was dull white in *H. tibetana,* whereas it varied from light to dark brown in other two species. The shape of the seeds were oblong to ovate, oval, elongate and elliptical in *H. rhamnoides,* ovate in *H. salicifolia* and elliptical in *H. tibetana.* The moisture content in the fresh fruits varied from 72.9 to 85.0%. The total oil content varied from 3.2 to 4.6% in *H. rhamnoides* and 2.0% in *H. salicifolia.* The vitamin C content varied from 219-642 mg/100g in *H. rhamnoides* to 947 mg in *H. salicifolia* and 356 mg in *H. tibetana* (Singh *et al.,* 1995).

Among the 12 biotypes of seabuckthorn growing in different areas of Lahaul-Spiti, weight of 100 seeds varied from a maximum of 1.331 g in *H. tibetana*, to 1.142 g in *H. salicifolia* and 1.103-0.718 g in *H. rhamnoides*. The total oil content was maximum in the seeds of *H. tibetana* (16.9%), which was significantly higher ($P<0.05$) than the seeds of *H. rhamnoides* (8.8-14.6 %) and *H. salicifolia* (10.5%). There was no definite pattern of different characteristics of the seeds of different biotypes of seabuckthorn in relation to altitude of the region (Singh *et al.*, 2005).

Morphological changes during ripening of the fruits was studied by Singh *et al.* (2001) in a natural stand of seabuckthorn growing at 2,650 m amsl, on the riverside in Lahaul. It was observed that fruit size varied from 8.2-16.0 g/ 100 fruits with a maximum value of 17.7 g during the 4th collection on 8th September (1995). Vertical diameter varied from 5.4 to 7.8 mm with a maximum of 8.0 mm. Cross diameter varied from 5.5 to 6.2 mm with a maximum value of 6.9 mm. The length of fruit stalk varied from 2.9 to 3.2 mm. Colour of the fruits was initially light yellowish green with both ends light red. After one month, fruits were yellow with both ends red. After 2 months, fruit were reddish yellow. Initially the fruits did not break easily during plucking of the fruits. After one month, fruit skin damaged, if not collected with extra care. After two months, fruit released the reddish white paste like material during plucking. The total soluble solid varied from 8.7 to 13.8% with a maximum values on 8th September, 1995.

The mineral composition (Zn, Cu, Fe, Ca, Mg, Na and K) of seabuckthorn fruit pulp and seed components were studied by Singh *et al.* (2001) in 8 populations of seabuckthorn (*H. rhamnoides* ssp. *turkestanica* and *H. salicifolia*) growing in Lahaul valley (2,600-3,200 m amsl), a dry temperate Himalayan region. In the fruit pulp of *H. rhamnoides*, Zn varied from 8.7 to 12.5 ppm; Cu, 3.1-4.9 ppm; Fe, 427-800 ppm; Ca, 2,310-3,325 ppm; Mg, 500-585 ppm; Na, 55-137 ppm and K, 8,575-10,025 ppm. In seed, Zn varied from 20.7 to 27.8 ppm; Cu, 4.1-8.4 ppm; Fe, 18.1-74.7 ppm; Ca, 480-1,380 ppm; Mg, 1,080-1,725 ppm; Na, 60-133 ppm and K, 6,225-9,350 ppm. The values of Zn, Cu, and Mg were higher in the seeds as compared to pulp, whereas other minerals (Fe, Ca and K) were higher in pulp, with the exception of Na, which was almost similar in both parts. The values for *H. salicifolia* were in general in the range of *H. rhamnoides* populations.

Gupta and Singh (2005) studied the mineral accumulation dynamics during fruit maturation, in 8 years old strands of *H. rhamnoides* ssp. *turkestanica* from Lahaul (semi-arid) and Spiti (Cold arid), *H. salicifolia*, also from Lahaul and HI-1 exotic form of *H. rhamnoides*, raised at High Land Agriculture

Research and Extension Center of CSK Himachal Pradesh Agricultural University, at Kukumseri (2,730 m amsl), in Lahaul, a dry temperate region of Himalayan state of Himachal Pradesh. For the present study, the fruits were collected during first week of September (I collection), third week of September (II collection), and first week of October (III collection) in the year 2003. In fruit soft part (pulp), K ranged from 4,200 to 7,000 ppm, Na (200-1,000 ppm), P (300-1,175 ppm), Fe (80-420 ppm), Zn (5-36 ppm), Mn (13-34 ppm), Mg (220-650 ppm), S (250-1,700 ppm) and B (8-90 ppm). In pulp, content of K was maximum of 7,000 ppm during II collection in HI-1. Na was maximum (1,000 ppm) both in Lahaul and HI-1 forms during III collections. P was maximum (1,175 ppm) in I and II collections in Lahaul and *H. salicifolia* forms and I collection of HI-1. Fe was maximum (420 ppm) in I collection of Lahaul form. Mn was maximum (34 ppm) in I collection of HI-1. Mg was maximum (650 ppm) in I collection of HI-1. S peaked at 1,700 ppm in II collection of HI-1. Content of B was found to be maximum (90 ppm) in the III collection of *H. salicifolia*. In seeds, K ranged from 2,200 to 4,000 ppm, Na (100-600 ppm), P (1,425-3,725 ppm), Fe (50-160 ppm), Zn (15-57 ppm), Mn (20-45 ppm), Mg (640-960 ppm), S (1,200-2,250 ppm) and B (37-102 ppm). In seeds, K was maximum (4,000 ppm) in II collection of Spiti form. Na was maximum (600 ppm) during III collection in HI-1. P was maximum (3,725 ppm) in II collection of Spiti form. Fe was maximum (160 ppm), in I collections of both HI-1 and *H. salicifolia*. Mn was maximum (45 ppm) during III collection of Lahaul form. Mg was maximum (960 ppm) in I collection of Spiti form. S was maximum (2,250 ppm) in I collection of Spiti form. B was maximum (102 ppm) in I collection of Lahaul form. Zn was found to be almost similar in both pulp and seeds. In general, with fruit maturation, K, Mn, Mg, S and B increased in *H. salicifolia* fruit pulp. In all the four forms studied, P and Fe decreased with the fruit maturation. In seeds, the general trend with fruit maturation was an increase of K, Fe, Zn, Mn and Mg in Lahaul form. Na, P, Zn, Mg, Mn and B increased in HI-1. In Spiti form Fe, Zn and Mn increased with the fruit maturation. In *H. salicifolia* K, P, Mg and B increased. Overall, the fruit pulp was more rich in K, Na and Fe, whereas seeds were found to be more rich in P, Mg, Mn, S and B.

4.3 INTRODUCTION OF EXOTIC ACCESSIONS

Five exotic improved forms of seabuckthorn have been introduced from source country by Singh (2003) at Kukumseri in Lahaul. The first exotic form "H1-1" introduced in 1995, after 7 years of growth, had a diameter of 2.8 cm at base, 178 cm of height, a narrow canopy spread of 29 cm. This form had the fruit weight of 30.5 g/100 fruits. The two other exotics HI-2 and HI-3, after 4 years of growth had the heights of 37 and 52 cm and canopy spread of

6.5 and 12 cm, respectively. The other two exotics HI-4 and HI-5 introduced in 2001, have also adapted to the local conditions. No decease has been recorded in any exotic form introduced at the center. All the forms were nearly thornless, except HI-4. Further 4 exotics have been introduced in 2004 from source country at the Kukumseri farm. These 9 exotic forms have quite high potential in raising promising orchards and developing hybrid forms for the region.

4.3.1 Evaluation of Genotypes

Singh and Singh (2001) found that genotypes of seabuckthorn from Spiti (cold arid region) showed more vigorous growth (133-167 cm) than those of Lahaul, a semi-arid region (132-158 cm) after 5 years of growth. Growth of exotic form 'HI-1' was medium (145 cm). Conopy of exotic form was significantly narrow (23.6 and 20.6 cm), as compared to the genotypes of Lahaul (34.1-53.8 cm and 24.8-63.0 cm) and Spiti (23.6-36.4 cm and 20.6-47.5 cm). Exotic form was significantly ($P<0.05$) lesser thorny (1.3 thorns/10 cm of 2 yr old branch) than the genotypes of Lahaul (2.7-3.5 thorns) and Spiti (3.3-4.2 thorns). Fruit size was maximum in case for exotic form (35.1 g/100 fruits) followed by Spiti (12.6-17.8 g) and Lahaul (12.4-15.2 g). Fruits stalk was also maximum in case of exotic form (3.4 mm) than local forms (1.1-2.5 mm). In most of the local genotypes, fruit colour was reddish-orange, whereas it was yellowish-orange for the exotic form. Total soluble solid was significantly ($p<0.05$) higher in the genotypes of Spiti (10.6-12.1%) than those of Lahaul (8.8-9.8%) and exotic form (9.5%). The HI-1's fruit also had the maximum oil content (5.2 %), which was significantly higher than local genotypes (1.7-4.3 %).

Recently, another study on the biochemical evaluation (vitamin C, carotenoids, vitamin E and flavonols) was carried out in the 2 elite forms (one each from Lahaul, a semi-arid region and Spiti, an arid region) of *H. rhamnoides* ssp. *turkestanica* and one elite form of *H. salicifolia,* also from Lahaul and HI-1, an exotic form of *H. rhamnoides,* all raised at the farm area of High Mountain Agricultural Research and Extension Centre of the University at Kukumseri (2,730 m amsl) located in district Lahaul-Spiti of Himachal Pradesh. Bioactive substances were analyzed in the fruits of 4 elite forms during first week and third week of September and first week of October 2003 and 2004. *H. salicifolia* is a tree form (4-9 m) and other forms are shrubby in appearance (1.5-3 m). Fruit colour ranged from reddish orange in Lahaul and Spiti forms of *H. rhamnoides* to yellow in *H. salicifolia* and yellowish-orange in HI-1. Fruit weight varied from a maximum of 32.8 g/100 fruits in HI-1, followed by 26.8 g in *H. salicifolia*, 12.8 g in Lahaul form and 12.1 g in Spiti form. HI-1 was nearly thornless, whereas *H. salicifolia* was mild thorny and other two forms were highly thorny. In the fruits collected and analyzed for fatty acids in 2003, among the saturated fatty acids, Palmitic acid is the most prominent and values varied from a maximum of 30.7-35.8% of total pulp oil in Spiti form, values being lesser and equal in Lahaul and *H. salicifolia* forms. Palmitoleic acid is the most dominant unsaturated fatty acid and its value being much higher in *H. rhamnoides* forms (45.6-52.9% of total pulp oil) than *H. salicifolia* (32.9-41.0%). Among the other unsaturated fatty acids, Linoleic acid was maximum in *H. salicifolia* pulp oil (12.7-16.1%), and minimum in Spiti form (3.4-8.2%). Linolenic acid was maximum in Lahaul form (0.8-2.6%) and minimum in Spiti form (0.8%). Arachidonic acid was also found in *H. salicifolia* (0.7-1.0%) (Singh *et al.*, 2004).

Fruits were again collected in 2004 and analyzed for physico-chemical characteristics. The saponification value of oil was 203. Iodine value was 65 and acid value was 0.496. FFA in terms of oleic acid was 0.249%. UV-Visible absorption was found to be 449 nm, 283 nm and 223 nm. The oil colour was 26R+20Y+6B (through Lovibond Tintometer) (Singh and Sawhney, 2005). Vitamin C value of seabuckthorn fruits ranged from 1,658 to 27,400 mg/kg. Vitamin C varied from 2,305 to 4,877 mg in Lahaul form. In Spiti form, it varied from 1,444 to 1,866 mg. In case of HI-1, its value varied from 2,110 to 4,423 mg. It varied from 24,670 to 29,840 mg/kg in *H. salicifolia.* Vitamin C was maximum in *H. salicifolia* (27,400 mg/kg), followed by 3,368.7 mg in Lahaul form, followed by 2,982.7 mg in HI-1 form, and a minimum of 1,658.7 mg in Spiti (Singh and Sawhney, 2005).

In Lahaul form, carotenoids' content varied from 118 to 171 mg. In Spiti form, it varied from 108 to 248 mg. In HI-1 its value varied from 220.9 to 446.5 mg. In case of *H. salicifolia* the value of carotenoides varied from 7 to 20 mg/kg. It was maximum in HI-1 (3237 mg/kg), followed by 163.67 mg in Spiti, 156.7 mg in *H. salicifolia* form, and a minimum of 145.3 mg in Lahaul form (Singh and Sawhney, 2005). Vitamin E value ranged from 33.1 to 1,467 mg/kg. In Lahaul form, its value varied from 100.5 to 109.9 mg. In Spiti form, it varied from 73.4 to 108.2 mg. In HI-1, its value varied from 1,347 to 1,536 mg. In *H. salicifolia,* it varied from 31.9 to 34.8 mg/Kg. Vitamin E was maximum in HI-1 form (1,467 mg/Kg), followed by Lahaul form (106 mg), 88.4 mg in Spiti form and minimum of 33.1 mg in *H. salicifolia.* á T1 was major tocopherol, followed by á T3 (Singh and Sawhney, 2005).

Value of flavonols ranged from 153 to 350.3 mg/kg. In Lahaul form, it varied from 208 to 308 mg. In Spiti form, its value varied from 122 to 180 mg. It varied from 270 to 428 mg/kg in *H. salicifolia.* In HI-1 it varied from 210 to 270 mg. Flavonols was maximum in *H. salicifolia* (350.3 mg), followed by

Lahaul form (265 mg), 238.3 mg in HI-1, and minimum of 152.7 mg in Spiti form (Singh and Sawhney, 2005).

H. salicifolia form has been selected due for high vitamin C values, large fruit and few thorns. Some exotics of *H. rhamnodes* have also been found promising for the cultivation programme in the region. As the local forms are highly thorny with low fruit yield and susceptible to pests, in order to find out suitable forms of *H. rhamnoides* ssp. *turkestanica* for cultivation, further studies are being carried out for the selection of suitable genotypes from the seabuckthorn resources growing in Ladakh and Uttranchal and other Himalayan regions of India.

4.3.2 Germination

The germination behaviour of seeds of 11 natural populations of *H. rhamnoides*, the most dominant and widely distributed species and one each of *H. salicifolia* and *H. tibetana* also growing in some sites of district Lahaul-Spiti and border area of adjacent district of Kinnaur was studied by Singh *et al.* (2005). The seeds of seabuckthorn growing in arid region of Spiti showed better germination rate (98-100%) than those of semi-arid region of Lahaul (21-97%) (2,600-3,200 m amsl). Among the species, *H. tibetana* showed a maximum rate of seed germination (100%) followed by *H. rhamnoides* (21-100%) and *H. salicifolia* (37%). Among the species, it is also clear that after 14 days, biotype of *H. tibetana* attained the maximum length of 8.71 cm, which is significantly higher ($P<0.05$) than those of *H. rhamnoides* (4.3-6.5 cm) and *H. salicifolia* (3.7 cm). Seedlings of all the biotypes had longer shoots than roots, with the exception of *H. tibetana*. Seeds with brown colour, oval, ovate and oblong shapes showed better growth than the cylindrical, elliptical, with the exception of *H. tibetana*. In general, seeds of medium to large size had better growth than the small seeds.

4.3.3 Storage of Fruit

Studies on post harvest storage of fruit was carried out by Paul *et al.* (2004) in 4 containers, i.e., paper bags, cloth bags, polythene bags (air filled and punched) and under 3 storage conditions, i.e., room temperature, incubator (25°C) and refrigerator (4°C). After 20 days of experiment, seabuckthorn fruits in small amount in paper bags under refrigerator showed better storage for longer duration, followed by polythene bags (punched and air filed). A total 29 fungi were recorded growing on the fruits of seabuckthorn. Among them, *Fusarium moniliforme, Alternaria tenuis, Phoma, Aspergillus* and *Penicillium* species were found associated with shrunk and spoiled fruits.

4.4 FODDER

In the month of July, crude protein content in the leaves of 5 biotypes of *H. rhamnoides,* varied from 17.5 to 20.5%, which were significantly ($P<0.05$) lesser than *H. salicifolia* (21.6%). Fat content also varied from 3.5–4.8% in *H. rhamnoides* to 4.6% in *H. salicifolia.* In September month, while there was no definite pattern in CP content, in general, fat content increased. NDF values varied from 27.1-31.6% in the leaves of *H. rhamnoides* to 26.2% in *H. salicifolia* in July month. The ash content also varied from 5.3-8.1% in *H. rhamnoides* to 5.1% in *H. salicifolia.* In general, the NDF content decreased considerably in September, whereas the ash content increased (Singh *et al.*, 1999).

In another study by Singh *et al.* (2001), crude protein content in the leaves of Indian and exotic form (HI-1) was similar in May month (30.4% and 29.3%, respectively) and it declined in October month to a minimum of 22.8% and 22.3%, respectively. EE value increased with time and it was initially higher in local form (3.5%) than the exotic form (2.5%), however it increased more significantly ($P<0.05$) in the exotic form (9.6%) than the local form (5.6%) in October month. High contents of crude protein and EE in the leaves make both forms of seabuckthorn a useful fodder for the livestock of cold desert Himalayas. Seabuckthorn leaves should be preserved as quality feed for livestock during winter. Dwivedi *et al.* (2003) also estimated 18.7-22.8% crude protein, 1.4-5.3% fat, 8.8-11.7% crude fibre, 6.7-8.2% total ash, 0.16-0.23% P, 1.5-2.2% Ca, 10 mg/kg Cu, 23.5 mg Zn, 8.2 mg/100 g vitamin E, 1.8% flavonoids and 496 mg/100g Fe in the leaves of seabuckthorn growing in Ladakh. These studies show that leaves of seabuckthorn are a quality fodder with high nutritional values. The leaves are also rich source of flavonoids (1.81%).

Singh *et al.* (2001), working at IVRI regional center at Palampur, found that nitrogen bound to NDF was 40.79% of the total nitrogen while the fraction of nitrogen linked to ADF (considered to be unavailable) was 9.21% indicating that around 30% of the cell wall bound nitrogen was available for ruminal degradation. The content of total tannins, hydrolysable tannins, condensed tannins and fiber bound condensed tannin was 9.9, 8.8, 1.1 and 0.54 per cent, respectively. Degradation kinetic parameters as studied by the *in vitro* gas production technique showed that the potential gas production from dry matter of leaves was 74.15 ml/200 mg while the rate of degradation was 0.0614. Gas production at 6, 12, and 24 hr was 31.5, 44.6 and 64.93 respectively, which increased to 30.0, 48.6 and 76.3 on addition of PEG (a tannin complexing agent) indicating the adverse effects of tannins present in seabuckthorn, which needs further studies.

4.5 INSECT-PESTS

Studies carried out by Bhagat *et al.* (2001) at Kukumseri in *vis-à-vis* Lahaul valley revealed that, the Death Hawk Moth, *Acherontia* styx (western) caused considerable damage to this plant at larval stage besides an unidentified mite species. However, some defoliating beetles belonging to sub family Melolonthinae viz. *Brahmina cariacea* (Hope), *Brahmina* sp. too have been found attacking seabuckthorn. Although the attack of these beetles was sporadic however, they reduced plant vigour and fruit bearing. *Holotrichia longipennis* Br. also caused damage to this plant at early stage of growth. It has been observed that the Indian Meal Moth, *Plodia interpunctella* (Huber) of Phycitidae family caused damage to berries of seabuckthorn in storage conditions.

4.6 MYCORRHIZAE

Extensive colonization by arbuscular mycorrhizal fungi (AMF) was studied by scientists of Institute of Himalayan Bioresources Technology (IHBT), Palampur (Gulati *et al.*, 2001) in the young feeder roots of seabuckthorn, with typical mycorrhizal structures, such as hyphal coils, arbuscules, vesicles, extra material hyphae, and chlamydospores. The extrametrical spore production was very copious. Noticeably formation of a large number of endospores was also observed, more frequently so in the root samples collected shortly after the melting of winter snow. The population densities of AMF spores were by far markedly higher in the rhizosphere soils as compared to the coil collected from non-rhizosphere/ uninhabited sites. *Glomus* spp. were found to be predominant amongst the AMF associated with the plant. Ectinomycorrhizae were not detected in the plant. Physico-chemical analysis of soil showed that the soils were silty loam to silty clay loam in texture, alkaline in reaction, high in organic carbon, and low in available N, P and K. Heavy root colonization and high rhizospheric population densities of AMF suggested a role for these fungi in the establishment of seabuckthorn in its natural ecosystems. Very intense binding of the soil particles by extrametrical hyphae also indicated a definitive function for AMF associated with seabuckthorn in maintaining soil stability in the natural habitats of the plant.

4.6.1 Soil Microflora

Grover *et al.* (2001), also working at IHBT, carried out the studies undertaken on the bacterial and fungal flora present in the soil samples collected from natural habitat of seabuckthorn growing in Lahaul and Spiti valley. Population enumeration in rhizospheric and non-rhizospheric soils, done by Soil Dilution Technique, showed positive rhizospheric influence on bacterial and fungal population. The differences obtained in population densities of these microorganisms were highly marked between rhizospheric and nor-rhizospheric soil samples, considerable differences were also obtained in population counts of both bacteria and fungi in the soil samples collected form various locations. Population densities of bacteria were substantially higher than the population densities of fungi in these soil samples. The majority of bacterial isolates cultured form the soils under cover by seabuckthorn contained rod shaped, gram positive or Gram negative, and motile or non-motile forms. A large number of the rod shaped gram positive, motile and endospore forming bacteria have been tentatively designated as Bacillus forms. A small number of coccoidal and coccobacillus forms were also detected in the soil samples. The fungi which were identified, amongst the large number of fungal isolates raised in spure culture from the soils excavated from natural growths of seabuckthorn, belonged to *Acremonium, Aspergillus, Aureobasidium, Candida, Chrysosporium, Colletotrichum, Geotrichum, Gliocladium, Paecilomyces, Papulospora, Penicillium, Torula, Trichoderma, Tricothecium, Mucor* and *Fusarium* spp.

Paul (1997) also found a new fungus *Phyllactinia miracula* from the leaves of *H. rhamnoides* growing at Kukumseri in Lahaul.

4.7 AGROFORESTRY SYSTEMS

Studies on traditional agroforestry practices were carried out by Singh (2001) in seabuckthorn (*H. salicifolia* D. Don.) stands growing in Tinu Village (3,150 m altitude) of Lahaul. Out of total area of 56 ha, cropped area is 25 ha, seabuckthorn forest 12 ha, *Salix daphanoides* 4 ha and rest is pasture land. Peas and potato are the important cash crops in the village. Area under seabuckthorn forest vegetation varies from 0.08 ha to 0.91 ha per family. Farmers cultivate seabuckthorn trees as windbreaks, shelter belts or pure stands, which meet their fuelwood requirement (180-1,500 kg because is being used for 'per' through out the text). During the green fodder's scarcity days of early winter (September to November), farmers drive their livestock to feed on the green and nutritious foliages of seabuckthorn stands. Under shade conditions of its canopies, seabuckthorn severely affects the pea and potato crops up to 4 and 2 m, respectively.

Agroforestry systems were also studied by Kuniyal *et al.* (2001) in Khoksar (3,200 m), Jahlama (3,000 m) and Hinsa (2,700 m) villages of Lahaul valley. Seabuckthorn (*H. rhamnoides*) was very important tree/shrub under indigenous agroforestry systems at Jahlama and Hinsa. In Jahlama, seabuckthorn showed 60.9% of the total density and 33.5% of the total importance value indices (IVI). Under agroforestry system, this plant species was much important, however, IVI of willow (*Salix daphnoides*) was relatively higher (49.9%). In Hinsa area, it was important after willow (in terms of density and IVI). The study reveals that seabuckthorn has pivotal role in the traditional agroforestry systems. The people are using seabuckthorn extensively in agroforestry practices, though the farming community was not aware of its economic potential. Seabuckthorn has also been a major scarcity of other sources of energy.

The allelopathy investigation was carried out in HPAU, Palampur by Upadhyaya *et al.* (2001) under laboratory conditions to assess the allelopathic effects of seabuckthorn on the germination percentage and seedling growth of pea, wheat and buckwheat. The experiment was laid out in RBD with various levels of concentrations viz. 0, 1, 2, 5, and 10% along with three replications. Water extract of seabuckthorn leaves significantly inhibited the germination percentage and seedling growth of pea, wheat and buckwheat at concentration between 1 and 10%. The effect was more pronounced with increase in level of concentrations. The root and shoot growth of pea, wheat and buckwheat were affected significantly by 2 per cent and higher concentrations of leaf extracts. The maximum inhibitory was observed in buckwheat followed by wheat and pea.

Singh *et al.* (2001) studied the pattern of root distribution in 1-2 years old nursery plants and 10-12 years old stand of seabuckthorn. Horizontal roots in 1-2 year old extended upto 65 cm, and vertical roots were 47 cm deep. 92 % of root biomass was confined in top 10 cm of soil horizon. Root system of 10-12 year old plants was well developed and extended horizontally up to 537 cm and vertically up to 127 cm. Strong root system of seabuckthorn makes it a valuable soil binder on fragile mountainous land. Surface lying horizontal roots need to be cut to alleviate the competition under the intercropping system. However, special care is to be taken during weeding under nursery conditions and pure seabuckthorn culture.

4.7.1 Biomass and Fruit Yield

Above ground biomass in a stand of seabuckthorn, selected for genetic improvement studies, varied from 0.13 kg per plant with a 4.6 cm girth and 78 cm height, to 6.6 kg in the plant with a 19 cm girth and 360 cm height. Fruit yield also varied from 10 g to 950 g per plant. Root seedlings were estimated on the horizontal roots of seabuckthorn plants growing on the river side. Number of sister plants of several generations varied from 9 in a 245 cm long root to 29 plants in 682 cm long root (Singh and Dogra, 1996). Plant showed a high rates of biomass production for fuel as well as fruit production.

4.8 PROPAGATION

Best time for seed collection was second half of September to first half of October. Seed treated with cold water for 7 days and sown at the depth of 1-2 cm soil in the furrows during the second to third week of April showed high rates (more than 50%) of seed germination. Cuttings of 1 cm diameter and 20 cm length, collected from sun facing young plants, when treated with 500 ppm Naphthalene Acetic Acid (NAA) and planted in the nursery bed with a soil mixture of soil+sand+FYM (2:1:1), gave a high rate of plants survival (Singh, 1995). Singh *et al.* (1997) also found the best results of propagation in the cuttings of 20 cm length. Awasthi *et al.* (2001) in Dr. Y.S. Parmar University of Horticulture and Forestry, Solan, H.P., also favoured 500 ppm NAA treatment. Two-year-old cutting with top bud showed a success of 61.6 per cent rooting in the second year. There was very less difference in rooting percentages (2%) between two-years-old cuttings with top buds of different lengths (6″ and 9″). Polybag experiments showed that one-year-old cuttings gave the highest rooting percentages in soil + sand +FYM (2 :1:1) media during first year, while the rooting percentages increased by 37 per cent when two-year-old cuttings were taken. Three-year-old cuttings, were still higher by 17.80 per cent over one-year-old cuttings. 500 ppm NAA solution treatment to two-year-old cuttings gave the highest rooting percentages (61.10%) even after two years.

4.8.1 Micropropagation

In order to fast propagate the quality plant materials of seabuckthorn, we have been working on the development of micropropagation protocol at our university center at Bajaura (Kullu). As a prerequisite first step in the development of micropropagation protocol in seabuckthorn (*H. rhamnoides* ssp. *turkestanica*, Lahaul form), aseptic cultures from the dormant buds have been established. For getting the aseptic cultures from buds, a sequential scheme of sterilization procedures have been standardized, wherein the explant (dormant buds) was sterilized by thorough washing in 0.1% Tee-pol detergent, treatment with tetracycline solution (250 mg/100 ml DDW), 70% EtOH, and final treatment with 0.1% $HgCl_2$. This sterilization procedure resulted in contamination-free cultures without any toxicity to the explant (Singh and Sawhney, 2005).

In general, in BAP + NAA combinations, the survival rate was 80 to 90%, with multiple shoot frequency (10 to 60%) and range of multiple shoots/ explant (1-3 shoots) and average shoot length range (5.2 to 6.1 mm) on various strengths of MS (Murashige & Skoog, 1962) medium. In general in BAP +

IAA, the overall explant survival rate was 80%, with average shoot length range of 3.35 to 6.3 mm. In majority of BAP + IAA combinations, there was no multiple shoot except with multiple shoot frequency (30%) and range of multiple shoots/explant (1-2). In BAP + IBA combinations, explant survival was 90 to 100%, with multiple shoot frequency (10-40%), and range of multiple shoots/explant (1 to 2-3), and average shoot length range (4.38 to 9.6 mm). Across the different hormone combinations, overall rate of explant survival ranged between 80 to 100%, (average survival of 85.83%) with multiple shoot frequency range of 10 to 60% (average multiple shoot frequency of 21.67%) and range of multiple shoots/explant varied from 0 to 3, and average shoot length/explant of 5.79 mm. Further study is in progress on increasing multiple shoot formation in explants, rooting and hardening under green house conditions (Singh and Sawhney, 2005).

4.9 ETHNOBOTANY

An ethnobotanical survey was conducted by Lal *et al.* (2001) at IHBT, Palampur, who found that the fruits of seabuckthorn are of great medicinal value and are widely used by the *Amchis* (traditional doctors) for treating different types of ailments (like stomach disorders, lung troubles, skin diseases etc.). It is used in the form of single plant drug as well as polyherbal formulation also. The medicine is prescribed as fresh or in the form of powder/jelly. Among three species of *Hippophae* (*H. rhamnoides*, *H. salicifolia* and *H. tibetana*) found in Himalayan region, *H. rhamnoides* is dominating one and widely used by the *Amchis.* Uniyal and Uniyal (2001) discovered that Aurveda, evolved during 5000-500 BC, have described medicinal properties of seabuckthorn as Amlavetas. *Amlavetas* is well-known Ayurvedic plant in *Samhita* and *Nighantu.* During last many years, Pharmaceutical industries are using many different species in the name of Amlavetas. In North India, petioles of *Rheum emodi* and other species are being traded in place of Amlavetas. Tibetan doctors in India are using pills named as Kyuru-25, Tsowo-25, Duetse-Chesoe and Seru-25 etc. made from seabuckthorn for the treatments of lung and breathing disorders and blood related diseases (Dorjee, 2004).

4.10 CONCLUSIONS

Forms of seabuckthorn in Himachal Pradesh (India) have been found quite rich in vitamin C and oil etc. However, they are thorny and low yielding. Therefore, systematic surveys need to be conducted in other parts of Himalayas like Ladakh for the study of genetic diversity and selection of promising forms of seabuckthorn to meet the requirement of plantation programmes. Some progress has been made on the standardization of methods for the propagation and cultivation of seabuckthoirn, however techniques for the management practices, i.e. spacing, application of irrigation, fertilizers, especially organic farming, pruning, need to be done urgently to increase the productivity of seabuckthorn plantations and also to convert the wild forests into productive stands. Research studies are also required to identify new molecules in seabuckthorn fruit, leaves and bark and their mode of action. Identification of pests and their control is another area, which need special attention of researchers.

References

Awasthi, R.P., Sankhyan, H.P. & Bawa, R. 2001. Research on seabuckthorn in Spiti valley of cold desrt Himalayas. In: Seabuckthorn-A Resource for Health and Environment in Twenty First Century (V. Singh and P.K. Khosla, Eds.). *Proceedings of International Workshop on Seabuckthorn* p. 14-19, February 18-21, 2001, New Delhi, 274p.

Bhagat, R.M., Kahsyap, N.P. & Singh, V. 2001. Insect-pests associated with seabuckthorn (*Hippophae rhamnoides*). In: Seabuckthorn-A Resource for Health and Environment in Twenty First Century (V. Singh and P.K. Khosla, Eds.). *Proceedings of International Workshop on Seabuckthorn* p.133-135, February 18-21, 2001, New Delhi, 274p.

Grover, Y., Mediretta, V., Chand, R., Kaur, P., Sharma, B.M., Gulati, A. & Singh, V. 2001. Soil microflora associated with seabuckthorn. In: Seabuckthorn-A Resource for Health and Environment in Twenty First Century (V. Singh and P.K. Khosla, Eds.). *Souvenir of International Workshop on Seabuckthorn* February 18-21, 2001, New Delhi.

Gulati, A., Maitra, A., Leelaveni, A., Bhardawaj, S.K., Prasad, R. & Singh, V. 2001. Mycorrhizal Associations in Seabuckthorn (*Hippophae rhamnoides L.*) from or Laboul and Spiti Valley in Himachal Pradesh, India. In: Seabuckthorn-A Resource for Health and Environment in Twenty First Century (V.Singh and P.K.Khosla, Eds.). *Souvenir of International Workshop on Seabuckthorn* February 18-21, New Delhi.

Gupta, R.K. & Singh, V. 2005. Minerals dynamics during fruit ripening in seabuckthorn (*Hippophae* L.) growing in dry temperate Himalayas. In: The 2nd International Seabuckthorn Association Conference on "*Seabuckthorn–A Multipurposes Plant, Solution to Nutrition, Health and Environment*" August 26-29, 2005, Beijing, China.

Kuniyal, C.P., Vishvakarma, S.C.R., Kuniyal, J.C. & Singh, G.S. 2001. Seabuckthorn (*Hippophae* sp.): A promising crop for Land-restoration in the cold deserts of the western Himalayas, India. In: *Seabuckthorn-A Resource for Health and Environment in Twenty First Century* (V. Singh and P.K. Khosla, Eds.). Proceedings of International Workshop on Seabuckthorn, p. 1-6, February 18-21, New Delhi, 274p.

Lal, B., Ahuja, P.S. & Gupta, A.K. 2001. Application of seabuckthorn in Amchi System of Medicine. In: *Seabuckthorn-A Resource for Health and Environment in Twenty First Century* (V. Singh and P.K. Khosla, Eds.). *Proceedings of International Workshop on Seabuckthorn* p. 239-242, February 18-21, New Delhi, 274p.

Paul, Y.S. 1997. A new species of *Phyllactinia* from India. *Indian Phytopath.* 50 (2), 291-292.

Paul, Y.S., Thakur, B.R. & Singh, V. 2004. Post-harvest spoilage of seabuckthorn fruits under different storage conditions. *Global Seabuckthorn Research and Development* 2 (4), 1-3.

Prach, K. 1994. Vegetation succession on river gravel bars across the Himalayas, India. *Arctic and Alpine Research* 26 (4), 349-353.

Singh, B., Bhatt, T.K. & Singh, V. 2001. Nutritional evaluation of seabuckthorn (*Hippophae rhamnoides* L.) leaves as cattle feed. In: Seabuckthorn-A Resource for Health and Environment in Twenty First Century (V. Singh and P.K. Khosla, Eds.). *Proceedings of International Workshop on Seabuckthorn* p. 175-181, February 18-21, 2001, New Delhi. 274p.

Singh, R.P., Vaneet, J., Nege, D.V. & Jishtu. 1997. Studies on nursery techniques of *Hippophae rhamnoides* L. in cold desert of Himachal Pradesh, India. *Annals of Forestry* 5 (1), 35-38.

Singh, V. 2001. Fodder values of exotic seabuckthorn (*H. rhamnoides* L.). In: *Seabuckthorn-A Resource for Health and Environment in Twenty First Century* (V. Singh and P.K. Khosla, Eds.). *Souvenir of International Workshop on Seabuckthorn* February 18-21, 2001, New Delhi.

Singh, V. 2001. Traditional agroforestry practices of seabuckthorn (*H. salicifolia* D. Don.). In: *Seabuckthorn-A Resource for Health and Environment in Twenty First Century* (V. Singh and P.K. Khosla, Eds.). Proceedings of International Workshop on Seabuckthorn p.7-13, February 18-21, 2001, New Delhi, 274p.

Singh, V. 2003. Performance of some exotic forms of seabuckthorn. In: *International Congress on Seabuckthorn* p. 47-53, 14-18th September, 2003, Berlin.

Singh, V. & Nayyar, H. 2001. Germination behaviour of different biotypes of seabuckthorn growing in dry temperate Himalayas. In: Seabuckthorn-A Resource for Health and Environment in Twenty First Century (V. Singh and P.K. Khosla, Eds.). *Souvenir of International Workshop on Seabuckthorn* February 18-21, 2001, New Delhi.

Singh., V. & Sawhney, R.C. 2005. Progress Report (2004-2005) of the project "Biochemical characterization and micropropagation of Seabuckthorn (*Hippophae* L.) – A multipurpose plant of dry temperate Himalayas". CSK Himachal Pradesh Agricultural University Hill Agriculture Research and Extension Centre, Bajaura-175125 (Kullu) H. P. and Defence Institute of Physiology and Allied Sciences, (DRDO), Delhi-110054. Funded by Department of Biotechnology, Ministry of Science and Technology, New Delhi.

Singh, V. & Dogra, K.K. 1996.. Characteristics, distribution, biomass, regeneration and nutritional values of seabuckthorn. *Indian Forester* 122 (6): 486-491.

Singh, V. & Singh, R.K. 2001. Evaluation studies on the Indian and exotic forms of seabuckthorn (*Hippophae* L.). In: Seabuckthorn-A Resource for Health and Environment in Twenty First Century (V. Singh and P.K. Khosla, Eds.). *Proceeding of International Workshop on Seabuckthorn* p.107-111, February 18-21, 2001, New Delhi, 274 p.

Singh, V. & Singh, R.K. 2004. Morpho-biochemical variations in seabuckthorn (*Hippophae* L.) populations growing in Lahaul valley, dry temperate Himalayas. *Indian Forester* 130 (6), 663-672.

Singh, V. 2005. Effects of mulch, soil media and hormone on the rooting rates of hard wood cuttings of seabuckthorn. *Indian Forester* (Submitted).

Singh, V., Singh, R.K. & Awasthi, C.P. 2001. Morpho-biochemical changes during ripening of the fruit of seabuckthorn (*Hippophae rhamnoides* L.). *J. Non-Timber Forest Products* 8 (1/2), 114-116.

Singh, V., Singh, B. & Awasthi, C.P. 1995. Distribution, taxonomy and nutritional values of seabuckthorn growing in dry temperate Himalayas. In: *Proceeding of International Workshop on Seabuckthorn* p.52-59, Dec.12-17, 1995, Beijing, China, 206p.

Singh, V., Sharma, K.B., Singh, S.P., Awasthi, C.P. & Singh, B. 2001. Mineral composition of seabuckthorn (*Hippophae* L.) populations growing in Lahaul valley, dry temperate Himalayas. In: Seabuckthorn-A Resource for *Health and Environment in Twenty First Century* (V. Singh and P.K. Khosla, Eds.). Proceedings of International Workshop on Seabuckthorn p.170-174, February 18-21, 2001, New Delhi, 274p.

Singh, V., Singh, R.K. & Singh, B. 1999. Fodder values of foliage of seabuckthorn growing in Lahaul valley, dry temperate Himalayas. In: *Proceeding of International Workshop on Seabuckthorn* p. 55-59, Beijing, China, 97p.

Singh, V., Uppal, R. & Awasthi, C.P. 2001. Morpho-biochemical changes during ripening of seabuckthorn fruits. *J.Non-Timber Forest Products* 8 (1/2), 114-116.

Singh,V. 1994. A report on arboreal seabuckthorn (*Hippophae salicifolia*) from dry temperate Himalayas. *Journal of Tree Sciences* 3 (1), 67-68.

Singh,V. 1995. Rooting rates of hardwood cuttings of seabuckthorn (*Hippophae rhamnoides*). *Journal of Tree Sciences* 14 (2), 87-88.

Singh,V. 2003. Performance of some exotic forms of seabuckthorn (*Hippophae* L.) in dry temperate Himalayas, India. In: *Seabuckthorn-A Resource of Health, a Challenge to Modern Technology*. Proceedings of the 1st Congress of the International Seabuckthorn Association, p. 47-53, September 14-18, 2003, Berlin, Germany, 214p.

Singh,V., Gupta, R.K., Uppal, R., Singh, B. & Lata, S. 2005. Morphological and biochemical variations in the seeds of different biotyes of seabuckthorn (*Hippopahe* L.) growing in dry temperate Himalayas. *Non-timber Forest products* (In press).

Singh,V., Gupta, R.K., Tandon, S. & Sawhney, R.C. 2004. Fatty acid composition of fruit pulp and seeds oils of some selected forms of seabuckthorn (*Hippophae* L.). *Global Seabuckthorn Research and Development* 3 (2), 1-8.

Singh,V., Singh, R.K. & Singh, B. 1999. Fodder values of foliage of seabuckthorn (*Hippopahe rhamnoides* ssp. *turkestanica*) growing in Lahaul Valley. In: *Proc. of International Workshop on Seabuckthorn*, p. 55-59, August 29-September 2, 1999, China Science and Technology Press, Beijing, China, 97p.

Upadhyaya, R.G., Bansal, G.L., Kaushal, S.K. & Singh, V. 2001. Allelopathic effect of seabuckthorn (*Hippophae* L.) on germination and seedling growth of some important crops of Lahaul-Spiti (H.P.). In: Seabuckthorn-A Resource for Health and *Environment in Twenty First Century* (V. Singh and P.K. Khosla, Eds.). *Souvenir of International Workshop on Seabuckthorn*, February 18-21, New Delhi.

□□□

5.1

Seabuckthorn in Uttarakhand

V.K. Yadav, S.K. Sharma, V.K. Sah, V.K. Rao and Rohit Bisht

CONTENTS

Mountainous regions of the developing countries share broadly common characteristics in terms of rapid population growth, accelerated degradation of the environment and depletion of the natural resource base. Inappropriate ways of the current resource use pattern further accelerates deterioration in these areas. The state of Uttarakhand which comprises of 13 districts, 10 districts are hilly, also face similar problems, thus, under such conditions, the conservation of environment, soil and natural resources becomes utmost important. Similarly, nutritional and improving the socio-economic status of the people living in such remote and inaccessible areas is also equally important.

Among various underutilized plant genetic resources, Seabuckthorn possesses a very important place with respect to the species of the dry temperate regions of high Himalayan region. Seabuckthorn is also called Amesh, Chuk, Amil, Tarwar in different dialects of Uttarakhand state. Due to its wider adaptability, drought resistance capabilities, deep root system and capacity to fix atmospheric nitrogen through *Frankia* symbiosis, the plant has been widely planted in China to control soil erosion and reclamation of degraded lands. The leaves, berries and seeds of seabuckthorn have high nutritional and medicinal value and are excellent source of vitamins C, B_1, B_2, E, F, K, P, pro-vitamin A, sugars and organic acids. High content of vitamin C and E and other bioactive substances makes it a suitable species for pharmaceutical, cosmetic and food industry (Yadav, *et al*., 2005 and Rao, *et al*., 2006).

Although seabuckthorn grows naturally in the state of Uttarakhand but the local inhabitants are still not well-verse with this wonder plant. The fruit is utilized on a limited scale for the preparation of chutney, pickle and juice in some areas of the state. The local residents are aware about the ecological significance of plant as they plant it in controlling soil erosion and in landslide prone zones.

The followings are the extract of the various studies conducted by the authors during recent years in Uttarakhand.

5.1.1 SPECIES

Out of three species found in India, only two species, i..e., *Hippophae salicifolia* D.Don and *Hippophae tibetana* S. are reported so far from Uttarakhand. Among them *H. salicifolia* is the most common and widely distributed species. Seabuckthorn is reported to exist in abundance in three districts of Garhwal division viz., Uttarakashi, Rudraprayag and Chamoli and in Pithoragarh district of Kumaon division. *H. salicifolia* is reported from Gangotri, Harsil, Yamunotri valley and Har-Ki- Dun area of district Uttarakashi; Hanumanchatti, Badrinath area, Jhelum- Malari area of Niti valley in district Chamoli and Rambara-Gaurikund area in district Rudraprayag of Garhwal division. In Kumaon division this species is reported from Kali valley, Bogdiar, Gori valley, Budhi, Byanse, Darma area in district Pithoragarh. However, *H. tibetana* was reported from Raini, Niti, Gomukh area of Garhwal, while, Gori valley, Milam and Shin-La are other areas of Kumaon division of the state (Yadav *et al.*, 2006 c). As such, *H. rhamnoides* L. which is common in other parts of the seabuckthorn growing areas of the country was not at all reported by our team during the surveys conducted through ground truth, during 2004-07. Dwivedi *et. al.* (2006) also reported the above two species form various parts of the state.

5.1.2 AREA AND DISTRIBUTION

At present seabuckthorn is growing wild in different parts of the state and as such no systematic domestication work of any kind is available on the species. The survey of naturally habitat seabuckthorn areas by the researchers started in 2004 and the available information reveal that the estimated area in different parts of the state is about 3750 ha which occurs naturally between 1700-3000 m amsl (Table1). The figures may further increase on the availability of data from the inaccessible and remote areas.

Table 5.1.1: Estimated areas of seabuckthorn in Uttarakhand

District	Location	*Approximate area (ha)
Uttarkashi	1. Har-ki- Dun	100
	2. Yamunotri valley a. Hanumanchatti- Bhandaligad (On the bank of Yamuna) b. Kharsali- Digdalagad c. Hanuman chatti-On the bank of Hanuman Ganga (3 Kms) 3. Gangotri valley a. Dabrani- Lanka (on the bank of Bhagirathi and both side of National highway and adjoining areas)	1200 600
Rudraprayag	Gaurikund- Rambana	300
Chamoli	1. Mana Valley a. Lambagad- Badrinath b. Gobind ghat - Ghagharia 2. Niti Valley a. Suraithota- Malari (on the bank of Vishnu Ganga and Dornagiri hill in village Jhelum	100 200 600 200
Pithoragarh	Munsiari (Kali and Gori river valleys), Dharchula	450
	Total	**3750**

*Ocular estimation of surveyed area only

5.1.3 GENETIC DIVERSITY

The diversity available in the naturally growing populations in Uttarakhand was estimated and documented. On the basis of preliminary study, wide variations were observed for different morphological traits viz., plant form, number of fruits per branch, fruit size, fruit weight, fruit shape, juice content, leaf area, seed size, presence of thorns etc. (Yadav, *et al.*, 2006 a). The variations were also recorded for different biochemical traits viz., TSS, acidity (%), ascorbic acid content and sugar content (Table 2) in fruits (Yadav, *et al*, 2006 b). However, more comprehensive survey of seabuckthorn in Uttarakhand needs to be required to select some promising forms.

Table 5.1.2: Diversity in plant morphological characteristics of seabuckthorn growing in Garhwal Himalayas

Location	No. of leaves/ 10 cm branch length	Leaf area (cm^2)	Fruit Length (mm)	Fruit Dia. mm)	100 fruit weight (g)	N. of fruits/ 10 cm branch length	Fruit yield/ 10 cm branch length (g)
Gangotri	10.84	6.26	7.20	7.04	13.53	12.61	1.72
Bhairon Ghati	11.83	9.35	7.07	5.92	12.57	10.69	1.34
Dharali	17.79	4.55	6.95	5.94	13.34	34.26	4.57
Harsil	11.50	6.01	6.59	6.49	16.44	19.00	3.12
Harsil A	11.83	6.65	6.84	6.78	18.87	68.22	12.87
Chormi	24.17	2.28	6.95	6.07	14.07	38.00	5.35
Mukhawa	16.57	3.48	7.53	6.92	18.34	87.14	15.98
Jhala	19.83	5.23	6.41	6.44	15.21	22.89	3.48
Chhormee	21.92	3.43	7.92	6.68	18.62	44.03	8.19
Kachora	28.42	3.44	6.38	6.63	17.19	45.14	7.76
Jaspur	21.58	4.42	7.51	6.30	15.59	99.40	15.49
Jaspur A	20.57	7.81	7.14	6.55	17.35	84.27	14.60
Sukhi	16.75	3.32	7.05	6.51	14.89	76.88	11.45
Sukhi Top	13.67	2.88	6.10	6.51	17.22	34.07	5.86
Sukhi Down	29.83	2.88	5.78	5.55	13.94	48.68	6.79
Sonagad	22.17	6.03	6.54	7.24	18.64	76.37	14.24
Sonagad A	20.90	3.83	7.01	5.51	11.53	49.83	5.74
Dabrani	23.58	3.40	6.30	6.25	14.08	52.86	7.44
Gagnani	13.00	3.72	6.69	6.35	14.17	34.40	4.87
$CD_{0.05}$	1.05	0.22	0.26	0.54	0.61	9.27	1.43

Adapted from (Yadav *et al.*, 2006 a)

Table 5.1.3: Diversity in fruit physical characteristics of seabuckthorn growing in Garhwal Himalayas

Location	Fruit shape	Fruit colour	Juice (%)	Seed (%)	Residue (%)	TSS (° Brix)	Acidity (%)	TSS : Acid ratio
Gangotri	Round	Yellow	78.23	7.87	13.90	9.80	7.93	1.24
Bhairon Ghati	Round	Yellow	71.45	7.77	20.78	10.00	5.68	1.76
Dharali	Round	Yellow	74.30	7.60	18.10	8.93	5.38	1.66
Harsil	Round	Greenish yellow	77.43	6.35	16.22	9.13	5.62	1.63
Harsil A	Round	Greenish yellow	78.50	7.14	14.36	9.83	6.53	1.51
Chormi	Round	Yellow	74.95	8.25	16.80	9.00	5.66	1.59
Mukhawa	Round	Yellow	77.05	7.74	15.21	8.13	5.83	1.40
Jhala	Round	Yellow orange	70.70	6.78	22.52	9.87	5.84	1.69
Chhormee	Ovate	Yellow orange	73.33	7.66	19.01	9.20	6.28	1.47
Kachora	Round	Yellow orange	74.95	6.15	18.90	7.80	5.49	1.42
Jaspur	Ovate	Greenish Yellow	74.82	7.15	18.03	7.67	5.59	1.37
Jaspur A	Round	Yellow	79.23	7.20	13.57	8.23	6.09	1.34
Sukhi	Round	Yellow	77.96	7.14	14.90	8.33	5.90	1.41
Sukhi Top	Round	Yellow	74.50	5.90	19.60	6.93	4.89	1.42
Sukhi Down	Round	Yellow	70.94	6.63	22.25	7.93	5.31	1.50
Sonagad	Ovate	Greenish Yellow	70.83	7.77	21.40	8.87	6.43	1.38
Sonagad A	Round	Yellow	75.34	7.87	16.79	9.07	6.02	1.52
Dabrani	Round	Yellow orange	73.87	7.12	19.01	7.07	4.92	1.44
Gagnani	Round	Yellow orange	73.58	7.22	19.20	8.13	6.15	1.32
$CD_{0.05}$	-	-	0.86	0.28	0.44	0.22	0.14	0.04

Adapted from (Yadav *et al.*, 2006 a)

5.1.4 SELECTION OF PLUS TREES/ STANDS

On the basis of morphological and biochemical traits viz., fruit size, thorns, juice yield, fruit colour, chemical quality etc. ten plus trees/stands which were selected out of many different marked stands of seabuckthorn in various portion of the areas. Areas surveyed and stands marked include all three districts, i.e., Uttarkashi, Chamoli and Rudrapyrag of Grahwal division. Out of the ten 10 plus tree/ stands,4 are located in Gangotri valley, two each in Yamunotri area, Hanuman Chatti area of Badrinath and 2 in Niti valley area.

Table 5.1.4: Chemical characteristics of selected plus stand/trees

S. No.	Sample Name	TSS	% Acidity	Ascorbic Acid (mg/100gm)	Sugar %
1.	Gangotri Selection 1	8.0	5.57	409.20	1.42
2.	Harsil Selection 1	7.5	5.38	228.80	1.50
3	Dharali, Selection 1	8.0	5.70	484.00	1.61
4	Badrinath Selection 1	6.0	4.42	365.20	0.88
5	Badrinath (Mana) Selection 2	8.0	6.46	391.60	1.25
6	Yamnotri Selection 1	10.0	6.72	462.00	1.02
7	Jhelum Selection 1	9.5	6.72	400.40	0.97
8	Suraithota Selection 1	6.0	4.67	316.80	0.62
9	Yamunotri Selection 2	6.5	4.74	330.00	1.19
10.	Badrinath Selection 3	7.5	5.57	224.40	1.58

5.1.5 PROPAGATION

A considerable amount of work has been done in parts of Jammu and Kashmir and Himachal Pradesh on various aspects of propagation (Singh and Gupta 2003). Propagation through vegetative means is easy in seabuckthorn and we can achieve good success if the work has to be carried out in its natural habitat. Since, the natural habitat of seabuckthorn in Uttarakhand are generally inaccessible areas that remain devoid of human population for almost 6 months during winters, the work done on propagation and domestication of this plant in the state is negligible. Further, it was found that when the semi hardwood cuttings which were collected in the month of November, and planted elsewhere i.e. at Ranichauri (2000 m amsl), the survival achieved was low 20 per cent. The success percentage was higher, i.e., 60 per cent, when the cuttings were taken during March. On the basis of results obtained so far, it was observed that treatment of cuttings with IBA @ 400 ppm gives more rooting as compared to control. Best time for the collection of cuttings was found to be March (Yadav, *et al.,* 2006 b).

The propagation by seeds can be successful, if the seedlings can be utilized for the purpose of propagation by asexual methods. Highest germination percentage of seeds was recorded when seeds were treated with IAA @ 200 ppm followed by the treatment with GA_3 @ 400 ppm (Bisht *et al* .2007b). The survival percentage was found highest in stratification treatment although was at par with some of the hormonal treatments, i.e., GA @ 200, 400 ppm, IAA @ 100, 200 ppm and IBA 200, 400 ppm (Table 5).

5.1.6 HARVESTING

The importance of the fruits of seabuckthorn is unquestionable but the collection of small sized berries poses a great difficulty. Although, there are some methods reported for the harvesting of berries in Mongolia (Avdai and Chimed-Ochir, 2003), but most of the methods used are not very much applicable under our conditions. The natural stands in the state are located on steep slopes in inaccessible terrains and as such it is not possible to use any machine or spread any cloth or net below the stands for the collection of fruits. It is not even possible to comb the branches and collect the berries without damaging the plants. Further, the presence of thorns on the plants makes harvesting more difficult.

Under such condition, the most common practice that is followed for harvesting of berries is cutting back the branches from the natural stands and bringing them to the laboratory, home or processing unit and removing the berries by hand. The harvesting is difficult when the berries are ripe and they are rupture while their removal from the branches. Harvesting of fruits before sunrise, when the atmospheric temperature is low, was also tried and it gave slightly better results. By this method a person can harvest about 4 to 5 Kg fruits /hour.

Since, most of the natural stands of seabuckthorn are present in over steep slopes, the problem of harvesting can solved through only be obtained in the plantations on a flat surface where mechanization is possible.

5.1.7 VALUE ADDITION

At present this fruit is not being utilized for the preparation of any value added products in Uttarakhand except for the preparation of chutney and pickles that too on a very limited scale in some of the areas of the state. Unlike the state of Jammu and Kashmir, no fruit processing units have been established so far in the state for the commercial production of juices and other value added products. The important reasons for the above may be unawareness of the people about the benefits and uses of this wonder fruit. Therefore, there is an urgent need to impart demonstrations and trainings to the local residents, in order to increase the awareness about this fruit. The

Table 5.1.5: Effect of different seed treatments on germination (%), survival (%) and number of leaves of seabuckthorn seedlings

Treatments		Germination (%)				Survival (%)			No. of leaves		
		Days after sowing				Days after germination			Days after germination		
		10	20	30	Mean	90	120	Mean	90	120	Mean
T_1	GA_3 100 ppm	18.33	26.67	56.00	33.67	80.23	40.68	60.45	9.66	17.47	13.57
T_2	GA_3 200 ppm	22.33	52.00	65.33	46.56	90.58	62.96	76.77	9.46	22.13	15.80
T_3	GA_3 400 ppm	29.00	58.33	66.67	51.33	92.00	67.68	79.84	9.60	23.13	16.37
T_4	IAA 100 ppm	26.00	46.00	60.00	44.00	83.93	57.30	70.62	10.07	24.73	17.40
T_5	IAA 200 ppm	33.33	61.67	71.33	55.44	87.09	58.37	72.73	11.47	22.27	16.87
T_6	IAA 400 ppm	8.33	33.67	54.00	32.00	81.00	45.19	63.10	9.00	18.47	13.73
T_7	IBA 100 ppm	8.66	46.33	50.00	35.00	88.80	44.28	66.54	7.60	19.80	13.70
T_8	IBA 200 ppm	25.00	50.00	50.00	41.67	85.55	61.30	73.42	9.33	22.60	15.97
T_9	IBA 400 ppm	22.67	51.33	51.33	41.78	89.34	65.99	77.66	10.00	18.27	14.13
T_{10}	Kinetin 100 ppm	17.67	40.67	41.67	33.33	88.40	52.15	70.27	8.93	26.47	17.70
T_{11}	Kinetin 200 ppm	20.33	44.67	44.67	36.56	90.84	67.76	79.30	8.50	24.27	16.38
T_{12}	Kinetin 400 ppm	5.00	18.00	23.33	15.44	92.85	46.46	69.66	8.00	23.27	17.60
T_{13}	KNO_3 1%	12.00	25.67	43.33	27.00	89.82	57.19	73.50	8.80	16.70	12.75
T_{14}	$MnSO_4$ 1%	18.00	36.00	45.00	33.00	96.87	19.51	58.19	11.07	21.93	16.50
T_{15}	KCl 1%	24.67	40.00	40.00	34.89	74.51	54.02	64.26	11.13	22.40	16.77
T_{16}	Lukewarm water	25.67	37.67	43.00	35.44	97.66	27.93	62.79	12.27	21.33	16.80
T_{17}	Stratification	20.00	29.00	33.33	27.44	96.25	69.76	83.00	10.80	24.40	17.60
T_{18}	Control	9.33	16.67	16.67	14.22	87.47	13.18	50.32	7.26	18.00	12.63
Mean		19.24	39.69	47.54	-	88.51	50.65	-	9.60	21.53	-
		$CD_{0.05}$				$CD_{0.05}$			$CD_{0.05}$		
Treatment (T)		5.80				17.22			NS		
Days after sowing (I)		2.37				5.74			1.21		
T x I		10.06				24.35			NS		

Adapted from Bisht *et al* .(2007b)

technology for the preparation of ready to serve and squash beverages was standardized using *H. salicifolia*. It was found that 5 % juice and 14° Brix total soluble solids were found to be the best on the basis of sensory quality for preparation of RTS beverages and 25 % juice / pulp and 50° Brix TSS were optimum for the preparation of good quality squash (Table 6 & 7). Also, both types of beverages were shelf consumable up to six months of storage both at room as well as refrigerated conditions showing only slight changes in sugars and sensory quality. (Bisht *et al.* 2007 a & c). However, the ascorbic acid of the prepared products underwent more than 50 per cent reduction during 6 months of storage, although these changes were slightly lesser in the products stored in refrigerated conditions as compared to their counterparts stored at ambient conditions.

Table 5.1.6: Sensory evaluation (overall acceptability) of different seabuckthorn RTS beverages on 9 point hedonic scale for standardization of recipe

Pulp (%)	**T.S.S. (°Brix)**			**Mean**
	10	**12**	**14**	
5	6.12	7.25	8.00	7.12
7.5	5.50	5.75	6.25	5.83
10	4.00	4.50	5.75	4.75
Mean	5.20	5.83	6.66	
			$CD_{0.05}$	
		Pulp (P)	0.89	
		TSS (T)	0.89	
		P x T	NS	

Table 5.1.7: Sensory evaluation (overall acceptability) of different seabuckthorn squash beverages on 9 point hedonic scale for standardization of recipe

Pulp (%)	**T.S.S. (°Brix)**					**Mean**
	40	**45**	**50**	**55**	**60**	
25	6.17	6.78	7.45	7.21	7.14	6.95
27.5	5.92	6.88	7.12	6.68	6.25	6.57
30	5.28	5.81	6.84	7.05	6.54	6.30
Mean	5.79	6.49	7.14	6.98	6.64	
				$CD_{0.05}$		
			Pulp (P)	0.45		
			TSS (T)	0.58		
			P x T	NS		

Table 5.1.8 : Effect of storage temperature on total sugars, ascorbic acid and overall acceptability of seabuckthorn RTS beverage

Storage Interval (months)	Total sugars (%),			Ascorbic acid (mg/100 ml)			Overall acceptability		
	Ambient	Refrigerated	Mean	Ambient	Refrigerated	Mean	Ambient	Refrigerated	Mean
Initial	11.83	11.83	11.83	24.00	24.00	24.00	8.10	8.10	8.10
2	10.65	10.65	10.65	15.00	24.00	19.50	7.53	7.53	7.53
4	10.60	10.61	10.61	12.80	14.72	13.76	6.80	7.47	7.14
6	10.01	10.40	10.20	5.19	6.12	5.65	6.31	7.01	6.66
Mean	10.77	10.87	-	14.25	17.21	-	7.18	7.53	-
		$CD_{0.05}$			$CD_{0.05}$			$CD_{0.05}$	
Storage condition (T)		0.02			0.62			0.04	
Storage Intervals (I)		0.01			0.87			0.06	
T X I		0.02			1.23			0.08	

Table 5.1.9: Effect of storage temperature on total sugars, ascorbic acid and overall acceptability of seabuckthorn squash beverages

Storage Interval (months)	Total sugars (%),			Ascorbic acid (mg/100 ml)			Overall acceptability		
	Ambient	Refrigerated	Mean	Ambient	Refrigerated	Mean	Ambient	Refrigerated	Mean
Initial	45.45	45.45	45.45	137.50	137.50	137.50	8.10	8.10	8.10
2	44.44	44.44	44.44	50.00	80.00	65.00	7.61	7.61	7.61
4	44.40	44.44	44.42	40.20	68.80	54.50	7.26	7.20	7.23
6	44.04	44.24	44.14	27.20	40.26	33.73	6.30	7.10	6.70
Mean	45.58	44.64	-	63.72	81.64	-	7.31	7.50	-
		$CD_{0.05}$			$CD_{0.05}$			$CD_{0.05}$	
Storage condition (T)		0.008			0.44			0.05	
Storage Intervals (I)		0.010			0.62			0.06	
T X I		0.017			0.88			0.09	

5.1.8 INSECT PESTS AND DISEASES

During a survey of nurseries in 2005-06 in the hilly region of Uttarakhand, seabuckthorn plants with root rot caused by *Rhizoctonia solani* Kuhn. were found in Gangotri valley and in nurseries at Ranichauri. Severe mortality of seabuckthorn plants (20-25 %) was observed during July -August at Forestry block of University campus. The infected plants appear sick which gradually turn yellow leading to defoliation of all leaves within 2 to 3 weeks. When uprooted, collar region of dead plants revealed a decay of the bark and secondary roots. Roots were shredded and filled with white mycelial growth on the surface of freshly infected areas. *Rhizoctonia* was consistently isolated from the lesions below the bark on PDA, and on water agar amended with streptomycin sulfate. Similarly aphid attack on seabuckthorn was also reported for the first time from Garhwal Himalayas during 2007 (Yadav, *et al.*, 2007, Singh, *et al.*, 2007).

5.1.9 CONSTRAINTS IDENTIFIED

1. Natural range of occurrence in remote and inaccessible areas: Unlike Ladakh and Lahul Spiti areas, in Jammu & Kashmir and Himachal Pradesh, respectively, no human population resides in the seabuckthorn growing areas of Uttarakhand state after October-November, due to snowfall and cutting off of these areas from rest parts of the state. These people remain away till April of the following year. As a consequence of this, the actual work on the utilization and propagation is hampered due to inaccessibility.

2. Sparse natural vegetation on steep slopes: Since, seabuckthorn has still not been domesticated in Uttarakhand state, the harvesting and other operations are very difficult on steep slopes and inaccessible terrains. The use of any mechanical method is difficult and spreading of net or cloth below the stands during harvesting and collecting the fruit in it is also not possible. Therefore, cutting of branches, taking them to some distant place and then harvesting the berries is practised. Harvesting methods for our conditions are destructive and may pose future threats for existence of the species. An improvement in harvesting methods is therefore, required.

3. Presence of thorns: Most of the natural stands in the state are having thorns on the branch that makes the harvesting difficult. Although thornless populations have also been found and need to be multiplied.

4. Quality planting material: In order to domesticate the species in the state, the first and fore most requirement is the availability of quality

planting material which is not available at present. Under such conditions there is a need to propagate and multiply the superior natural stands and plant them in such locations wherein we can find the natural human population through out the year, who can take care of the plants and harvest the benefits of this fruit.

5. Low success rate with vegetative methods of propagation: Propagation of the species if done out side of its natural range, results in very low success rates with various propagation methods.
6. Poor documentation of available genetic diversity: Since a little work is reported to be conducted on this plant in Uttarakhand, the documentation of the available diversity with respect to various plant growth, fruit morphological and fruit bio-chemical parameters is scanty in literature. Such information is very much beneficial for the success of any breeding programmes.
7. Unawareness about the use and methods for utilization: People are also not aware regarding the potential uses and benefits of the fruit. Therefore the fruit is being wasted in the wilds of the state. As such no domestication of the species has been done. Although some trainings that were conducted around the natural habitat of the species were successful in increasing the level of awareness of people.
8. No work on introduction of elite germplasm has been done: The aspect of the introduction and plantation of superior germplasm from other Indian states and from abroad is still untouched in Uttarakhand state.
9. Dioecious nature of the plant: It is very difficult to differentiate between the male and female plants before flowering and fruiting. Therefore, this poses problems while multiplying the germplasm for plantation.
10. Short harvesting duration: The duration of harvesting of the crop is very short. If the fruits ripen on plant itself, the juice is squeezed out of the fruits during harvesting and is wasted. Visiting all the natural habitats of the state, simultaneously, during the maturation of the crop is a great problem.
11. Slow rate of growth: The rate of growth is very slow, so it is not possible to get immediate results.
12. Needs for standardization of methods for value added products: A number of new value added products and neutraceuticals can be prepared from the fruit, but this needs extensive research on the subject.

13. Absence of private players. Unlike Laddakh, where industries have come up for commercial processing and value addition of the fruit, such interventions are altogether missing in Uttarakhand.

5.1.10 FUTURE INTERVENTIONS REQUIRED

A. Suggested areas for plant improvement

- Estimation of available area under seabuckthorn
- Estimation of genetic diversity
- Biochemical and molecular analysis of available diversity
- Standardization of Agro-techniques for cultivation
- Efforts for accelerated domestication (thornless plants)
- Development of markers for distinguishing between male and female plants
- Selection of elite genetic material
- Standardization of vegetative propagation techniques
- Development of protocols for micro-propagation
- Intra and Inter-specific hybridization
- Mutation breeding for specific target trait
- Introduction of elite germplasm
- Estimation of genetic diversity available in *Frankia*
- Development of Agro-forestry modules by involving seabuckthorn
- Development of methodology for enhancing nitrogen fixation
- Utilization of seabuckthorn for reclamation of degraded cold barren areas

B. Suggested areas for product improvement

- Development/ standardization of novel food products
- Bioactivity profiling of available diversity
- Development of suitable technologies for harvesting of fruits
- Study on pharmacological uses of Seabuckthorn oil
- Development of pharmaceuticals and neutraceuticals
- Development of cosmetic products
- Training of local residents on preparation of different value added products
- Joint ventures with industry

Future Prospects of seabuckthorn in Uttarakhand

Seabuckthorn can play an important role in sustainable development of the existing fragile and unstable eco-system of temperate and dry temperate regions of the state. The 'wonder plant' can be utilized in fighting from malnutrition, conservation of natural resources, diversification of horticulture and social and economical empowerment of inhabitants of the seabuckthorn growing areas. Moreover, if seabuckthorn cultivation is started over large areas in the state it can also take lead in the supply of raw material to seabuckthorn industries. A vast potential of seabuckthorn in the state lies in following areas:

1. Nutritional Security

Similar to other hilly states of the country, Uttarakhand also shares large inaccessible remote areas. This seems is misfortune of the people that they are not aware about nutritional and medicinal power of this important plant growing in their vicinity. The proper utilization of this extraordinary fruit, which is found in these areas, possesses great potential for meeting out the nutritional demands of people especially for vitamins, minerals and other bio active compounds.

2. Rehabilitation of degraded land

In different parts of the world, seabuckthorn has been used successfully for the rehabilitation and reclamation poor and problematic lands. In India seabuckthorn is found most useful in greening of cold arid regions of Ladakh, where no other species thrives well. In temperate and dry temperate region of Uttarakhand major portion of the land is under waste land (Table 10), where only selected species can be grown. Due to its proven ability for fixation of atmospheric nitrogen and addition of organic matter, seabuckthorn seems to be ideal plant for the purpose.

Similarly, establishment of various Hydro-Electric Projects in the state also caused massive losses to natural resources and areas become more prone to land sliding and erosion. In such areas restoration of natural resources is possible by reforestation by an appropriate species such as seabuckthorn.

Table 5.1.10: Details of degraded barren land in Seabuckthorn habitated districtS

District	Chamoli	Uttarakashi	Pithoragarh
Total wasteland (in sq. Km)	4[illegible]92.64	4564.33	4141.05
Per cent of total geographical area	49.23	56.94	58.40

Source : www.dolr.nic.in/wasteland

3. Self employment to un-employed youths

One of the major advantages in Uttarakhand as compared to other seabuckthorn growing areas of the country is that majority of the natural seabuckthorn populations are located on the *Yatra Routes*, i.e., on way to Gangotri, Yamunotri, Badrinath, Kedarnath and Hemkund Sahib. This offers a great opportunity to the local residents around these *yatra routes* to process the fruit and sell their value added products to the pilgrims and tourists.

4. Seabuckthorn based industry

Other seabuckthorn areas in the country like Ladakh region of Jammu and Kashmir, Lahaul-Spiti area of Himachal Pradesh are located more distant from the plain areas of the country. Contrary to them, most of the natural seabuckthorn populations in Uttarakhand exist in near to national highways/ roads going to Gangotri, Yamunotri, Mana valley, Niti valley etc. Collection of fruit from these areas and establishment of pulping/ post harvest unit is easy. Similarly, fruit maturity time for seabuckthorn also vary, depending on the altitude of the area, thus it can supply fruits for more time than other parts of the country. At present, there are no industries operating on the processing and value addition of seabuckthorn in the state. If any organization comes forward for the commercial utilization of this fruit for the development of neutraceuticals, cosmaceuticals, pharmaceuticals etc., it would be profitable venture.

5. Eco-tourism

Uttarakhand is also known as "Dev Bhoomi" and many spiritual and religious places are located in the state. The state is also a place for many natural beautiful and scenic places viz. Valley of flowers, *bughyals* , biosphere reserves (Nanda devi) where tourist visits for their spiritual and mental relief. Plantation of seabuckthorn in these areas can attract more tourists and will be increasing the income of the state.

6. Reduction of pressure on native forest

Being an excellent fodder and fuel plant, seabuckthorn, if planted on larger area it will be greatly reduce the pressure on other species, as has been the case of Ladakh and Nepal.

7. Research and Development

More than 70 National and state research and development organizations are located in the state, seabuckthorn can prove to be an ideal plant for research and development of different products. Although lots of work has been done in other parts of the world but it is still felt there is lot to be researched upon on this wonder plant.

There is no doubt that seabuckthorn possesses a great scope and potential for meeting out the nutritional requirement of people in the state, increase their income and support to the tourist industry in the state. The need of the hour is the due support from government agencies and sincere efforts from local residents, private organizations and government bodies for the planned, systematic and profitable utilization of the fruit.

References

Avdai Ch. and Chimed-Ochir, G. 2003. Harvesting of Seabuckthorn Fruits in Mongolia. In : Seabuckthorn (*Hippophae* L.) :A Multipurpose Wonder Plant. Virendra Singh (Editor-in-chief), Vol. I, Indus Pub. Co. New Delhi. pp 443-447

Bisht, R.; Sharma, S.K. and Yadav, V.K. 2007a. Studies on preparation and storage evaluation of Seabuckthorn (*Hippophae salicifolia* D Don.) ready to serve beverage. *Asian J. Hortl. Sci.* (In press)

Bisht, R.; Sharma, S.K. and Yadav, V.K. 2007b. Effect of different pre sowing treatments germination and plant growth characteristics of Seabuckthorn (*Hippophae salicifolia* D Don.). *Asian J. Hortl. Sci.* (In press)

Bisht, R.; Sharma, S.K. and Yadav, V.K. 2007c. Studies on preparation and storage evaluation of Seabuckthorn (*Hippophae salicifolia* D Don.) squash beverage. *Indian Food Packer* (submitted)

Dwivedi, S.K.; Singh, R and Ahmed, Z.2006. The Seabuckthorn. FRL (DRDO), Leh Ladakh, Jammu and Kashmir, India.57 p

Rao, V.K, Yadav, V.K., Sharma, S.K., Sah, V.K and Singh, V. 2006.Seabuckthorn (*Hippophae spp*) : The Wonder plant for higher Western Himalaya. National Symposium Production, Utilization and Export of Underutilized fruits with commercial potential. Held at BCKV, Kalyani, W.Bengal, w.e. 22-24, November, 2006, Book of Abstract, Pp11-12.

Singh, K.P., Prasad, D. and Yadav, V.K.2007. The First Report of *Rhizoctonia solani* Kuhn. on Seabuckthorn (*Hippophae salicifolia* D. Don) in Uttaranchal Himalayas. *J. Mycol. Pl. Pathol*, 37(1) :126-12

Singh, V.and Gupta, R.K.2003.Propagation of Seabuckthorn (*Hippophae rhamnoides* L.). In : Seabuckthorn (*Hippophae* L.) :A Multipurpose Wonder Plant. Virendra Singh (Editor-in-chief), Vol. I, Indus Pub. Co. New Delhi. pp 315-333.

Yadav, V.K., Sah, V.K., Singh, A.K. and Sharma, S.K. 2006a.Variations in morphological and biochemical characters of Seabuckthorn (*Hippophae salicifolia* D. Don)

populations growing in Harsil area of Garhwal Himalaya in India. *Tropical Research & Extension, 9 : 1-7*

Yadav, V.K., Sharma, S.K, Sah, V.K, Rao, V.K. and Singh, V. 2006c. Morphological and Biochemical diversity in Seabuckthorn (*Hippophae salicifolia* D. D.Don) and identification of suitable berries for processing. Book of Abstract (Eds. B.K. Yadav, Athapol Noohorm, S K Rakshit) International Conference on Innovations in Food and Bioprocess Technologies, held at Asian Institute of Technology Bangkok, Thailand, hed 12-14 December, 2006, p 25.

Yadav, V.K.; Sah, V.K., Rao, V.K. and Sharma, S.K.2006b. Annual Progress Report: Selection, Multiplication and Utilization of Seabuckthorn. Department of Genetics & Plant Breeding, GBPUA&T, Hill Campus, Ranichauri, Tehri Garhwal, Uttarakhand.

Yadav, V.K.; Sah, V.K., Rao, V.K. and Sharma, S.K.2007. Annual Progress Report: Selection, Multiplication and Utilization of Seabuckthorn. Department of Genetics & Plant Breeding, GBPUA&T, Hill Campus, Ranichauri, Tehri Garhwal, Uttarakhand

Yadav, V.K.; Singh, A.K. and Tripathi, S. 2005. Nutritional and medicinal importance of Seabuckthorn. Book of Abstract, National Consultative Workshop on Medicinal and Aromatic Plants, held at GBPUA&T, Pantnagar, Uttarakhand, w.ef. June 25-27, 2005.

□□□

5.2

Seabuckthorn in Uttarakhand

H.B. Naithani and N.S. Bist

CONTENTS

5.2.1 INTRODUCTION

Seabuckthorn is a general term used for a medicinal, deciduous and dioecious, shrub-tree, *Hippophoae* Linn., which belongs to the family Elaeagnaceae. The name may be related to the fact that in England this spiny shrub and trees (it can grow either way depending on soil and climate) used to grow on sand dunes along the sea beaches. Similarly, the German name of Seabuckthorn - Sanddorn may be translated as sand spine. Legends about seabuckthorn tell that the Greeks used it in the diet of race horses; another legend mentions that its leaves were the preferable food for flying horses-*Pegasus*. The name *Hippophae* is also derived from *Hippophae* means *hippos,* a horse and *phao,* which means to destroy. Another legend refers to the custom in some ancient kingdoms to execute convicts by dropping them into a barrel of boiling oil. The legend tells that if the oil in the barrel was substituted by the seabuckthorn oil, the convict had a chance to survive.

Seabuckthorn generally grows on riverbanks. It is distributed in the Hindu Kush Himalayas, adjoining areas of China, parts of Europe and the former USSR as well as in the Scandinavian Region. Russian immigrants brought it to America in the beginning of 20th Century. It is widely distributed throughout the temperate zones up to an elevation 5,200 m. It can resist low temperature up to - 43^{0}C and can withstand temperature as high as 40^{0}C. The total distribution of seabuckthorn in three of the main producing countries, namely China, Mangolia and the former USSR is approximately 8,10,000 ha (natural growth) and 3,00,000 ha plantations (Koykov, 1985; Huang Quan *et al.*, 1990; Lu Rongsen, 1992). In India as, in the cold desert areas of Himachal Pradesh, it constitutes about 24.84% of geographical area of the state (Singh *et al.*, 1991).

The mankind has used seabuckthorn for at least 12 centuries. The plant is of great utility and is sometimes referred as "*Kalpataru*", the legendary tree that provides the fruits of desire (Duhoons *et al.*, 1996). Its fruits are supposed to be a "storehouse of nutrients". Many items like jam, soft drink, sauce, pickle, jelly and biscuit are made from it and exported by China. Its medical uses attributed to Theophrastis and Dioskorid has been described in the ancient Greek texts and in other classic Tibetan literature, "the Rgyud Bzi" [the four books of Pharmacopoeia of Tang Dynasty (618-907 AD)]. The clinical tests on its medicinal uses were first initiated in Russia during 1950 (Gurevick, 1956). Its oil was formally listed in the pharmacopoeia in 1977 and clinically tested in Russia and China (Xu, 1994). The pharmacological committee of the Mimispy of Public health of the former USSR approved the seabuckthorn oil in clinical use in Hospitals (Rongsen, 1992).

5.2.2 GEOGRAPHICAL

The Uttarakhand State is located in the north of India in the Himalayan ranges, the State has 13 districts and a geographic area of 53,483 km^2 (1.6% of the country's geographic area (Fig.5). The population of the state is 8.48 million (0.8% of country's population, Census 2001), of which 25.6% is urban and 74.4% rural. The average population density is 159 persons per km^2. The tribes constitute 3.0% of the state population. The total forest area is 34,661.52 sq. km., which is 64.81% of the total geographical area of the state. The per capita forest area in the state is 0.49 ha.

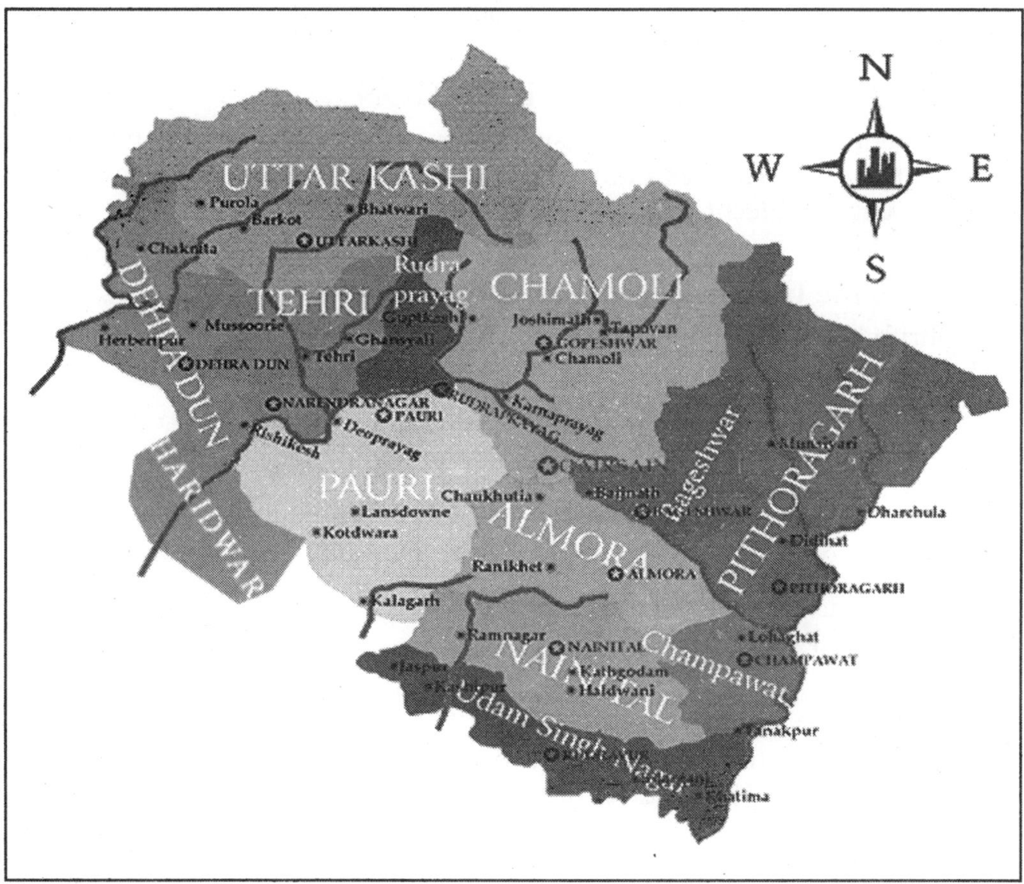

Fig. 5.2.1. Map of Uttarakhand

The Uttarakhand Himalaya is extremely rich in various forms of plant life and abounds in genetic diversity of medicinal plants, timber species like Sal and Deodar, wild relatives of crop plants and other economically important species (Rao, 1994). Around 427 species of flowering plants have been listed under various categories of Indian Red Data Book of which 30 species occur in the Uttarakhand. Altitudinally the forest can be classified into the following categories:

• Sub-tropical	(<1500 m)
• Warm temperate	(1500 m-2500 m)
• Cool temperate	(2500 m-3000 m)
• Sub-alpine	(3000 m–3500 m)
• Alpine	(3500 m-5500 m)
• Nival	(>5500 m)

5.2.3 CLASSIFICATION / DISTRIBUTION

The genus *Hippophae* Linn. consists of four species among which the most important is *H. rhamnoides* Linn. The other species are *H. salicifolia* D. Don and *H. tibetana* Schlecht and *H. gyantensis*.

The key to the Indian species is mentioned below:

(i)	Shrub less than 1 m; leaves in clusters of 3, resembling whorls.	*H. tibetana*
	+ Shrub 2-3 m; leaves alternate, or alternate and opposite.	2
(ii)	Scales on the lower leaf surface without hair.	*H. rhamnoides*
	+ Scales on the lower leaf surface with hair.	3
(iii)	Bark white; leaves narrow, 3.5-5.5 x 0.3-0.5 cm.	*H. gyantensis*
	+ Bark not white; leaves 4-12 x 0.6-1.4 cm.	*H. salicifolia*

(A) H. tibetana Schlecht. in Linnaea 32:296. 1863. *H. rhamnoides* subsp. *tibetana* (Sch.) Ser. In Beih. Bot. Cetrab. (Abt. 2) 25.18. 1909. *H. rhamnoides* auct. Non L. sensu. Hook. f., Fl. Brit. Ind. 5:203. 1886.

Vernacular Names: *Ames, Chuk* (Kumaon); *Tarwar* (Garhwal).

Shrub up to 0.5 m high, bushy, broomy. Stem hard, tortuosus, knobby; branches often ending in arms. Leaves 8-20 mm long, 2-4 mm broad, whorled, normally three at nodes, margins regular, base cuneate, apex obtuse, upper surface green with silvery rusty coloured, scales, brownish beneath; petioles 1 mm long. Fruit 8-11 mm long, 6-9 mm broad, peduncle 1-2 mm long. Seeds 4-5 mm long, 3-4 mm broad, winged, flat, with partly longitudinal furrow.

Distribution: In Uttarakhand, *H. tibetana* has scattered distribution and is found in Ranimimi, Niti, Nilang and Gomukh (Garhwal), Kali valley, Bagutiyal Gori valley, Buddhi, Byans and Dharma (Kumaon).

Phenology: Flowering: May-June; *Fruiting:* September.

(B) H. salicifolia D. Don, Prodr. Fl. Nepal 68. 1825. *H. rhamnoides* subsp. *salicifolia* (D. Don) Serv. in Bot. Centrabl. Beih. 25:18. 1909. *Elaeagnus salicifolia* (D. Don) A. Nelson in Amr. J. Bot. 22:682. 1935. *E. fasciculata* (Wall. ex steud.) Nelson in Am. J. Bot. 22:682. 1935. *H. conferta* Wall. Num. List. 4032. 1831.

Vernacular names: Uttarakhand: *Ames, Chuk* and *Chu* Sikkim and Bhutan: *Tare, Tari-caum, Tarbu.*

A small deciduous tree up to 2.5 m girth and 10 m high with rounded spreading crown, often spinescent when young; bark reddish-brown with deep longitudinal furrows, very rough; blaze 2.5 cm, fibrous, pale-yellow or yellowish-brown; twigs clotted with rusty-brown scales. Leaves petiole, 4-12 cm long, 0.6-1.4 cm broad, oblong-lanceolate, stellately pubescent above when young, becoming bluish-green and with varying amount of pubescence when mature, canescent with dense short stellate tomentum beneath, the midrib beneath pale rusty brown with a few scales mixed with the hairs. Fruit 6.5 cm long, ovoid, yellow with scattered scales, very acid to the taste, remaining long on the trees.

Phenology: Flowering: April-May. Fruiting: October.

Distribution: It is distributed along river banks within an altitudinal range of 2000-3000 m. In Garhwal, it occurs along Rupin-Supin, Yamuna and Bhagirathi rivers in Uttarkashi district, along Alaknanda in Chamoli, Bhuyander river in Valley of Flower, Osla, Har ki Doon, Mallakaliphat, Garhwal (9000), Daoli, Chamoli (6500), Pandukeshwar (Badrinath), between Dyara and Dodital (Uttarkashi), Harsil and on way to Kedarnath and Yamontri. In Kumaon, region is occurs along Kali and Gori rivers in Pithoragarh District.

5.2.4 PLANTATION TECHNIQUES

It grows best in deep, well drained, sandy loam soil rich in organic matter. It spreads through suckers, which can be separated and multiplied. Plantations are established with suckers, seedlings, cutting, layering, grafting, and other propagated material such as the seedlings from root fusion or the plantlets produced through tissue culture. Seedlings can also be raised in polytubes, which are easy to transport. It has profuse suckering habit and yields 12-13 suckers/plant. These suckers can be separated and transplanted in the 2nd

fortnight of the March. One to two years old, pencil thick stems are used as cuttings and are planted in the month of March. These are immersed in running water for 48 hours, subsequently treated with 500 ppm IBA and then planted in the field. This results in 80-90% successes in rooting. These are transplanted at a spacing of 2-3 m after one-year (Dwivedi *et al.*, 2001). Rows should be oriented in a North-South direction to ensure maximum light. It is sensitive to soil moisture deficits, especially in spring when flowers and fruits are in the developing stage. (Li, 2002). It responds well to phosphorus fertilizer, however, nitrogen fertilizers can adversely affect root nodulation and it delays the development of nodules after inoculation with *Frankia* (Akkermans *et al.*, 1983). Seed germination of *H. rhamnoides* is 100% Singh *et al.* (1991) indicated that seeds do not have any dormancy and remain viable for one year. Germination in *H. rhamnoides* is completed between 17 and 21 days. Germination tests were also conducted without removing the pulp from the seed to see its effect on germination. Seeds with pulp have 20.6±7.83 per cent germination and take 35 to 62 days. The ratio of male to female plant is important for maximizing the number of fruit bearing trees. Recommendation for male: female ratio varies with the plant density and region. In British Columbia, with an orchard planting of 4000 trees per hectare, 1:6 male female ratio is considered adequate (Li and Schroeder, 1999). A natural seabuckthorn forest can yield 750 to 1500 kg of berries per hectare. Moderate pruning increases the yield. The crown should be pruned annually to remove overlapping branches and long branches should be cut to encourage development of lateral shoots. Weed control is very important especially for promoting growth of newly planted seedlings (Li and Schroeder, 1999). The fruits have a mean weight of 34.6 mg, measured 9.4 x 7.8 mm and on an average contain 82.2% moisture and 11.5% soluble solids.

5.2.5 USES

- Fruits are good source of vitamin A, other carotenes, vitamin E and several other tocopherols, flavonoides. Vitamin E content (202.9 mg/ 100 g fruit) is higher than wheat embryo, safflower, maize and soybean. Berries of seabuckthorn are second only to Aonla in vitamin C content. It is 3 to 16 times higher than that of the Kiwi fruit (*Actinidia sinensis*), which is very famous for the high vitamin C content. It is also rich in several other vitamins such as B_1, B_2, K and P. Zn content varies from 1.8 to 6.1 ppm; Cu 2.6-5.2 ppm; Fe 100.6-301.8 ppm; Ca 1,695-3,270 ppm; Na 17.5-160 ppm and K 5,825-11,275 ppm. (Anon., 1959).

- Twenty fatty acids have been identified in the oil of *H. rhamnoides*. The oil from all the studied varieties contains 67-75% palmatic and palmitolic acids.
- Leaves are used for making tea.
- The fruits act as an anti-aphrodisiac in female.
- Its bark is an effective blood purifier and tumor inhibitor.
- Fruit juice is used in cardiac trouble.
- Regular intake of fruits has a prophylactic action against cold, cough and provides resistance to other diseases.
- Seabuckthorn oil is used to promote the healing of skin, such as burns sunburns, eczema and psoriasis.
- The emulsion is used to prevent wrinkles of eye region.
- The paste of the ripe fruits, turmeric and grain floor is externally applied to beautify the face and other parts of body of the women.
- Its oil is approved as "Green Food" in China.
- It also improves the mucous membranes of gastro-intestinal tract and acts as a natural dietary supplement.
- The preparations from the fruits are utilized to prevent gum bleeding.
- Help to coordinate the function of liver, spleen and kidney.
- Traditional medicinal practitioners particularly in Mayar and Bhanga villages in Lahul make medicines from the fruits for the treatment of lung disorder, blood pressure and indigestion.
- Bhutanese use the fruits of *H. tibetana* as a medicine for stomach ache.
- Fruits of *H. salicifolia* contain 84% antioxidant activity.
- Ladies and children often consume the fruit, locally called '*tsastalulu*' in Ladakh.
- Leaves are good nutritive fodder for cattle and sheep/goats as they are rich in protein (18-22%), fat (4-5%) and other micronutrients.
- Paste of the bark is used to heal wounds and ulcers.
- The dried berries are stored and are supposed to be an effective cure for preventing cracking of lips in dry climate of Ladakh.

Apart from medicinal uses, it has other values also. The wandering livestocks browse on its foliage particularly during early winter (October-November) when other sources of fodder dry up. For example, the endangered double humped camel found in the Nurba valley, Ladakh, survives on seabuckthorn for its food requirements. The fruits are also used for polishing gold and silver. The thorny branches are used to fence the cultivated land. Its wood provides fuel and charcoal. It is also used for making handles for household implements (Dwivedi et al. 2001). The calorific value of stem was recorded as (4430.3 ± 87.8), branches and twigs (4,361.0 ± 73.3), roots (410,837.9±values) and leaves (3577.1 ± 61.4). The plantation of one ha area can meet fuel wood requirement for twenty families in the Spiti Valley (Singh *et al.*, 1991). Tannin has also been obtained from the leaves. The pulp remaining after removal of juice is used as a food colouring material. Its root system contains nitrogen-fixing nodules, thus it improves soil fertility.

5.2.6 CONCLUSIONS

H. salicifolia is the most common species in Uttarakhand (India) and up till now no emphasis has been given on extraction and utilization of its products. It has big fruits with high content of vitamin C and contains rich crude protein in its leaves. It has fewer spines; therefore, the extraction of seeds is easy. In recent years the juice of Burans (*Rhododendron arboreum*) has become very popular in Uttarakhand, which is selling at good prices. Similarly, developing suitable products from seabuckthorn can raise economic status of the local people, at the same time conserving and extending the area under it with the involvement of people. It occurs at sub-alpine region which are covered by snow during winter. These areas do not have much vegetation, therefore, if these areas are developed into seabuckthorn plantations, both people of these areas as well as the state government can earn from such activity. Following actions are suggested:

- To make scientific exploration in the state to assess the extent of area under these species and explore possibilities of managing and improving them with the involvement of local people.
- Standardize cultivation practices, pruning, and application of fertilizers (organic), harvesting and post harvest care.
- To develop local value addition and some branded products and promote its marketing.

References

Anonymous. 1959. Wealth of India - V: 100. CSIR, New Delhi.

Akkermans, A.D.L., Roelofsen W., Blom J., Huss. D. & R. Harkink. 1983. Utilization of carbon and nitrogen compounds by *Frankia* in synthetic media and root nodules of *Alnus glutinosa, Hippophae rhamnoides,* and *Datisca cannabina. Can.J. Bot* 61, 2793-2800.

Duhoon, S.S., M.N. Koopar & U. Chandra, 1996. Seabuckthorn (*Hippophae* spp.) - A less known wonder plant of Ethno-Micro-Botanical importance in cold desert of India. *J. Econ. Taxon. Bot. Add. Ser* 12, 43-45.

Dwivedi, S.K., Attrey, D.A. & Chaurasia, O.P. 2001. Seabuckthorn: A hope for greening of Ladakh. - Ext. Folder No. 3 pp. 1-5-DRDO - Leh (Ladakh).

Gurevick, S.K. 1956. The application of Seabuckthorn oil on ophthalmology. *Vestin. Ohamologu,* 2, 30-33.

Haung Quan 1990. The development and utilization of Seabuckthorn plants in Mongolia. In *Hippophae* No. 2, p. 43-46.

Koykov, N.T. 1985. In Seabuckthorn. Moscow: Forest Industry, Translated into Chinese by Zhang Zheming.

Li, T.S.C. 2002. Product Development of Seabuckthorn, Trends in new crops and new uses, p.393-397.

Li, T.S.C. & Schroeder, W.R. 1999. A growers guide to Seabuckthorn. Agriculture and Agri-Food Canada. 70 p.

Rongsen, Lu. 1992. Seabuckthorn. A multipurpose species of fragile mountains. ICIMOD No. 20. 1-50-Kathmandu, Nepal.

Rao, R.R. 1994. Biodiversity in India (Floristic Aspects). Bishen Singh Mehendra Pal Singh, Dehra Dun, 313 pp.

Singh, R.P., Bahar, N., Prasad, B. & Gupta, M.K. 1991. Ecological studies of *Hippophae rhamnoides* Linn. In Spiti Valley of Himachal Pradesh. *Annal. Ard. Zone* 20(2),119-122.

Xu, M. 1994. The medical research and exploitation of Seabuckthorn. *Hippophae* 7, 32-34.

□□□

6

Seabuckthorn in Sikkim Himalayas

B.C. Basistha

CONTENTS

6.1 INTRODUCTION

Himalayas of Sikkim by virtue of its altitudinal variations, steep slops and different aspects of mountains and heavy monsoon lead to wide range of climatic variations within short distances, is also known to be a hotspot of biodiversity (Lachungpa, 2000) with highest density of vegetation in the eastern Himalayas (Srivastsva, 2000).

Seabuckthorn mostly grows in and around riverbank and sea store, bearing fruits in clusters and with thorny branches. In Nepali the plant is named as *Ghangaru,* in Hindi the plants is popularly known as *Charma* or *Kalabis,* in Sherpa language it is called as *Khurpu* and locally in North Sikkim the Bhutia called the plant as *Tarobo.* For last few decades, the plant has been gaining wide attraction from researchers and ecologists world around particularly after its economic potentialities were exposed in nutritional and medicinal values. Serial trial and experimental works leading to research programmes were done in the last decade particularly in Russia, China, Canada and India. Its ecological advantages were also explored through research programmes in many Himalayan and temperate climatic zones of South Asia, Russia and China.

Studies have revealed that seabuckthorn could be regenerated through vegetative propagation using cuttings, and through micro-propagation (Yao and Tigerstedt, 1995; Rongsen, 1992; Basistha *et. al,* 2001), which would benefit our further approaches. European papers and literatures from China reveal that the plant has a wide range of adaptability from 1,000 m to 4,100 m amsl and in the temperature ranges of –40 to + 40^{0}C. However, this may not be as suitable in or Indian region because of the geographical locations of the region, aspects, and differing latitudinal characteristics of Indian Himalayas to that of European ones, where snowfall comes to as lower as to 1,000 m amsl (Rongsen, 1992).

6.2 USES

Seabuckthorn (*Hippophae salicifolia* D. Don) is amongst the eight species of *Hippophae* Linn. (Rousi, 1971) and has a multipurpose uses in food and pharmaceutical industries, medicinal and aromatic factories, to high altitude farmers and researchers (Rongsen, 1992) and may be called as wonder plant of the dry temperate Himalayas (Singh, 1998) due to its multilateral potentialities and utilities in ecological and economic aspects. Its importance is mentioned in the writings of ancient Greek scholars such as Discorid and Theophrastus. Wherein, seabuckthorn was known as a beneficial dietary supplement for domestic animals and is equally liked by beers, birds etc.

Seabuckthorn, besides having above ecological advantages, have multifarious use in drugs and aromatics industries and also have wide range of essential elements in its fruit pulp, seed, leaves and bark. It contains high amount of Vitamin C (300-1600 mg/100 g) about 4-100 times higher than other vegetables/fruits. The fruit oil contains Vitamin E (200 mg/100 g) and has anti-cancer activity. The β-carotene, an important bioactive compound, is present in sufficient (250 mg/100 g) in its fruit oil. Lycopene, flavonoid (0.2% in juice), amino acids, and about more than 100 kinds of bioactive elements have been identified in various components of seabuckthorn leaves, fruits, which help improve the micro-circulation of bloods capillaries and nourish other skin/hair ailments. (Singh *et al.*, 1995; Singh, 1998; Singh, 2001) and thus has high values in development of medicine, herbal and health care products. During 1998-2001 studies were conducted in the Sikkim Himalayan region at Lachen and Lachung region of North Sikkim (Basistha *et. al.*, 2001).

Seabuckthorn has ability to fix atmospheric nitrogen into soil (Andreeva *et al.*, 1982; Akkermans *et al.*, 1983; Cireasa *et al.*, 1983; Basistha, 2001) and thus help enrich the soil. The plant helps in binding soil with its long root networks (Rongsen, 1992; Singh *et al.*, Singh, 1998; Basistha, 2001). The plant persist over a wide range temperature (+40°C to -43°C) and annual precipitation ranging from 250 mm to 800 mm and can even survive at low (15%) moisture level promoting better growth of other high altitude species; poplars, pines etc. (Lei *et al.*, 1983; Rongsen, 1992). *H. salicifolia* (Anonymous, 1998) has been found growing in the sandy and riverine terrain of North Sikkim over 2,800 m to 3,100 m amsl sustaining with the fragile and slide prone slops and erosion affected riverine areas. With the increasing human interferences in these areas, like elsewhere in the Hindu Khush Himalayan region (Jodha *et al.*, 1992; Rongsen, 1992) the vegetation of this area have been subjected to decreasing density. Most of the older seabuckthorn plants are either found to be chopped or destroyed in order to meet the domestic purposes. The local people however, use the plant parts as timber, firewood, fodder and fencing purposes, preparation of edible items like pickles, jam and dying ink, etc.

In China and some European countries, the plant has been much diversely used in different fields. A wide range of products are being synthesized (Zhang, 1987; Rongsen, 1992; Singh,1998; Xiao *et al.*, 1979) form its leaves, bark, fruits, seeds, roots etc. and hence the plant is designated as wonder plant.

6.3 DISTRIBUTION

The natural habitat of seabuckthorn in Sikkim lies in the dry temperate hills especially in the north district ranging from altitude of 2,800 to 3,300 m amsl. The plants were observed in areas of 250-300 ha of riverine, and non-riverine and torrential areas as natural orchard. The plant for its regeneration favoured the landslide areas of sun facing slopes and riverbank areas. The total distribution of seabuckthorn in Sikkim is approximately 500 ha spreading in the dry temperate region over altitudinal ranges of 2,800-3,300 amsl. They are widely distributed in different slopes and riverbank areas in and around Lachen, Zema I, Zema II, Zema III and Lachung regions and mainly in the North-eastern aspects of the slopes. *Hippophae* in Sikkim is mostly observed to be of *salicifolia* species with small patches of *H. rhamnoides* in upper belt of the natural habitat which extends from 2,800 to 3,300 m amsl. The plant is observed to be colonized at either side or river banks of River Zema, River Lachen and sandy slopes of River Lachung and near by Thangu in North Sikkim. There is distinct variation of plant size and density of population of *Hippophae* in different altitudes and slope facet. The soil type and climatic features of the region are dominant factors for plant density and production characteristics.

The ethnic population of the native area love the plant for they found some qualities in it. They use the plant as a cheap source of firewood, as per local respondents the wood of seabuckthorn burns with almost blue flame and gives a good heat with minimum of smoke and ash. People prepare jam and pickles out of the fruits, which, they say, is a good digestive agent also. The local women weavers, craft-men use the juice from the fruit as dying ink for their traditional blankets, bedcovers and sweaters etc. As also cited in the literatures from China and European countries (Rongsen, 1992) the native birds of high altitude are seen hovering over the seabuckthorn trees. Local people informed that bears and other wild animals also like the ripened fruits. However, due to excessive deforestation for firewood and cultivation of other agricultural/ horticultural crops the plant seems to be in danger, for it the preventive measures are to be initiated on time. The present study on the plant has revealed that the seabuckthorn of Sikkim Himalayas although growing in different ecological and micro-climatic situations possesses many beneficial characteristics, if properly explored may serve for the socio-economic purpose of the local community of the region and to the economy of the state. Sikkim, having about 24% of its geographical area as snow clad land and whereas no other vegetation virtually grows economically, development of seabuckthorn orchards, may add to the sustainability of the socio-economic status and may also help fulfill domestic needs of hill area tribes. There is an urgent need of research as regards to agro-techniques for commercial cultivation of this plant and may have to be standardized at local level and further a need based, phase wise programmes is to be sorted out for the effective commercial scale seabuckthorn cultivation in Sikkim Himalayas.

6.4 CONCLUSIONS

The above discussion thus clearly indicates that sebuckthorn, which is presently growing in a concentrated slopes of North Sikkim, dominantly in Lachen and Lachung areas, may be further popularized so as to derive at the beneficial aspects in a manner such that the local people could be benefited from its identified advantages. It is optimistic that seabuckthorn cultivation may help improve reforestation, conserve fragile hills from recurrent landslides and soil erosion and also add to the economic sustainability of the marginalized hill tribes.

References

Akkermans, A. D. L., Roelofsen, W. J., Bolm, K., Hussdanell & Harkink, R. 1983. Utilization of carbon and nitrogen compounds by *Frankia* in synthetic media and root nodules of *Alnus glutinosa, Hippophae rhamnoides* and *Dastica cannabina. Can. J. Bot.* 61, 2793-2800.

Andreeva, I. N., Fedorova,E. E., Il'yasova, V.B. & Tibilov, A. A. 1982. Ultra structure of nitrogen fixing and wintering nodules in one year seedlings of *Seabuckthorn* and oleaster *Hippophae rhamnoides, Elaegnus argentea. Soviet Plant Physiol.* (USA) 29, 109-116.

Anonymous. 1982. State Report on Agriculture Census, 1980-81, Vol. I, Bureau of Economic and Statistics, Gangtok, Sikkim.

Anonymous. 1998. Environment Impact Assessment Report, 510 MW Teesta Hydroelectric Project, Stage V, Vol. I, EPC, Forest Department, Government of Sikkim.

Basistha, B.C. 2001. Vivipary in Seabuckthorn (*Hippophae salicifolia* D. Don), Sikkim Science Society, Gangtok. *J. of Hill Research* 14 (1), 67.

Basistha, B.C., Adhikari, I. M. & K. K. Thapa. 2001. A case study of *Hippophae* Linn. with special references to agro techniques in the Sikkim Himalayas: In *Proceedings of International workshop on Seabuckthorn*, New Delhi, February.

Cireasa, V.I., Rosu, F., Drobata, M.A. & Drobata,G. 1983. Research on techniques for production of *Hippophae rhamnoides* planting material, Journal's article, Romania.

Dobrista, S.V. & Novik, S.N. 1992. Feedback regulation of nodule formation in *Hippophae rhamnoides, Plant and Soil* 144, 45-50.

Jodha, N. S., Banskota, M. & Pratap, T. 1992. Sustainable Mountain Agriculture, Vol.1, ICIMOD and IBH Oxford publication, New Delhi.

Lachungpa, U. 2000. Rare and endangered animals in Sikkim, Their conservation status today. In: Sikkim Herald, Vol, XXLIV (112), Sikkim Government Press.

Lie, Q., Zhong, Y. & Yang, H. 1983. Effects of nitrogen supply from Seabuckthorn in artificial woods of Simon poplar. Forest Science and Technology.

Rongsen, Lu. 1992. Seabuckthorn: A multipurpose plant species for fragile mountains, In Occasional paper no 20, ICIMID, Kathmandu.

Rousi, A. 1971. The Genus *Hippophae* L. A taxonomic study. *Ann. Bot. Fennica* 8, 177-227.

Schroeder, W.R. 1988. Planting and establishment of shelterbelts in humid severe-winter region. *Agr. Ecosystems Environ.* 22/23, 441-463.

Singh,V., Singh, H. S. & C.P. Awasthi. 1995. Studies on Distribution, Taxonomy and Nutritional values of *Seabuckthorn* growing in dry temperate Himalayas. *Proceedings of International workshop on Seabuckthorn* pp. 52-59, 1995, Beijing.

Singh, V. 1998. *Seabuckthorn* a wonder plant of dry temperate Himalayas. *Indian Horticulture*, 43 (2), 6-8.

Srivastava, R.K. 2000. Landslide prevention and control measures in Sikkim. In: Sikkim: prospective for planning and development. Rai, S.C.; Sundheryal, R.C.; Sharma, E. (eds.), Sikkim Science society and Bishan Singh & Mahendra Pal Singh, Dehradun, pp. 597-604.

Thomas, S.C., Li., W.R., and Schroeder P. 1996. *Hort. Technology*. 6 (4).

Wahlberg K. & Jeppsson, N. 1990. Development of cultivars and growing techniques for Seabuckthorn, black chokeberry, *Lonicera* and *Sorbus*. Severiges lantbruks-universited Balsgard-Avdelningen for hortikulturell vaxiforadiing Verksamhetsberatteise 1988-1989 pp 80-93.

Xiao Zhuyin, 1979. Seabuckthorn a multipurpose plant species for fragile mountains. ICIMOD occasional paper No. 20, Lu Rongsen ed., pp. 14. Kathmandu, 1992.

Yao Y. & Tigerstedt, P.M.A. 1995. Geographic variation of growth rhythm, height and hardiness and their relations in *Hippophae rhamnoides*. *J. Amer.Soc.Hort. Sci* 120 (4), 691-698.

Zhang Fushun 1987. Content of Seabuckthorn Oil and its Physical-chemical Properties, composition of fatty acids and biological active substances in Northwest China. In: *Chinese Oil and Fat (Shaanix)* No.2, pp. 25-30.

□□□

7

Seabuckthorn in SOWA-RIGPA (Tibetan Medicine)

Padma Gurmet

CONTENTS

7.1 INTRODUCTION

"Seabuckthorn", is also known as *tSer-Mang, tSer-Sa-Lu-Lu, Shib-Shu-Lu-Lu* or *sTar-Bu,* in Ladakh(India), is one of the dominant plant species of the Trans-Himalayan regions of Ladakh and Lahaul-Spiti. It is abundant in all eight major valleys of the Trans-Himalayan cold desert region, namely Lahoul, Spiti, Changthang, Indus, Nubra, Daras, Suru and Zanskar. Although multifarious traditional uses of this plant can be widely seen, in the form of the bio-fences that protect gardens and agriculture lands, fuel stocks for winter, sheep and goat enjoying the leaves and Amchis (traditional doctors) collecting the fruits for medicine, Seabuckthorn grows largely as an unwanted weed and its full potential has been not been acknowledged in an appropriate manner by the natives of this region. Due to the harsh climatic conditions and their remote locations, Ladakh and Lahoul-Spiti have been largely isolated from the rest of the world until recent decades. The inhabitants of these regions have thus had to depend to a large extent on the surrounding wild plants for their basic commodities, including food, fuel, fodder and medicine. *Hippophae* species are among the most commonly used of these plants, with a particularly wide range of applications.

The earliest written information about the medicinal uses of Seabuckthorn is reported in the *rGyud-bZhi,* which is the classical medical text book of Sowa-Rigpa (Amchi / Tibetan medicine). It is one of the most important and widely-used ingredients in various Sowa Rigpa formulations. Three different species of Seabuckthorn are mentioned in the classical medical texts, the *dTrimed Shelgong* and the *Shelthring,* namely *Sa-sTar* (*Hippophae tibetana*), *Bar-sTar* (*Hippophae rhamnoides* ssp *turkestanica*) and *Nam-sTar* (*Hippophae salicifolia*). Of these, *Bar-sTar* (*Hippophae rhamnoides* ssp *turkestanica*) is the most commonly used species in Tibetan medicines. Recent developments and scientific inputs have added new dimensions and directions for Seabuckthorn and the plant has now entered a phase of transition, from traditional uses to modern scientific products and the scope is widening day by day.

7.2 MYTHOLOGICAL ORIGINS

The mythology of Seabuckthorn goes back to the famous period of the ocean churning, which is considered as the beginning of the universe in Hindu mythology. When nectar came out of ocean, after many days of effort from the gods and *Asuras* (demigods), the clever Rahu (king of *Asura*) ran away with the nectar pot and started drinking it. When Brahma saw this, he threw *Chakara* on Rahu and cut him from the neck into two pieces. Rahu started bleeding and since he had had a sip of nectar, a drop of it fell from his throat onto the earth. It is believed that Seabuckthorn and a number of other plants grew out of the nectar that dropped from Rahu's throat and thus that the plant contains some of the properties of nectar (Phuntsog 1983). Hence Seabuckthorn is a very potent plant and has a high medicinal value.

In many villages of Ladakh, Seabuckthorn is also regarded as the residence of *Nagas* (Lhu). These are elemental forces or spirits, often associated with water sources and specific components of the environment. For this reason, large and old Seabuckthorn trees are regarded as holy, people must be wary of them and will not dare to cut them down or to pollute near to them, as they might well be *Lu-Tser* (residence of *Nagas*).

7.3 TRADITIONAL CLASSIFICATION

The three verities of Seabuckthorn mentioned in the *dTrimed Shelgong* and *Shelthring* texts are classified and named according to the variation in size between the different species. *Sa-sTar* is the smallest, with a height of less than 2 ft., so it is given the prefix *Sa* (ground). *Bar-sTar* is given the prefix *Bar,* meaning medium, and has plants growing up to 8 ft. in height. Finally, *Nam-sTar,* (*Nam* meaning 'sky') is the tallest, shaped like a tree. On the basis of this classification, the botanical identification of the different spices can be established as: Sa-sTar -*Hippophae tibetana;* Bar-sTar- *Hippophae rhamnoides* sp. *turkestanica;* and Nam-sTar - *Hippophae salicifolia.* While all three species are found in Ladakh & Lahoul-Spiti, *Bar-sTar* (*Hippophae rhamnoides*) is the most common species and is found in most parts of Trans-Himalaya and is particularly abundant in the Indus, Nubra, and Suru valleys. *Hippophae salicifolia* is not found in Ladakh region but it is fairly widespread in the Lahul valley. *Sa-sTar* (*Hippophae tibetana*) is the rarest species and is found in only in a few area of Spiti and in the Rangdum area of the Zanskar valley (Gurmet, Chaurasia, & Ballabh 2000; Chaurasia & Gurmet 2003).

7.4 TRADITIONAL USES

Hippophae is a multi-utility plant for the local population and many of its uses can be easily seen throughout trans-Himalayan regions. Due to its spiky thorns, the most common use is as bio-fencing for plantation areas, orchards and agriculture lands which are effectively protected with Seabuckthorn bio-fences. Being one of the coldest regions of the world, fuel is a major problem in Ladakh and Lahul-Spiti. Seabuckthorn is one of the most useful fuel plants, as it grows without any effort and produces a very warm and long-lasting fire. During community gatherings, social ceremonies and festivals in monasteries, Seabuckthorn is widely use as fuel for cooking food, as it is also in most households. Hippophae is also a wonderful fodder plant for livestock and is particularly enjoyed by sheep, goat and camels, for whom the leaves are considered very healthy. In the Nubra valley of Leh district, a large number of semi-wild Bactrian Camels survive largely on Seabuckthorn and their health is significantly better than the domestic ones.

Due to its hardness, the wood of Seabuckthorn is also occasionally used for building roofs, to make handles for agriculture appliances and so on. The use of Seabuckthorn as a food substance is not widespread in the region, apart from children making juice out of it and picking the berries to eat. Some sources report that a few people use this plant to make *Chang* (traditional wine), but this has not been verified. Of all of the many uses of Seabuckthorn, the most important is as a medicinal plant in the Sowa-Rigpa traditional medical system.

7.5 SEABUCKTHORN: AN IMPORTANT MEDICINAL PLANT

The use of Seabuckthorn as medicine is described in the text books of Chinese medicine, Ayurveda and Sowa Rigpa. The earliest reference concerning the medicinal uses of Hippophae has been described in detail in the *"rGyud bzi"*, which is the fundamental text book of Sowa Rigpa. The recently published Chinese Pharmacopoeia of 1977 also describes the medicinal use of Seabuckthorn, and it is identified as the Ayurvedic medicinal plant *Amlich* or *Badriphal* (Sharma & Chunekar 1998). According to these medical systems, Seabuckthorn has a wide range of uses including: treatment of respiratory problems (especially asthma); purifying the blood; strengthening digestion; coordinating the balance of functions among various organs (liver, stomach, spleen, kidney and heart); assuaging pain caused by wounds; promoting tissue healing and regeneration; invigorating blood circulation; combating old age; improving strength; lowering blood fat; and preventing cancer. Seabuckthorn is rich in vitamin 'C' and contains a unique combination of other important vitamins such as A, B1, B2, E & K. It is also a rich repository of proteins, organic acids, carotenoids, flavonoids and minerals like iron, cobalt, molybdenum etc (Dwivedi, Singh, Ahmad 2006).

7.6 "SOWA-RIGPA" HIMALAYAN ART OF HEALING

It has been already mentioned that the earliest reference to the medicinal uses of Seabuckthorn is found in the *rGyud bzi* (Four Medical Tantras) of Sowa Rigpa, therefore it is important to give a brief introduction to this medical system. Sowa-Rigpa is commonly known as Tibetan or Amchi medicine and is among the well-documented, codified classical medical traditions of the world. Originating in India, the system was enriched in Tibet and has a living history of more then 2500 years. Sowa-Rigpa has been popularly practiced throughout the central and south Asian countries of Tibet, Mongolia, Bhutan, China, Nepal, in the Himalayan regions of India and in some parts of the former Soviet Union. In India, it has been popularly practiced in Ladakh, Himachal Pradesh, Arunachal Pradesh, Sikkim, Darjeeling and now in Tibetan settlements all over India (Gurmet 2005).

Sowa-Rigpa is based on the Ayurvedic theory of the *Panch Mahabhuta* (Jung-wa-lna /five elements) and *Tri-dosh* (Nespa Sum / three humors). According to this theory all the animate and inanimate phenomena of this universe are composed of *Panch Mhabhuta* (Jung-wa-Lna/ five elements). It is on the basis of this theory of five basic elements that the sciences of physiology, pathology and pharmacology are established. Being taught by Buddha, philosophically Sowa-Rigpa is based on Buddhism (Anonymous 2004). The pharmacopoeia of this medicine is based on the skillful use of plants, minerals, animal parts, precious and semiprecious stones and metals. In particular, the competent use of a large number of Himalayan medicinal plants is a unique feature of this medicine.

7.7 SEABUCKTHORN AND SOWA-RIGPA

Seabuckthorn is one of the most important and widely used medicinal plants in Sowa Rigpa. Although *sTar-Bu* is the classical name for this medicinal plant there are many synonyms like *gLawa-tsirma, Lai-tsir, Tsar-tsir, Mai-ldan, Stin-ge-sochan, Yalga-gyur* etc. Mostly it is the fruit of Seabuckthorn that is used in medicine. The berries are collected during October-November and after cleaning, are left to dry in the sun, before being used to prepare medicines or being made into concentrated decoctions (*Khanda*). *sTar-Bu* is used in compound form and is mixed with other plants and minerals.

Ro (taste), *nus pa* (potency) and *Nowo* (nature) are three important factors that define the efficacy of ingredients in Sowa-Rigpa. Seabuckthorn is sour (*sKur-wa*) in taste (*Ro*), sharp (*rno*) and light (*Yang*) in potency, and neutral (*nyom*) in nature (*Nowo*). Though there are only minor differences in medicinal uses and potencies among the three species of Hippophae, *Hippophae rhamnoides* is by far the most commonly used. According to *Dri med sel gong* and *sel thring* (the main Pharmacognosy and *materia medica* texts in Sowa-Rigpa), *Hippophae salicifolia* (*Nam-sTar*) and *Hippophae tibetana* (*Sa-sTar*) are more effective for treating lung and throat disorders, while *Hippophae rhamnoides* (*Bar-sTar*) is better for dealing with blood disorders.

There are around two hundred popular Seabuckthorn-based formulations in the various pharmacopoeias of Sowa Rigpa. These formulations are mixtures of between two and fifty different substances. Out of these, most of the Seabuckthorn-based drugs are used in the treatment of lung and phlegm disorders, blood disorders, menstrual problems, throat infections, liver problems, spleen and stomach disorders, peptic ulcer, joint pain, dysentery, appetite stimulant, skin disease, toothache, asthma, against tumors, stone and cancer, diabetes, coughs and colds, as well as being used as a general tonic. Some important formulations are given below, along with the major uses (Anonymous 2004, Dorjai 1995, Dawa, Namgail & Phuntsog 1990, Phuntsog, 1983, Lama, Suresh & Thomas 2001).

Table 7.1: Some important Seabuckthorn formulations and their major uses:

Name of the medicine	Major Disease/ Disorder
sTar-bu 3 sByor	Digestive stimulant and against Neoplasm
sTar-bu 19	Gynecological problem like vaginatetis, salpingitis.
sTar-bu 5	Throat infections
sTar-bu 18	Cervicitis, vaginatetis, salpingitis, menstrual disorder.
Si-dru 8	Antacid, peptic ulcer and gastritis.
sTar-chang	Traumatism and antitussive
sTar-bu pan tsa	Against vomiting and bronchitis.
Chong- bZi 21	Peptic ulcer, acidity, bilious vomiting and diarrhea, phlegm disorders.
sMug-poi-ninpo- bDun-sByor	Peptic ulcer and antacid.
Srolo 11	Lungs disease
Gar-nag 10	Indigestion, gastric neoplasm, gallstone, gastrointestinal polyps, uterine polyps.
Dai-Nyug	Appetite stimulant, peptic and duodenal ulcer, bleeding peptic ulcer, indigestion and stomach problem.
Skuru-25	Hypertension, blood disorders, fever, acidity, liver and stomach problem.
sMug-po nYoms ldan jor-wa	Chronic peptic and duodenal ulcer, acidity and phlegm
Tan-nad chi-sman	Tumors, gastric neoplasm, gallstone, gastrointestinal polyps, uterine polyps.
Tan-joms Zla sil	Gallstone, blood tumor
Or-sman dai chaid	Asthma, stomach and liver disorder, anemia, hiccups, edema and piles.
Nul-byor	Edema
Lo nad chi Joms Chen po	Lungs disease, lungs tonic.

contd...

contd...

Dud rTsi chi sos	Chronic cough and cold, bronchitis.
Dud rTsi ril kar	Digestion stimulant and acute colicky pain
Gun dum 10	Bronchitis
Udpal 25	Used against poison, food poison, peptic ulcer, liver and gall disorders.
Udpal 8	Liver and blood disorders.
Gyam-tsa 10	Spleen disease.
Usu 6	Gastric, diarrhea, dysentery.
Kar po 6 sByor	Irritation in rectum
Skad sman jamling drug da	Problem in vocal cords
sNon boo 6 thang	Irritation of skin, eczema and other skin problems.
Mo nad chi Joms chen mo	For general gynecological problem.
Ol- Si 25	Menstrual problems,
Zi-ched 11	Reduce labour pain, dissolve haematoma.
Chang khog rTsa shal	Against wounds, tumors and neurological problem.
Zla shail dud rtsi thigpa	Chronic fever, dysentery, vomiting blood, edema.
Gag sail rgyadpa	Chronic constipation, problem of delivery & placenta.

(Note: The name of medicine is written as per pronunciation and the uses of medicine are tentatively translated from the classical texts)

7.8 CONCLUSIONS

Seabuckthorn is one of the most important plant species of the Trans-Himalayan region, with remarkable scope to become a plant of great economic importance. It has tremendous potential to improve the economic and environmental conditions of poor and environmentally harsh desert areas of this region. The vast ethnobotanical knowledge of the local population is an added advantage for the development of Seabuckthorn as an economic plant. Particularly Amchis' (Sowa Rigpa practitioners) knowledge about the medicinal uses of Seabuckthorn is an important basis for exploring the potential development of new medicines to treat some of the chronic diseases of the modern age. Thus, if the rich ethnobotanical knowledge of the local population could be explored within modern scientific dimensions, Seabuckthorn could indeed prove to be the 'golden bush' for the economically backward regions of Ladakh and Lahoul-Spiti. At the same time, it is very important to take concerted action for the conservation and cultivation of Seabuckthorn, to ensure its sustainable development and utilization. Keeping in view the vastness and barrenness of the Trans-Himalayan region, *Hippophae* should in particular be adopted for widespread soil conservation and afforestation projects.

References

Anonymous, 2004. rGyud-bzi, Tibetan Medical and Astro Institute, Dharamshala, HP. India

Chaurasia & Gurmet, 2003. A Checklist on Medicinal & Aromatic Plants of Trans-Himalayan Cold Desert (Ladakh&Lahaul-Spiti), Field Research Laboratory, Leh and Amchi Medicine Research Unit, Leh

Dorjai, Gawai .1995.Thungs-Pai-Te Maid-Shel-Gi Melong, Me-Regs Pai Tun Khang Lhasa, Tibet

Dawa. Pod-Ki-Sowa-Rigpa-Las Sman Zas Jor zoi Lag Lain Sang Sgo Baid Pai Ldai-mig, Rig Drag Publication, New Delhi

Dwivedi, S.K., Singh, Ranjit. Ahmad, Zakwan.2006. The Seabuckthorn.Field Research Laboratory, Leh

Gurmet, Padma., Chaurasia, O.P.& Ballabh, Basant 2004. Studies on some important medicinal plants growing along Indus river of Ladakh Himalayas, *Bull. Medico-Ethno-Botanical Research* (CCRAS), Vol.XXV,No 1-4 (2004) pp.1-11 New Delhi.

Gurmet, Padma 2005. "Sowa-Rigpa" Himalayan Art of Healing, Indian Journal of Traditional Knowledge (CSIR), Vol.3 (2) April 2004 pp-212-218 New Delhi.

Lama, Yeshi C., Ghimire Suresh K., Thomas, Yildiz A. 2001. Medicinal Plants of Dolpo, People and Plants Initiative & WWF Nepal Program Kathmandu, Nepal.

Namgail, Gurmet & Phuntsog, S.T 1990. Amchi Pharmaco-Therapeutics, Central Council for Research in Ayurveda, New Delhi.

Phuntsog, Dilmar Gai Shes Stanzin 1983. Dri Med Shel Gong/ Shel Taing, M/S D. W. Tashigang, BPO Nimo Leh-Ladakh.

Sharma, S.K & Chunekar, K.C 1998. Rastriya Ayurveda Vidyapeeth, New-Delhi.

□□□

8

Seabuckthorn for Food and Health

Farhath Khanum and A.S. Bawa

CONTENTS

8.1 INTRODUCTION

Since ancient times seabuckthorn berries are utilized as dietary supplement, source of herbal medicines, skin treatments and remedies both in Europe and Asia. A legend says that before 12 century BC, the ancient Greeks were surprised to find that some sick horses let loose to die a natural death, became strong and energetic again. The source of this magic was traced to a kind of shrubberry. They named the shrub *Hippophae*, meaning trees that make horse shine. Recently, it has been extensively planted across much of northern China, and in other countries, to prevent soil erosion and to serve as an economic resource for food and medicinal products. One of the most well recognized properties of seabuckthorn, traditionally utilized by Eastern and more recently by Western medicine, is nourishing, revitalizing, and restorative action on the skin and mucous. Oil extract, obtained from the berries and called seabuckthorn oil are used to promote healing of heat burns, radiation burns, sun-burns, poorly healing wounds and eczema. High content of fat soluble vitamins (A, E) and nutrients (essential fatty acids, phytosterols) makes seabuckthorn an indispensable ingredient in restorative, anti-aging and revitalizing skin care. Seabuckthorn oil effectively helps combat symptoms of malnourished or prematurely aging skin. It is also taken as dietary supplement to improve the conditions of the mucous membranes and as a natural source of carotenes, phytosterols, and essential fatty acids.

8.2 COMPOSITION

Seabuckthorn has been reported to contain more than 190 compounds in the seeds, pulp, fruit and juice. These compounds include fat soluble vitamins (A, K, E), 22 fatty acids, 42 lipids, organic acids, amino acids, carbohydrates, vitamins C, B_1, B_2, folic acid, and flavanoids, phenols, terpenes and tannins (Katiyar *et al.* 1990; Yang and Kallio, 2002; Li and Schroeder, 1996). It also contains 20 mineral elements. Many of the substances that are found in seabuckthorn are known to have beneficial effects on health. The concentration of vitamin C in seabuckthorn fruit is higher than strawberry, kiwi, orange, tomato, carrot, hawthorn (Bernath and Foldesi, 1992; Lu, 1992) and many other fruits (Katiyar *et al.* 1990). Vitamin E content in seabuckthorn (202.9 mg/100 g fruit) is higher than wheat embryo, safflower, maize, and soybean.

8.2.1 Flavonoids

The flavonoids are an important component of seabuckthorn. Zhang *et al.* (1988 a,b) have reported isorhamnetin as the major flavonoid constituent. The other flavonoids detected include quercetion, kaempferol, myriatin and their glucosides.

Table. 8.1: Constituents of Seabuckthorn Fruit (per 100 grams fresh berries)

Component	Content	Reference
Vitamin C	200-1,500 mg (typical amount: 600 mg)	Rousi and Aulin (1977)Yao *et al.* (1992), Yao and Ticherstedt (1992)
Vitamin E (mixed tocopherols)	Up to 180 mg (equal to about 270 IU)	Lu (1992)
Carotenoids, including beta carotene, lycopene, zeaxanthine	30-40 mg	Kudritskaya *et al.* (1984), Lu (1992)
Fatty acids (oils); the main unsaturated fatty acids are oleic acid (omega-9), palmitoleic acid (omega-7), palmitic acid and linoleic acid (omega-6), and linolenic acid (omega-3); there are also saturated oils and sterols (mainly β-sitosterol)	8-12% (3-5% in fruit pulp, 8-18% in seed): fatty acid composition and total oil content vary with sub species	Franke and Muller (1983). Lu (1992),

8.2.2 Carotenoids

A large group of fat-soluble compounds, are responsible for many of the yellow, orange and red colours of the fruits of seabuckthorn. Carotenoids are also powerful antioxidants that protect plants from oxidative damage. Carotenoids protect humans by hindering free radical chain reactions caused by aging, smoking, pollution and other stresses. Carotene is dominant among the carotenoid pigments in seabuckthorn, the concentration of which varies from species to species and the place of origin (Zhang, 1987).

8.2.3 Seabuckthorn Berry Oil

Seabuckthorn oil has multiple benefits in the area of restorative and anti-aging skin care. Natural antioxidants and essential fatty acids help reverse damaging effects of sun radiation, reduce skin inflammation, promote natural restorative processes, minimize long term effects of sun exposure, like wrinkles, dryness, dark spots etc. The oil is well tolerated by any type of skin and helps combat acute sunburns and provide long term anti-inflammatory, restorative and revitalizing action.

8.3 HEALTH APPLICATIONS

Seabuckthorn berry was the common medicine used in ancient China by Tibet and Mongolian doctors, it had pharmacological functions like making expectoration easy, good for lungs and stomach, invigorating the function of the spleen, removing blood stasis and promote blood circulation etc., it was already recorded in The Rare Medicine Book of the King of Moon, the four medical dictiondries, in Tang dynasty, and a book on Chinese herbal medicine in Qing dynasty. In 1977, the Chinese Department of Public Health officially listed *Hippophae* in the The Dictionary of Chinese Medicines. Seabuckthorn has been shown to have a potent antioxidant activity, mainly attributed to its flavonoids and vitamin C content. Both the flavonoids and the oils from seabuckthorn have several potential applications. Some of the significant medical applications are listed below.

8.3.1 Mucous Membranes

Seabuckthorn seed and pulp oils combine high levels of beneficial fatty acids, natural antioxidants and vitamins as well as plant sterols. Clinical studies and animal experiments have shown multiple beneficial effects of the oils on mucous membranes (Erkkola and Yang, 2003). Mucous membranes cover the respiratory, digestive, uro-genital tracts and the inner surfaces of the eyes. Mucous membranes are important channels for interactions and substance exchanges between the human body and the environment. They are often the major routes for pathogens and external toxins and allergens to enter the body. The health condition of mucous membranes plays an important role in the general well-being of the whole body. Mucous membranes are constantly under the challenges of genetic deficiencies, diseases, stress, aging, side effects of medical treatments and environmental factors such as air and water pollutants. As a result people often suffer from troubled mucous membranes. Medical solutions for problems of mucous membranes are often unsuccessful.

8.3.2 Oesophagytes

Oesophagytes is often referred to as Radiation Esophagytes because it usually afflicts radiation patients. Chemotherapy is also suspected of causing esophagytes. It is an inflammation of the oesophagus lining. The experts call it "temporary discomfort because of pain, swelling, or dryness of the throat." Forty patients were diagnosed with oesophagytes. Thirty-four were males and six were females. 3-5 ml (0.96 mg/ml) of seabuckthorn seed oil were given to patients 3 times per day. The average treatment lasted 14 months. The effective rate was 97% (Li Minghzhong *et al.*, 1989).

8.3.3 Cancer Therapy

Most of the work done in this area has been with laboratory animals. Anticancer properties of seabuckthorn seed oil are effected in several ways. It is the combined action of anticancer factors, β-carotene, flavonoids, polyphenols, terpene, alkaloid and trace elements. Together they exert concerted effects that are far more potent than just the sum total of their individual effects. Anticarcinogenic factors in seabuckthorn seed oil can cut short the cancer process. For instance, it blocks carcinogenic effects of NDMA, hepatitis C virus, and aflatoxin, thus preventing growth of cancer cells. Antioxidants also effectively prevent damages done by free radicals, and cancer cell formation. Seabuckthorn seed oil stimulates the immune functions and activities of globulin C, increases antibodies, and enhances phagocytosis of macrophage, thus strengthening the body's resistance to cancerous mutations.

Seabuckthorn seed oil plays a very important role even for those who had major operations, undergoing chemotherapy or radiation. These are all drastic measures and the body will have to suffer great loss making the patient feeble and depleted. Chemo and radiation necessarily hurt the digestive system, blood generation, liver and kidney functions, and skin health. The patient may suffer nausea, vomiting, poor appetite, body weakness and bad mood, and many may experience complications such as uremia. The fact is many patients do no die of cancer, but of complications. Seabuckthorn seed oil not only has very good effects against cancer, it also can improve life quality of the patients by reducing the adverse effects of the above mentioned drastic measures. It provides the patient with a rich and comprehensive supply of nutrients that help improve conditions of the patient and resistance. Seabuckthorn seed oil also improves blood circulation and removes stasis, and eliminates waste, promotes tissue growth and avoids infection at the site of the operation. For those under going chemotherapy, taking seabuckthorn seed oil may help counteract the many side effects, improving gastrointestinal functions, increasing appetite, restoring liver and kidney functions, keeping the patient in good health. Used with other cancer drugs, it may improve the treatment and reduce recurrence and metastasis. Post-operation radiation may result in skin burn, ulceration, drastic drop of white blood cells and lowering of resistance to pathogens. Seabuckthorn seed oil increases the self-repairing ability of the traumatized organs and cells, promotes the functions of the spleen and bone marrow and blood generation, and protects the heart, spleen, liver, kidney against the damaging effects of radiation. This not only helps strengthen the body's resistance, but also prevents other problems radiation may lead to. Agarwal *et al.* (2002) have shown that alcohol extracts of seabuckthorn protect the bone marrow from radiation damage.

8.3.4 Cerebral-cardiovascular Health

Both the incidence and death rate of cerebral-cardiovascular disease are very high and currently it is the number one killer taking more than 16 million lives per year. The principal culprit of this disease is arteriosclerosis that is closely related to high blood fat. Seabuckthorn seed oil actively reduces blood fat level on the one hand and nurture the blood vessels and improves the quality of the vessels. Of the rich unsaturated fatty acids, oleic acid reduces cholesterol; linoleic acid regulates blood pressure and reduces serum cholesterol and prevents arrhythmia; and linolenic acid reduces blood pressure, dissolves accumulated fat and promote metabolism. Together they also inhibit platelet agglutination and prevent thrombosis. Vitamin E is a strong antioxidant, promoting metabolism, preventing oxidation of high density cholesterol in the serum, thus inhibiting and reducing cholesterol settlement along the blood vessel wall. It also prevents oxidation of unsaturated fatty acids and enhances immunity.

Seabuckthorn seed oil has tonic effects to cerebral-cardiovascular systems. Total flavonoids increase blood flow of the coronary artery and nutrients in the blood supply to heart muscles, lower oxygen consumption, strengthen muscle contraction, thus improving the heart function and to increase the anti-hypoxia under normal or below normal blood pressure. Sitosterol softens the blood vessels, and increases their tenacity, improves its elasticity, thus preventing hardening (arteriosclerosis). Vitamins E and other antioxidants remove wastes including peroxides, ailing and dead cells from the blood, and avoid damages the wastes may do to the artery walls. 5-serotonin and betaine protect the cerebral cardiovascular functions by comprehensive regulation and coordination of the nervous, endocrine and the immune systems (Xiao, 2003, Cheng *et al.,* 2003, Yang and Kallio, 2002). A herbal tea formulation has been made at DFRL Mysore with seabuckthorn leaves as the major ingredient. The formulation was fed to rats that were made hypercholesterolemic by feeding cholesterol at 2% level. The formulation was found to reduce serum cholesterol and LDL levels (Swamy *et al.,* 2004). The formulation was also found to increase the antioxidant defence enzyme activities *in vivo.*

8.3.5 Gastrointestinal Health

Importance of digestive health can never be over-emphasized as this is the provider of nutrients for the body to ensure its normal functions. Statistics show that 25% of adults, 32% of students, 65% of married women, and 90% of elderly people suffer from gastrointestinal disorders, and the incidents are on the rise. These include gastric and duodenal ulcer, acute and chronic gastritis, chronic diarrhea, colitis, etc. Containing numerous anti-inflammatory and anti-ulcer substances, the pharmaceutical effects of seabuckthorn seed oil is direct and clear (Zhou Yuannpeng *et al.*, 1998, Xing *et al.*, 2002). The concerted and mutual promoting actions of the more than 100 elements have positive effects on the endocrine, circulation, immune, and nervous systems and this in turn creates a most favorable condition for the digestive system. First, the oil provides a protective coating inside the stomach, intestines and duodenum, thus preventing pathogens from harming them, and keeping pathogenic changes from spreading. The anti-inflammatory and anti-ulcer properties of b-sitosterol-b-D-glucosides, ursolic acid and betaine promote ulcer healing, prevent spread and inhibit erosion. β-carotene, vitamin E, unsaturated fatty acids stimulate cell metabolism at site and repair of injuries. Meanwhile, seabuckthorn seed oil has remarkable inhibiting effects on gastric acid and abnormal increase of gastric proteinase, thus keeping the ulcer from spread and metastasis.

8.3.6 Liver Cirrhosis

Liver cirrhosis is a common chronic hepatic injury caused by chronic hepatitis B, ethanol consumption and metabolic disorders, etc. The patients often die of hepatic failure due to portal hypertension, bleeding of oesophageal and gastric varices. Recent studies have shown that fat storing cells called hepatic stellate cells (HSCs) are the main collagen producing cells in fibrotic liver. Under the influence of inflammatory cytokines, vitamin A-rich cells are activated, proliferating and transforming into myofibroblasts, producing extracellular matrix(ECM) (Gao *et al.*, 2003). Retinoic acid droplets and retinoic acid receptors (RAR) diminish. Recent studies have also shown that when retinyl esters and RAR contents are restored in HSC, HSCs would remain in the inactivated state. Hence, HSCs are regarded as the therapeutic targets for prevention and treatment of hepatic fibrosis Gao *et al.*, 2003. In this study, seabuckthorn (*Hippophae rhamnoides,*) was used in cirrhotic patients to determine its effect on the changes of fibrotic parameters, improvement of liver function and whether it could be used as a therapeutic antifibrotic agent. Zhao *et al.*, (1998) reported that seabuckthorn could protect the liver from damage by CCl^4. A combination of an antiviral drug and seabuckthorn in treating patients with chronic hepatitis B could shorten the duration for the normalization of serum ALT. A clinical trial demonstrated that seabuckthorn extracts helped normalize liver enzymes, serum bile acids, and immune system markers involved in liver inflammation and degeneration (Gao *et al.*, 2003). In addition, seabuckthorn oil protects the liver from damaging effects of toxic chemicals, as revealed in laboratory studies (Cheng *et al.*, 1990). Experiments show seabuckthorn seed oil reduces liver damage by alcohol, paracetamol and carbon tetrachloride, a strong poison to the liver. It also has a preventive property to fatty liver. This is because it increases metabolism of cholesterol and fat, thus extensively improves liver fat metabolism disorders. This in turn may reduce incidence of cancer and arteriosclerosis.

8.3.7 Seabuckthorn Seed Oil and Immunity

Immunity is the ability and process of resistance of the human body against pathogens harmful to our body. This is crucial to good health because it means not only freedom from diseases and pains, but also an optimal condition physically, mentally and socially. Continuous good health depends on two things: prevention and wholistic maintenance. In other words, to eliminate pathogens before illness strikes and to avoid further pathological changes and complications when it does. Seabuckthorn seed oil provides more than 100 nutrients and bioactive substances, and many of them are considered immune building factors, such as linolenic acid, vitamins E, A, carotene, polyphenols, flavonoids and trace elements including iron, copper, manganese, selenium etc. This group of well-coordinated and well-proportioned immune building factors regulates and improves the immune system in different processes and at different levels, restores disturbed immune system, strengthens the body's resistance to diseases, thus effectively preventing diseases and speeding up recovery. Experiments show that seabuckthorn seed oil can substantially enhance the immunity of cells and body fluids, promotes phagocytosis of macrophage, and markedly restore natural killer cell level. Since the later 1970s, scientific studies led to many important discoveries about the adverse effects of free radicals and oxidation on our health. It is found that more than 80 diseases are related to free radicals, the most common ones include stroke, myocardial infarction, infant anemia, diabetes, arteriosclerosis, tumors, cataract, radiation damage, burn, brown spots, wrinkles, aging, arthritis, lung disease, nephritis, hepatitis, etc. Seabuckthorn seed oil is a pure natural antioxidant store. Its rich content of flavonoids, glucosides, phenols, terpenes, vitamins E, A, C, β-carotene, and trace elements including iron, zinc, selenium, manganese etc., are all antioxidants with very low molecular weight to neutralize free radicals. With collective power they are even more effective. At the same time, seabuckthorn seed oil can also activate superoxide dismutase whose role in the body is to eliminate free radicals. As a natural immune enhancer, it maintains the stability of the immune system and keeps the supervisory role of the system normal, thus eliminating the mutant and dead cells caused by free radicals, and enhancing phagocytosis of macrophage and killing cancerous cells.

8.3.8 Sub-health Condition

The World Health Organization found that 10% of world adult population are in sub-health condition in which all physical and chemical indexes seems normal, but the person is suffering from all kind of discomfitures and problems. Sub-health condition includes (i) chronic fatigue complex as results of over taxing physical and mental work or long term stress, such as insomnia, forgetfulness, somnolence, profuse dreaming, dizziness, distention in the head, mental fatigue, cumbersome limbs, shoulder and back soreness, lack of concentration, proneness to flu and cold, etc; (ii) poor appetite, dizzy head and vision, depression and fright, vexation, agitation, irascibility, tidal fever, vacuity sweating, congestion, short of breath, general body weakness, irregular menses, poor sexual performance which are the results of endocrine disorders, menopausal complex and aging; and (iii) vacuity sweating, congestion, short of breath, palpitation, dizzy head and vision, general lack of strength which occur after a serious illness or chronic diseases. This is a critical state of health in which pathogens may invade and diseases take place. Seabuckthorn seed oil provides various nutrients and bio-active substances the body needs, such as unsaturated fatty acids, carotenoids, phytosterols, amino acids, vitamins and trace elements. They regulate and harmonize the endocrine, circulatory, autonomic nerve and digestive systems. This with antioxidant and free radical resistant properties, may effectively restore health and get rid of sub-health conditions.

8.3.9 Anti-ageing and Youthfulness

Experts compared seabuckthorn seed oil with five most frequently used skin creams and found that the content of free fatty acids, carbohydrates, phytosterol, phosphatide, vitamins E and A, carotenoids that are dermatopically effective are many times higher in seabuckthorn seed oil than their counterparts. They are organically combined together providing more protection and nutrition to the skin. The free fatty acids for an instance, is easily absorbed by the skin which it helps keep slightly acidic, and inhibits fungus and mould. Linoleic acid and linolenic acid promote repair of tissue damage, and the hydrocarbon prevents excessive moisture loss of the skin, softens and lubricates the skin and keep it smooth and fair. Sitosterol improves mirco-circulation in the skin, phosphatide forms a protective film to prevent moisture loss and keep invaders at bay, and acts both as a moisture regulator and an antioxidant. As an antioxidant, fighter and neutralizer of free radicals and peroxides, and immune builder, seabuckthorn seed oil is also an effective anti-aging agent, slowing down the aging process both internally and externally.

8.3.10 Skin

An ingredient of the oil, palmitoleic acid, is a component of skin. It is considered a valuable topical agent in treating burns and healing wounds. This fatty acid can also nourish the skin when taken orally if adequate quantities of seabuckthorn or its oil are consumed; this is a useful method for treating systemic skin diseases, such as atopic dermatitis (Yang *et al.*, 1999). It is an ingredient in sunblock-*Hippophae* oil has UV-blocking activity as well as emollient properties-and it is an aid in promoting regeneration of tissues (Ianev *et al.*, 1995). The fruit may also be used for benefiting the hair: the name *Hippophae*, means shiny horse, and refers to the good coat developed by horses feeding off the plant. The miracle properties of stimulation to tissue regeneration and epithelial healing with analgesic and anti-inflammation effects but without leaving scars, make seabuckthorn seed oil a most welcome remedy for burn, cuts, frostbites, bed sores etc.

8.3.11 Anti-radiation

Radiation is not limited to post operational radiation treatment of cancer. As a matter of facts, many people either due to their profession or special environment, may be subjected to radiation including computer operators, cell phone users, medical personal related to X-ray, electronic or nuclear workers, astronauts, pilots, stewards stewardesses etc. Vitamin E, carotenoids, flavonoids, unsaturated fatty acids, malic acid and selenium promote self-reparation of damaged tissues and cells. As a whole seabuckthorn seed oil stimulates bone marrow and blood generation and thus protects the heart, spleen, liver and kidney from radiation damage.

8.3.12 Woman's Health

Gynecological problems most frequently seen are menstruation pain, menopausal syndromes, anemia, cervical erosion, cervicitis, vaginitis, etc. Seabuckthorn seed oil provides comprehensive nutrients and has multi-effects on the endocrine, circulation, nerve, immune and reproductive systems by regulating and coordinating them. It also has antibacterial, antifungal, anti-infection and anti-inflammation properties and both internal and external use can effectively deal with menstruation pain, cervical erosion, cervicitis and vaginitis.

8.3.13 Children's Health

Children need a complete and well-balanced nutritional structure to grow strong and healthy both physically and mentally. Seabuckthorn seed oil provides vitamins A, E, D, K, etc. Carotenoids, trace elements including iron, zinc, calcium, copper, manganese, selenium and iodine which are easy to absorb without any hormones. At the same time, it contains various immune regulating factors that increase children's resistance against diseases and viruses without side-effects. Therefore, it is better to add seabuckthorn seed oil to the diet of children including infants.

8.3.14 Synergetic Effects

Seabuckthorn also contains many kinds of substances that are rare in other plants, such as terpene, alkoloid that are also useful for good health. Specialists and scholars are amazed that so many health building factors are combined organically in seabuckthorn seed oil. Research found that the ratio among these bioactive nutrients is so beneficial to human health that they are easily absorbed when taken. These bioactive components are not playing their own games but are well coordinated, supplementing and reinforcing each other to fight pathogens. This synergetic effect, or the integrative effect, can only be achieved by the miracle hand of Mother Nature. For an instance, there are many antioxidants such as vitamin E, vitamin A, vitamin C, polyphenols, flavonoids, etc. The antioxidant property of seabuckthorn seed oil is not simply the sum total of individual elements, but that of the synergetic effects of all working together, thus superceding the total quantity and achieving a new quality.

8.3.15 Burns and Seabuckthorn

Zhou *et al.*, (1998) has reported a study wherein thirty-two cases were treated, the oldest was 78 years old, the youngest was one year and two months old. Twenty-eight cases of scalds were caused by various reasons, and four cases were caused by gas burns; among them 12 cases were degree I burns, 18 cases were degree II light burns, 2 cases were degree II serious burns. In 28 cases, they were treated with *Hippophae* seed oil directly after burning, and in 4 cases, they were treated in other hospital first and then transferred and treated with *Hippophae* seed oil. The study reports that seabuckthorn oil had magic effects in treating burns and scalds, easy to use, reliable and effective.

8.4 MANUFACTURE OF SEABUCKTHORN PRODUCTS

Many agree that food and eating are an important part of their lives and that nutrition is a prime concern when making food and eating decisions. Most consumers now believe that food can have a potent impact on health. The movement towards better health through improved nutrition has spurred the development of new foods, food supplements, herbal remedies and other enhanced food products. The "natural products" industry, which includes fresh and organic foods and beverages, dietary supplements and "all natural" personal health care products, grew 25% in 1996 and continues to experience ongoing growth. Food trend futurists predict that by year 2006, a shift in focus from the health aspects of food to the entertainment and pleasure associated with food will occur. Future consumers will want specialized products and services. The consumer will expect a blend of pleasure and medicinal, customized to individual needs. Diversity will still matter, but self-care and the combination of the pleasure and medicinal, is likely to become very important in the future. Current and future trends for new foods such as seabuckthorn bode well. Seabuckthorn, as a crop and future food source, has been gaining attention because of its nutritional benefits.

Following is a list of products being developed at Defence Food Research Laboratory, (DRDO) Mysore.

Health drink: A number of health drinks have been developed from seabuckthorn pulp such as squash, spiced squash, squashes blended with other fruits such as mango, pineapple, apple, guava, grapes etc. Ready To-Serve (RTS) drinks (Kumaran *et al.*, 2004, 2005a).

Jam and Jellie: These products have also been developed from seabuckthorn pulp alone or in combination with other fruits such as mango, pineapple, apple, grapes and papaya (Kumaran *et al.*, 2003).

Yogurt: Seabuckthorn constitutes the major ingredient along with milk and bifido bacteria (Kumaran *et al.*, 2005b).

Fruit juice powder: Ready to mix fruit juice powder has been prepared after spray drying the seabuckthorn pulp.

Herbal tea: A tasty, refreshing, stress busting, thirst quenching tea has been developed from the dried leaves of seabuckthorn.

Bakery products: A number of baked goods such as biscuits, cakes, cookies, rusks, buns and breads have been developed from seabuckthorn leaves and leaf extracts.

Wine: Wine has been developed from seabuckthorn pulp alone and in combination with grapes.

Seabuckthorn oil: Oil has been extracted from the seeds and pulp portions of the fruit individually.

Food colourant: Seabuckthorn yellow has been isolated from the waste left over after the extraction of pulp from the fruits.

Animal feed: Residues after extraction of colourants has been utilized in the development of animal feed.

A number of food products have also been developed by other workers such as fruit juices (Beveridge *et al.*, 1999, Dwivedi, et al. 2001, Chuhan *et al.*, 2003).

8.5 CONCLUSIONS

It is no exaggeration that seabuckthorn is a great herb store of natural remedies by itself, and if you decide to take just one supplement, this is it. Cardiotonic, stomachic, dermatropic, antioxidant, anti-infection, anti-fungal, anti-inflammatory, anti-radiation, anti-constipation, healthy cholesterol maintaining, blood vessel softening, blood fat reducing, pro-blood circulation and pro-natural healing, seabuckthorn has a wide range of uses both internal and external, including the prevention and cure of cardiovascular disease, cancers, ulcers, burns, and immune disease, gastrointestinal disease, respiratory disease, urinary disease such as yeast infection, and gynecological disease, skin disease, etc. As seabuckthorn seed oil protects the white blood cell against damage by radiation and chemotherapy and speeds up regeneration of tissue cells and promoting natural healing, it can provide essential help to those under going radiation and chemotherapy for cancer patients.

- Seabuckthorn regenerates skin cells,
- Seabuckthorn is anti-aging,

- Seabuckthorn seed oil works on demodex mites that are said to cause rosacea. Seabuckthorn supports prostate health,
- Seabuckthorn supports hormone balance,
- Seabuckthorn seed oil promotes white blood cell health,
- Seabuckthorn seed and seabuckthorn pulp oils used to alleviate symptoms of redness, inflammation, pustules, blemishes, itching,
- Seabuckthorn seed oil has been used successfully in yeast infections, thrush, vaginal, foot problems,
- Seabuckthorn seed and seabuckthorn pulp protects liver functions,
- Seabuckthorn oils strengthen the immune system,
- Seabuckthorn protects against inflammatory itis', such as arthritis, polyarthritis and more due to the high levels of β-carotenes and anti-oxidants in the oils,
- Seabuckthorn seed oil inhibits gastric secretions,
- Seabuckthorn oil relieves pain and accelerates the repair process of gastic and duodenal epithelial tissue and mucosa,
- Seabuckthorn seed oil was topically used [3-4 times a day] on 60 children [4 months-12 yrs] with ulcerative stomatitis. Significant improvement in 2 days w/all 60 cases, 55 cases cured in 3-5 days, 2 severe cases cured in 8 days,
- Seabuckthorn seed and seabuckthorn pulp oils are anti-inflammatory with analgesic effects reported,
- Seabuckthorn seed oil sprayed on topically once a day, cured cervititis after 3 months of treatment,
- Seabuckthorn seed oil and seabuckthorn berry oil gets rid of inflamed skin, itching, pustules, demodex mites, bumps, dry, flaky skin,
- Seabuckthorn seed and seabuckthorn pulp oils rejuvenate & regenerate the skin and body, inside and out,
- Seabuckthorn seed oil, 1.5 teaspoons a day for 1 month for curing peptic ulcers,
- Seabuckthorn seed oil and seabuckthorn pulp oils mixed 1:1, orally for dry mouth, 5 g a day for 4 weeks, improves symptoms,

- Seabuckthorn seed oil promotes tissue granulation for burns, wounds and scalds,
- Seabuckthorn seed and seabuckthorn pulp oils reduce peroxidation of cell membranes,
- Seabuckthorn seed oil helps regulate immune functions, antagonizing immune suppressants,
- Seabuckthorn seed and seabuckthorn pulp, are used to reduce the immune suppressive and hematoxic effects of chemotherapy and irradiation therapy,
- Mix of seabuckthorn seed and seabuckthorn pulp improved oesophagitis caused by irradiation therapy,
- No side effects reported.

References

Agarwal, P.K. and Goel, H.C. 2002. Protective effect of RH-3 with special reference to radiation induced micronuclei in mouse bone marrow, *Indian J. Exptl. Biol.* 40 (5), 525-530.

Bernath, J. and Foldesi, D. 1992. Seabuckthorn (Hippophae rhamnoides L.): A promising new medicinal and food crop, *J. Herbs, Spices, Medicinal Plants* 1, 27-35.

Beveridge, T., Harrison, J.E. and Drover, J. 2002. Processing effects on the composition of seabuckthorn juice from Hippophae rhamnoides L. Cv. Indian summer. *J Agric Food Chem.* 50, 113-116.

Beveridge T., Thomas S.C, Li B., Dave Oomah and Smith, A. 1999. Seabuckthorn products: Manufacture and composition. *J Agric. Food Chem.* 47, 3480-3488.

Chauhan, A.S, Rekha, M.N, Ramteke, R.S. and Eipson, W.E. 2003. Seabuckthorn (*Hippophae rhamnoides* Lin.) berries: Harnessing its potential for processing. *J. Food Sci. Technol.* 40, 349-356.

Chen, T.G., Ni, M.K., Li, R., Ji, F. and Chen, T. 1991. Investigation of the biological properties of central Asian sea buckthorn growing in the province of Kansu, China, *Chem. Natural Compounds* 27, 119-121.

Cheng, J., 2003. Inhibitory effects of total flavones of Hippophae rhamnoides on thrombosis in mouse femoral artery and in vitro platelet aggregation, *Life Sciences* 72(20), 2263-2271.

Cheng, T., 1990. Acute toxicity of flesh oil of Hippophae rhamnoides and its protection against experimental hepatic injury, *Journal of Traditional Chinese Medicine* 15(1), 45-47, 64.

Erkkola, R. and Yang B. 2003. Seabuckthorn oils: towards healthy mucous membranes. *Agro Food Industry* May/June.

Franke, W. and Muller, H. 1983. "A contribution to the biology of useful plants. 2. Quantity and composition of fatty acids in the fat of the fruit flesh and seed of sea buckthorn," *Angewandre Botanik* 57, 77-83.

Gao, Z.L. 2003. Effect of sea buckthorn on liver fibrosis: a clinical study. *World J Gastroenterol.* Jul; 9(7), 1615-1617.

Goel, H.C, Gupta, D, Garg, A.P. and Bala, M. 2005. Protection of mitochondrial system by *Hippophae rhamnoides* L against radiation induced oxidative damage in mice. *J.Pharm Pharmacol.* 57 (1), 135-43.

Hirvi, T. and Honkancn, E. 1984. The aroma of the fruit of sea buckthorn, *Hippophae rhamnoides* L, *Zeitschrift fur Lebensmittel-Untersuchung and Forschung* 179, 387-388.

Ianev, E., 1995. The effect of an extract of sea buckthorn on the healing of experimental skin wounds in rats, *Dermatology* 48(3), 30-33.

Katiyar, S.K., Sharma, K., Kumar, N. and Bhatia, A.K. 1990. Composition of some unconventional Himalayan wild fruits. *J.Food Sci. Technol* 27, 309-310.

Kudritskaya, S.E., Zagorodskaya, L.M. and Shishkina, E.E. 1989. Carotenoids of the sea buckthorn, variety Obil' naya, *Chem. Natural Compounds* 25, 724-725.

Kumaran, M.S., Khanum Farhath, Bawa, A.S. 2003. Functional fruit jam from seabuckthorn pulp. *5th Intl. Food Convention* Dec 5-8, Mysor, India.

Kumaran, M.S., Khanum Farhath and Bawa, A.S. 2004. Spiced sebuckthorn mixed squash. *National Seminar on cultivation, Harvesting and Scientific Exploitation of seabuckthorn.* Aug 26-27, Leh, India.

Kumaran, M.S., Khanum Farhath and Bawa, A.S. 2005a. Development of seabuckthorn fruit yogurt. *National Seminar on Recent Trends in Biotechnology of fermented Foods.* July 13-15, Madurai, India.

Kumaran, M.S., Khanum Farhath , Vibhakar, H.S. and Bawa A.S. 2005b. Development of seabuckthorn mixed fruit squash using response surface methodology. *Intl. Symposium on Food Biotechnology, Biovision-* 2005, Oct 14-15, Thrissur, India.

Kumaran, M.S., Khanum Farhath, Jeevaratnam K. and Bawa, A.S. 2004. Optimization of conditions for preparing functional fruit jelly from seabuckthorn. *Intl.Conf. on Natural Products, Free radicals and Radioprotectors in Health* (NFRH-2004 and SFRR-India).

Li, Minghzhong 1989. Experimental researchers treating the white radiation oesophagytes cancer and injury with seabuckthorn oil compound. *Seabuckthorn* 2 (4), 37-40.

Li, T.S.C. and Schroeder, W.R. 1996. Sea buckthorn (*Hippophae rhamnoides*): A multipurpose plant, *Horticultural Technology* 6(4), 370-378.

Lu, R. 1992. Seabuckthorn: a Multipurpose Plant Species for Fragile Mountains. Intl. Ctr. Integrated Mountain Devleopment, Karmandu, Nepal. pp. 62.

Padmavathi, B., Upreti, M., Singh, V., Rao, A.R., Singh, R.P. & Rath PC. 2005. Chemoprevention by *Hippophae rhamnoides*: effects on tumorigenesis, phase II and antioxidant enzymes and IRF-1 transcription factor. *Nutr. Cancer* 51 (1), 59-67.

Rosch, D., Krumbein, A., Mugge, C., Kroh, L.W. 2004. Structure-antioxidant efficiency relationships of phenolic compounds and their contribution to the antioxidant activity of sea buckthorn juice, *J. Agric. Food Chem.* 51(15), 4233-4239.

Rousi, A. and Aulin, H. 1977. Ascorbic acid content in relation to ripeness in fruit of six Hippophae rhamnoides clones from Pyharanta, SW Finland, *Ann. Agr. Fenn* 16, 80-87.

Swamy, M.S.L, Khanum Farhath and Bawa, A.S. 2004. Hypocholesterolemic effects of seabuckthorn leaf based herbal tea. *National Seminar on cultivation, Harvesting and Scientific Exploitation of seabuckthorn*. Aug 26-27, Leh, India.

Wang, S. 1990. Studies on the chemical components in fruits of *Hippophae rhamnoides, Forest Res.* 3, 98-102.

Xiao, Z., 2003. The inhibitory effect of total flavonoids of hippophae on the activation of NF-kappa ß by stretching cultured cardiac myocytes, *Sichuan University Medical Journal* 34(2), 283-285.

Xing, J., 2002. Effects of sea buckthorn seed and pulp oils on experimental models of gastric ulcer in rats, *Fitoterapia* 73(7-8), 644-650.

Yang, B., Kalimo, K.O., Tahvonen, R.L, Mattila, L.M, Katajisto, J.K. and Kallio H.P. 2002. Effect of dietary supplementation with sea buckthorn (*Hippophae rhamnoides*) seed and pulp oils on the fatty acid composition of skin glycerophospholipids of patients with atopic dermatitis. *J Nutr Biochem* Jun;11(6), 338-40.

Yang Baoru, 1999. Effects of dietary supplementation of sea buckthorn oils on fatty acids in patients with atopic dermatitis, Proceedings of the International Sea Buckthorn Congress, ICRTS, Beijing.

Yao, Y., Tigerstdet P.M.A. and Joy, P. 1992. Variation of vitamin C concentration and character correlation between and within natural sea buckthorn (*Hippophae rhamnoides* L.) Populations, *Acta Agr. Scand.* 42, 12-17.

Zhang Maoshun, 1987. Treatnent of ischemic heart diseases with flavonoids of Hippophae rhamnoides, *Chinese Journal of Cardiology* 15(2), 97-99.

Zhang, T.Y., Hua, X., Xiao R. and Kong, S. 1988b. Separation of flavonoids in crude extracts from seabuckthorn by counter current chromatography with two types of coil planet centrifuge. *J .Liquid Chromatogr* 11, 233- 244.

Zhang, T.Y., Xiao, R. Xio, Z.Y., Pannel, Ito Y. 1988a. Rapid separation of flavonoids by analytical high speed counter current chromatography. *J.Chromatogr.* 445, 199-206.

Zhong, Fei, 1989. Effects of the total flavonoid of Hippophae rhamnoides on nonspecific immunity in animals, *Shanxi Medical Journal* 18(1), 9-10.

Zhao Yuanpeng, 1998. Study on the effect of hippophae seed oil against gastric ulcer, Institute of Medical Plants Resource Development, The Chinese Academy of Medical Sciences, Beijing.

□□□

9

Nutritive Value of Fruits of Seabuckthorn

Deepa H Dwivedi and Sanjai K Dwivedi

CONTENTS

Seabuckthorn (*Hippophae spp*), is a deciduous, dioecious, thorny shrub found growing wild in the cold arid climatic conditions of India, China/ Tibet, Magnolia, Russia, Finland, Germany, Nepal, Pakistan, Afghanistan etc. It bears small, soft, orange/yellow highly nutritive fruits with numerous foods, medicinal and cosmetic uses. The array of biomolecules synthesized by seabuckthorn has imparted a central role to the plant not only as health promoter but also for prevention and treatment of different diseases (Dwivedi *et al*, 2002). Seabuckthorn has been identified as *Amalavetas*, an ayurvedic plant. *Amaltas* name has been used for other plant species in various regions of India. However, *Amaltas* is the plant species which is real seabuckthorn in Himalayas, being used by traditional doctors for curing various health problems (Chauhan *et al*, 2003). In modern drugs development every plant part are being bioprospected for protection against cancer, radiation and use of its specific flavonoides in wound healing as an important antioxidant (Geetha & Sawhney,2004). Seabuckthorn is one of the most fascinating plant species in the world and can be termed as natural biofactory capable of synthesizing more than 200 biomolecules having immense medicinal, nutraceutical and cosmeceutical values.

9.1 CHEMICAL COMPOSITION AND NUTRACEUTICAL VALUE

Fresh seabuckthorn fruits are rich in nutrients such as carbohydrates, organic acids, amino acids and vitamins and the vitamin content of seabuckthorn is much higher than any other fruit or vegetable. Because of its high content of vitamin C, seabuckthorn is usually used to make soft drinks and other similar food products (Rongsen, 1992). The seabuckthorn juice is a good source of several amino acids and minerals (Table 9.2 and 9.3).

The fruits are very soft and juicy but perishable. Ripe fruits of Seabuckthorn are a rich source of various nutrients, minerals and biomolecules. Aqueous extract of fruits of Seabuckthorn are added to other products for decreasing their freezing points. Various products from seabuckthorn were used in the diet of Russian cosmonauts and Seabuckthorn cream was used for protection from cosmic radiation. Fruit yields about 33.5 per cent yellow pigment which can be used as natural food colour (Bawa, 2004). Fruit are rich in 5 hydroxytryptamine (5-HT) which is very rarely present in plant kingdom. Fruits also contain beteine compound which has anti ulcer activity (Geetha and Sawhney, 2004). The biochemical composition of ripe fruits of Seabuckthorn have been studies extensively by various researchers and have been presented in Table-9.1.Based on the various studies, brief of major nutrients of Seabuckthorn are discussed below-

Table 9.1: Physico-chemical characteristics of fruits Indian Seabuckthorn *(H. rhamnoides L.)*

Parameters	**Katiyal** *et al.* **(1990)**	**Chauhan,** *et al.* **(2003)**	**Dwivedi,** *et. at.* **(2006)**
Fruit size (mm)	5-7	5.38 - 5.42	6.62
Shape	ovoid	ovoid & Small	Round, oval, oblong
Colour	Orange Yellow	Orange Red	Yellow, Orange Orange, Red
Taste	Sweetish & sour	Sweetish sour with exotic flavour	Sour
Av. Fruit weight (mg)	-	210	140.7
Seed number	-	2-4	01
TSS °brix	-	10-12	14.3
Acidity as (% citric acid)	-	1.72 - 1.95	2.54 (as malic acid)
Total sugar (%)	-	6.2 - 6.95	1.03
Juice yield (%)	-	65-68	74.5
Residue (%)	-	25-28	19.04
Seed (%)	-	10-12	6.45
Moisture (%)	52.4	58.76	74.58
Ash (%)	1.76	1.8	1.8
Crude Protein (%)	2.23	-	2.64
Crude Fibre (%)	4.75	-	3.45
Total Carbohydrate (%)	32.37	-	20.56
Sodium (mg/100g)	69	-	41.28
Potassium (mg/100g)	622	30.08	1499.96
Calcium (mg/10 mg)	671	-	383.0
Iron (mg/10 mg)	16	-	11.68
Phosphorous(mg/10 mg)	74	-	0.02%
Ascorbic acid (mg/100g)	50	422-516	424.8
Carotenoides (mg/100g)	-	6.8-6.9	12839.6 (ug/100g)

Table 9.2: Contents of various amino acids (mg/100g) in seabuckthorn berry juice

Amino acids	Zhang *et al.* (1989),	Chen *et al.* (1988)
Aspartic acid	426.6	3.72
Threonine	36.8	6.24
Serine	28.1	5.31
Glutamic acid	19.4	2.65
Glycine	-	0.64
Alanine	21.2	2.50
Cysteine	3.3	0.82
Valine	21.8	2.85
Methionine	2.3	1.12
Tryptophan	-	0.51
Isoleucine	17.4	0.97
Leucine	-	1.94
Tyrosine	13.4	1.79
Phenylalanine	-	3.21
Histidine	13.7	1.06
Lysine	27.2	3.49
Arginine	11.3	0.47
Proline	45.2	12.28
Glycine	16.7	-

Table 9.3: Mineral content in seabuckthorn juice (mg/ml)

Parameters	Tong et al. (1989)	Zhang *et al.* (1989)	Liu and Liu (1989)
Potassium	100-806	0.147-0.209	-
Calcium	64-256	93.9-173	-
Phosphorus	-	82.1-206	-
Magnesium	53.3-165	39.8-103	-
Sodium	18.0-89.8	17.7-125	-
Cobalt	0.01-0.09	= 0.1	-
Chromium	0.47-1.00	0.108-0.287	-
Copper	-	0.158-0.653	<10.0
Manganese	0.81-3.86	1.17-2.60	-
Nickel	0.39-0.09	0.115-0.357	-
Strontium	0.08-0.45	0.19-0.616	-
Vanadium	-	0.002-0.009	-
Iron	5.93-16.1	4.13-10.9	-
Molybdenum	1.81	0.03-0.058	-
Zinc	2.09-6.13	0.431-1.25	-
Tin	-	0.045-0.259	-

Contd...

Table 9.3 Contd...

Selenium	-	7.96-11.3	-
Boron	-	0.43-1.38	-
Barium	-	0.168-0.362	-
Aluminium	-	2.2-16.7	-
Titanium	-	0.103-0.814	-
Lithium	0.06-0.15	0.132-0.33	-
Cadmium	0.002-0.015	<0.05	-
Arsenic	-	-	<0.5
Lead	0.06-0.27	0.431-0.761	<1.0

Vitamins: Fruits of seabuckthorn are a very rich source of vitamin 'C' and vitamin 'E' which indicates its strong antioxidant property. Therefore, fruits can be used in health protection juices and antitumor creams.

Organic acids: Fruit juice of seabuckthorn is rich in organic acids (2-3.5 % malic acid, oxalic acid, etc), which has been found to remit the toxicity of some antibiotics and barbitals, preventing teratogenesis and damage from X-rays. Presence of vitamin 'C', organic acid and tannic acids in the fruit of seabuckthorn make it an ideal source for the production of several beverages particularly health protection juices.

Carotenoids: Fruit of seabuckthorn is a very rich source of many carotenoids (total 39 identified), particularly ß-carotene which is one of the most bioactive substances and a precursor of Vitamin'A'and therefore promotes health of the body. Deficiency of carotenoides in human food leads to poor growth, weak eye sight, decreases tolerance to diseases and affects the mucous membrane of the gastrointestinal tract. Carotenoides, particularly ß -carotene and cantaxanthene, act as anti-carcinogens and anti swelling agents and therefore, cure different kinds of skin cancer. Carotenoides are an effective remedy against burn, frostbites, ulcers and various gynaecological problems.

Flavonoides and Phenols: Flavonoides isolated from seabuckthorn fruits and leaves have been found to improve the immunity of body, retard osmosis of the capillary wall and prevent vitamin C from breaking up. Many flavonoides of seabuckthorn have been found to neutralize free radicals; therefore they are anti-aging and anti-tumor. These flavonoides and phenols also control arteriosclerosis, lowering cholesterol level, eliminating the inflamation and preventing hyperthyroidism. The total flavonoides have been found to cure cardiovascular diseases and cancer. They improve the cardiac pump function, myocardial diastolic function and hemodyanamic performance and decrease myocardial oxygen consumption index and total peripheral vessel resistance. They exhibit the protective effect on myocardial ischemia. Leaf drug, containing flavonoides increase the wound healing after chemical burns and wounding. They inhibit the growth of tumors and exhibit a radioprotective action. They inhibit the growth of sarcoma 180, lymphosarcoma, Ehrlich tumor etc by 65-75 %. The most potent anti-tumour compounds are leucoanthocyanins. They also promote the anti-tumour effect during X-ray treatment and increase the resistance of human body to X-rays. Some flavonoides of seabuckthorn reduce the serum cholesterol level. Leaf catechins and flavonoides have also been proposed for the food industry.

Steroid: Among the sterols, b-sitosterol is considered one of the active compounds used to prevent and cure arteriosclerosis. The most studied sterols are b-sitosterol and stigmasterol, the biological activity of which includes anti-arteriosclerosis, healing of burns and stomach ulcer.

Betaine: There is a 0.09-0.36 % betaine compound (methylated product of glycine) which has very strong anti-ulcer activity and also curative effect on arteriosclerosis.

5-Hydroxy Ptamine(5-HT): This is rare compound of the plant kingdom found in seabuckthorn. It regulates human emotion, blood pressure, and body temperature and hormone level. It has been found to be anti-infection, anti-radiation and promotes coagulation of blood by transforming fibrinogen into fibrin.

Chlorogenic acid: Chlorogenic acid and other phenol compounds stimulate gastric secretion and take part in the diuretic function.

Coumarines: The leaves and fruits contain certain coumarines which strengthen the function of blood circulation in capillaries, are anti-vitiligo, anti-tuorigenesis and regulate the disorders of gall bladder.

Ursolic acid (trierpene) : It has the effect similar to adrenocortical hormone (ACH): It cures bronzed skin, health wounds, ulcers and inflammation.

ß –amyrinoleylcohol acid: Leaves of seabuckthorn contain this compound which dilates the cardiac and cerebral vessels, facilitates blood circulation and also lower blood pressure slightly.

9.2 AROMA VOLATILES

Seabuckthorn juice is immensely flavoured with exotic flavour and considered to be a good appetizer, which has health imparting attributes. A total of 60 volatile flavour components are identified in seabuckthorn fruit by GC-MS with concentration of about 36 mg volatile oil/kg .The volatile components of the seabuckthorn berries and characters impact components are listed in the Table 4.The aroma of seabuckthorn is generally due to the presence of certain aliphatic esters such as ethyl, 3-methylbutyli and cis-3-hexen-1-yl esters and the most important compounds are ethyl hexanoate, 3-methylbutyl, 3-methylbutanoate, 3-methylbutanoic acid, 3-methylbutyl benzoate and 3-methylbutyl octanoate whereas presence of terpene compounds are surprisingly low (Hirvi and Honkanen,1984).

9.3 PIGMENTS

The ripe fruits of Seabuckthorn have the potential for extraction of natural colouring substances for colouring edible food products. The rich yellow pigment of seabuckthorn is widely used as food additive since many synthetic chemical pigments have proved to be harmful to the health of human beings. The extraction of yellow pigment has been reported from pure juice and puree of seabuckthorn fruits and its yield was 33.5%(Bawa, 2004). The pure pigment is an orange-yellow powder that dissolves in water and alcohol. The colour varies significantly when the pigment is exposed to sunlight for 30 days at 80 °C for 40 min /day and it is stable at pH <6.5 but unstable at pH >.â -carotene is dominant among the carotenoid pigments in seabuckthorn, the concentration of which varies depending upon the place of the origin and the part analyzed. The seed oil from Chinese seabuckthorn varieties like Longxl, Shaanxl contain 159.8 mg/100 g â-carotene whereas the highest carotene content of 9059.6 mg/100g is reported in the pulp oil from Hetain and Xinjiang varieties of Seabuckthorn.

9.4 FLAVONOIDS, FLAVONES AND FLAVOLANS

Flavonoids are also important components of the fruit and various methods have been developed for its separation. Zhang et al. (1998) reported the major flavonoid constituent isorhamnetin in a crude ethanol extract of dried seabuckthorn fruit analysed by counter current chromatography with a 2-phase solvent system. Rapid separation of flavonoids by analytical high speed current chromatography is one of the best methods for the separation of seabuckthorn flavonoids.

Seabuckthorn oil has been reported to be the carrier of bio active substances. Out of total fatty acid, more than 80 per cent is unsaturated fatty acid. It contains about 60 per cent linoleic and linolenic acid. Pulp oil has more bio-active substances than seed oil. The oil is being extensively used for cosmetic and medicinal purposes. It has also been used for the protection against cosmic, UV and nuclear radiations.

9.5 VARIATION IN THE NUTRITIVE VALUES

The variations in the nutritive values/ bio chemical composition of fruits have been reported world wide. The variations in the nutritive values are primarily because of difference of the species of *Hippophae*. Different species have been reported to poses different compositions/ compound in fruit and plants. This variation is due to difference in the genetic trait/ make up of species which make one species different from others. However, variation in the same species can also be observed due to the effects of soil, climate, growing conditions, mophotype, stage of harvesting, method and time of collection, storage/determination etc. Various studies have been conducted to determine the variability in the ripe fruits of Seabuckthorn of the same species. The study conducted by Dwivedi *et al.* (2004, b) has shown the considerable variability in ripe fruits of Seabuckthorn (*Hippophae rhamnoieds ssp tukestanika* L). The colour of ripe fruits varied from pale-yellow, deep-yellow, orange-yellow, greenish-yellow, redish-yellow, yellow and light red. The shape of ripe fruits have also shown considerable variation as round, oval, oblong, ovate, elongate, elliptical and even capsular shape with flat and pointed end. The biochemical properties of fruits play vital role in the establishment of variability in fruit morphotypes. The total soluble solids content in the fruit juice plays important role in the processing of the juice. It has been observed that fruit juice contains 10.2 to 18.4 °B TSS with an average value of 14.3 °B. Low pH content of the fruit juice and high acidity(2-3 %) make the fruits of this plant unfit for fresh consumption/table purpose. Acidity varied from 1.46 to 3.62 percent with an average value of 2.54 percent. The fruits of seabuckthorn are known for its high vitamin 'C' content. However, the vitamin 'C' has been found to be most unstable and highly variable parameter in the fruit samples collected in the region. The highest Vitamin 'C' content was 1400 mg/100 g of ripe fruits while it was lowest with 138 mg/100g fruit. The average value of vitamin 'C' was 769 mg/100g fruit.(Dwivedi *et al.* 2006). The leaf samples collected from different places showed considerable variation in crude protein(18.1-22.84%), Ether extract(1.41-5.29%), Crude fiber(8.76-11.67%), total Ash(6.72-8.16%), total phosphorous(0.16-0.23%), calcium(1.52-2.23%) Vitamin E (8.19 mg /100 g), Iron (495.7 mg/kg) (Dwivedi *et al*, 2003,b). Seeds of seabuckthorn have shown variation in colour, light-dark brown to blackish brown and dull brown. Shape varied from oval to ovoid and oblong, elongate to elliptical.

Seabuckthorn products based on fruit pulp or juices are rich in vitamin-C, pro- vitamin-A and other vitamins (Dwivedi, *et al.*, 2004a), especially of the-B group and phytochemicals which are therefore considered to have therapeutic and chemopreventive activities (Li and Wang 1998; Li and Liu

1991). Seabuckthorn juice is known to block the endogenous formation of N-nitroso compounds more effectively than ascorbic acid and thereby prevent tumour production (Li and Liu 1991). It has been reported that concentrated seabuckthorn juice significantly inhibits the growth of sarcoma, lymphocytic leukemia and human gastric cancer cells. Seabuckthorn fruits are used for preparation of jelly and used against pulmonary complains and also in Siberia, the plant is used to prevent growth of tumor and sunburn protection, as emollient in prevention of eye region skin wrinkles and in other cosmetics preparation. In China, various parts/products of the plant utilized in medicines, skin creams, oil, jams, jellies, squashes and wines. Syrup prepared from it is used in lung complaints; a decoction is used for cutaneous eruptions. Fairly stable food products with high vitamin C potency have been prepared from the fruit (Bapalal 1968; Chauhan2003).

9.6 CONCLUSIONS

Seabuckthorn is a repository of bioactive compounds which have a restorat and regenerative function. The plant is able to tolerate abiotic stresses like high salt concentration, water, soil nutrient and low temperature stresses etc under the moist and hostile climatic condition of dry Himalayas.. Fruits of seabuckthorn are rich in flavonoides, polyphenols, vitamins, antioxidants, unsaturated fatty acids etc. Other plant parts viz leaf, stem, seed, bark, are also rich in medicinal properties which have been in use traditional system of medication for treatment of various ailments. A complete map of the bioactive agents in seabuckthorn need to be studied after extensive analysis and experimentation.

References

Bapalal 1968. *Nighantu Adarsha*, Vol-1. Chaukamba Bharati Academy, Varanasi, Pp128-130.

Bawa, AS. 2004. Seabuckthorn- A Potential functional ingredient for food industry. Souvenir and Book of Abstracts of National Seminar on seabuckthorn held at FRL, (DRDO), Leh, Aug, 26-27, 2004.Pp-5.

Chauhan, AS, Rekha MN, Ramteke, RS and Eipeson, WE 2003. Seabuckthorn Berries: Harnessing its full potential for processing. *J Food Sci. Technol.* 40(4).349-356.

Chen T et al. 1988. Studies of the biochemical composition of *Hippophae* and its quality assessment in Gansu Province. *Hippopahe.* 1: 19-26 (in Chinese).

Dwivedi S K, Paljor Eli, Attrey D P, Singh B. 2002. Propagation of common Seabuckthorn through hardwood cuttings in Ladakh. *The Indian Journal of Agricultural Sciences*.72 (4): 228-229.

Dwivedi, S.K., Kareem Abdul, Singh Ranjit and B Raut 2003. Studies on Evaluation of Seabuckthorn leaves for animal fodder in cold arid India. In: International Seabuckthorn Association meeting, Berlin (Germany), 14-18 Sep 2003. Abstract p-29-30.

Dwivedi S K, Attrey D P and Chaurasia O P. 2004 a. Studies on preparation and preservation of Seabuckthorn squash beverage. *Indian Journal of Horticulture*. 61(1): 78-80.

Dwivedi S.K, Singh Ranjit, Raut.B. 2004 b. Present status and future thrust on seabuckthorn research in Ladakh. Souvenir and Book of Abstracts of National Seminar on seabuckthorn held at FRL, (DRDO), Leh, Aug, 26-27, 2004.Pp-38-44.

Dwivedi S.K, Singh Ranjit, Ahmed Z. 2006. The Seabuckthorn. Field Research Laboratory (DRDO).Pp-8-24.

Geetha, S and Sawhney, R C. 2004. Seabuckthorn for health and medicine. Souvenir and Book of Abstracts of National Seminar on seabuckthorn held at FRL, (DRDO), Leh, Aug, 26-27, 2004.Pp-7-11.

Hirvi T, Honkanen E 1984. The aroma of the fruits of Seabuckthorn. *Food Res Technolo.* 179: 387-388.

Katiyar, SK;Sharma,K;Kumar, N and Bhatiya AK 1990. Composition of some unconventional Himalayan wild fruits. *J Food Sci. Technol.*27.309-310.

Liu J, Liu Z 1989. Research of proceedings of technology for seabuckthorn concentrated juice. Proceedings of International Symposium (*H. rhamnodies* L.) Xian, China , Oct. 19-23, pp314-317.

Li TSC, Wang LCH 1998. Physiological components and health effects of ginseng, Echinacea and seabuckthorn. In: Functionl Foods, Biochemical & Processing Aspects, Mazza G (ed). Technomic, pp 329-356.

Rongsen, Lu.1992. Seabuckthorn- A multipurpose plant for fragile mountains, ICIMOD, Kathmandu, Nepal.

Tong J, Zhang C, Zhao Z, Yang Y, Tian K 1989. The determination of the physical – chemical constitutes and sixteen mineral elements in Seabuckthorn raw juice. Proceedings of International Symposium on Seabuckthorn (*H . rhamnoides* L) Xian, China, Oct. 19-23. 132-137.

Zhang W, Yan J, Due J, Ren B, Guo J 1989. Preliminary Study of biochemical constitutions of berry of Seabuckthorn growing in Shanx province and their changing trend. Proceedings of International Symposium on Seabuckthorn (H. rhamnoides L) Xian, China, Oct/ 19-23 pp 96-105.

□□□

10

Fodder Values of Seabuckthorn

V.K. Sharma, J.S. Chauhan and Desy Rani

CONTENTS

10.1 INTRODUCTION

Seabuckthorn (*Hippophae* sp. Family Elaeagnaceae), unknown to many, grows widely in temperate regions of the world. Lahaul-Spiti district in Himchal Pradesh and Nubra valley in Ladakh (J&K) have very rich resources of Seabuckthorn. The forage and fruit production has been estimated at 15-25 tons and fruit products as 30-35 tons per hectares per annum. Seabuckthorn leaves and fruit have been used as natural fresh feed for livestock and poultry for many years. The significance has been well recognized in increasing weight gains, improvement of reproductive performance and increased resistance to various diseases of livestock. Seabuckthorn leaves, fruits and seeds are being used as tea after processing and fruit as ready to serve beverges, such as squash, syrup, jam and jelly. The industrial waste of these parts of plant available as a by-products, can be successfully used as animal feed. Market is bright for seabuckthorn and its by-products which can be used as various types of feed supplements and additives. The total green biomass and fruit production of seabuckthorn in Indian Himalayas is about 6,00,000 tons and 24,000 tones, respectively. During processing, the leaf-meal and fruit pulp as a left-over is available approximately to the tune of 60,000 and 2,400 tones, respectively and is available as by-products to be used as animal feeds. The leaves are very nutritious and as such can be fed to the animals. As per the information available, no systematic work, so far, has been carried out to study the effects of feeding seabuckthorn leaves, fruits, seeds and other industrial by-products available to the cattle and poultry to evaluate their nutrients utilization for various productive functions. Keeping in view the usefulness of seabuckthorn leaves, fruits, seeds and industrial by products, it will be of great importance to evaluate its foliage, various products and by products for nutritional values for livestock and poultry.

10.2 PRESENT STATUS

Seabuckthorn is a deciduous, thorny and nitrogen fixing shrub or a small tree, 2-4 m in height and grows abundantly in dry temperate Himalayas (Singh, 1994, 1995; Singh *et al.* 1995; Singh and Dogra, 1996; Singh 1998). When all the green forage species and grasses dry-up in August, the farmers drive their livestock to feed on the green foliage of seabuckthorn during the early winter (September-November). Singh *et al.* (1999) estimated fodder values of Indian seabuckthorn growing in Lauhal valley (Himalayan region). They concluded that crude protein (CP) contents varies from 17.5–21.5 per cent which is significantly higher than that of 17.1 per cent found in red clover. Ether extract (EE) was found to be 3.5 – 4.8 per cent. However, the proximate composition of other species of seabuckthorn (CP 23.9–30.3 per cent, EE 2.8 - 4.0 per cent, CF 15.5-23.3 per cent, NFE 30.1-45.1 per cent and ash 4.9-7.6 per cent) is comparable and even exceeds the nutrient contents of improved fodders like *Medicago sativa, Mellitus ablus*, millet and maize stalk.

The residues of seabuckthorn fruits accounting about 2 per cent of the total fruit weight are rich source of protein, fat, amino acids, vitamin A, C, D, and E, carotenoids, flavonoids and micro elements. Feeding on supplementary Seabuckthorn leaves and fruit residues weak pigs increased their weight by 9-13 per cent, goat milk increased by 6-7 per cent, emekens enhanced their grown rate by 9-13 per cent and laying out put by 23-29 per cent. Seabuckthorn leaves are very rich in crude protein (18-22 per cent) and fat (4-5 per cent), therefore, it is quality feed for livestock, particularly during early winter and can be preserved in the winter (Singh, 2001).

Although lot of work has been conducted in India on this plant species, but that lacks practical application in the feeding of livestock. Singh *et al.* (2001) estimated the nutritive value of the seabuckthorn (*H. rhamnoides* L.). They observed that the leaves are rich source of protein and are less fibrous, indicating their potential feeding value to the livestock. They also observed the adverse effects of tennins present in seabuckthorn . Chauhan *et al.* (2001) have pointed out that the fruits of the seabuckthorn is rich source of vitamin C, carotenoids and mineral matter. It contains substantial amount of vitamin P and vitamin K. Due to high acidic nature and exotic flavour, seabuckthorn berries are having good potential for producing various processed products like ready-to-serve beverages, squash, syrup, jam and jelly. The effect of seabuckthorn on the life of the people has been widely studied by Lat *et al.* (2001), Uniyal and Uniyal (2001) and Agarwal (2001). Patro *et al.* (2001), have studied the anti-oxygenic and anti-oxidant activity of seabuckthorn. Singh (1994) has published a report on arboreal seabuckthorn. Singh *et al.* (1995) have also evaluated the utilization of biomass and nutritional value of the seabuckthorn available in dry temperate Himalayas. Tiwari and Singh (2001) have reported the biomass production and nutritional values of seabuckthorn fruit and seed. The different experiments are being carried out in cattle for the utilization of industrial waste left-over, after processing of fruits, seed and leaves (for herbal tea preparation) for nutrient availability in animals (Sharma *et al.*, 2005).

In ancient Greece, leaves of seabuckthorn when added to horse fodder, resulted in weight gain and shiny hair, thus the Latin name *'Hippophae'* meaning shining horse (Rongsen, 1991). Besides browsing on the foliage, leaves and fruits residue can also be utilized as supplementary feed for livestock and poultry. The fruit residue is quite rich in protien, fat, amino acids, vitamin C and fat soluble vitamin contents which are as high as 426-503 mg/100 g has been reported in seabuckthorn foliage (Garonovich, 1980). The long term feeding with these supplementary additives promoted growth and improved immunity in animals. These results are supported by the experiences of the local herdsmen. In northern Ganehu province, a large area is covered by natural seabuckthorn forest where the herdsmen always graze their sheep on the seabuckthorn foliage, which fatten their livestock (Rongsen, 1994). Rongsen (1991) found that all the species of seabuckthorn were quite rich in protein. Among them, *H. rhamnoides.* ssp. *gyntsensis* contained the highest amount of crude protein 22.9 per cent and fat 6.1 per cent whereas, *H. neurocarpa* had the lowest amount of CP 11.5 per cent and fat 3.7 per cent.

Russian workers have also reported high values of water and fat soluble vitamins (vitamin P, E, C and carotenoids). The seabuckthorn cake also contains a wide range of minerals, which are optimals for an ideal fodder species. Thus the cake of seabuckthorn leaves is quite rich in biologically active nutrients and provides high energy, revealing the significance of seabuckthorn as a quality feed for the livestocks (Solonenko *et.al.*, 1993). Zhanmong (1992) compared the productivity of seabuckthorn with fodder shrubs and grasses and concluded that growing of seabuckthorn was more beneficial than other species of the fodders. Now, because it has a positive signification is accepted the etymdogy form Greek word: horse and shine. Romanian specialists demonstrated the Seabuckthorn which induces health for all animals, externalized in the aspect of their shine. The second significance is also from Greek, horse and poison, this is reflected in occident Europe tradition. Romanian specialists conceptional the beginning of the century was the same, it may be caudated by the plant's negative effect (Angel, 2005).

10.3 TECHNOLOGIES DEVELOPED

Nutritionally the leaves, fruits and seeds are very rich source of the protein, vitamins and minerals. It will be very useful to utilize and evaluate the leaves/ fruits/seeds and other industrial by products depending upon its availability and nutritive values for different production functions in different species of livestock. Many industrial houses are coming up and they are extracting syrup and oil, are preparing herbal tea by using the fruit, seeds and leaves of the seabuckthorn. The experiments are going on at H.P. Krishi Vishvavidyalaya to evaluate the feeding potential of left-over of the industrial waste after processing of fruit and leaves for the various production functions and nutrient utilization in cattle. After the completion of these experiments, the technology to use different parts of seabuckthorn, its fruit and other by products will be developed, making this plant a potential source in cattle production.

In Kelan Country, seabuckthorn beverage factories, poultry products processing factories, meat processing unit and refrigeration factories have been established, the credit for which goes to the exploitation of seabuckthorn resources, linking it to the development of rural economy.

10.4 OPPORTUNITIES

Seabuckthorn is a rich source of vitamins and other bioactive substances, which have great potential to provide health protective feed and besides many other pharmaceuticals. However, despite rich nutritional value, our country is yet to harness the rich potential of seabuckthorn in producing value-added food products and medicines. This plant is also an important source of fuel, wood, fodder and also serves as a soil binder in cold desert of Himalayas.

Nutritionally, the leaves and fruits are very rich source of the protein, vitamins and minerals. It will be very useful to use the leaves/fruits/seeds and other industrial by products depending upon its availability and evaluate the nutritive value for different production functions in cattle. The seabuckthorn leaves, fruits and seeds are being used for the production of tea, ready-to-serve beverages, squash, syrup, jam and jelly. Due to industrial processing, appreciable quantity of wastes in the form of leaf-meal, fruit-pulp and seed-waste remains unutilized which is rich source of protein, amino acids, fat, vitamin A, C, D, E, carotenoids, flavonoids and micro-minerals. The scientific and systematic use of these industrial by products in the feeding systems of different categories of the animals is to be exploited. It is therefore, presumed that if seabuckthorn and its industrial by products are used in animal feeding system, the animals may be benefited for the above mentioned nutrients. Therefore, an improvement in the economy of feeding, health status, production and reproduction in livestock will lead to the improvement in socio-economic status of small and marginal farmers.

10.5 LIMITATIONS

Now, this is well known fact that seabuckthorn leaves, fruits and its by products after extracting the oil can be used for feeding to the animals. As the seabuckthorn is growing on high hills and some inaccessible hilly areas and due to low population density of livestock in the hills, the consumption of leaves and byproducts in these areas is limited. It will involve a heavy transportation cost to use the leaves as fodder to a distant place. Primary industries are coming up in the hilly regions, which after processing of fruits, send the syrup and pulp to the heavy industries established in plains for further processing. This makes the primary and secondary products (left for animal feed industry) a costly affairs.

10.6 FUTURE STRATEGY FOR RESEARCH

Very limited and unclassified work has been done to evaluate the fodder (leaves) and industrial by products for various production functions and nutrients utilization to the different categories of livestock. This can be used as non-conventional source of animal feed and this can help to bridge the gap between the demand and supply of concentrates to the animals. Government of India is sponsoring projects to evaluate the feeding values of seabuckthorn to the animals and such one project has been sanctioned to the College of Veterinary & Animal Sciences, CSK HPKV, Palampur and the work has been started to evaluate the potential feeding value of seabuckthorn in cattle.

As per rough estimates the annual availability of leaf-meal and fruit processing industrial waste will be to the tune of 60,000 and 2,400 tones, respectively.

Table 10.1: Proximate Composition of Seabuckthorn and Clover

Species	Part	CP	EE	CF	Ash	NFE
H. rhamonoides	Female leaves	17.1	4.6	16.9	5.9	55.7
ssp. *sinensis*	Male leaves	16.2	4.2	13.0	4.1	62.5
Trifolium pretense	Stem + leaves	17.1	3.6	21.5	10.2	47.6
Trifolium repens	Stem + leaves	24.7	2.7	12.5	13.0	47.1

Table 10.2: Nutritional Values of Indian Seabuckthorn (*Hippophae rhamnoides* L.) and *H. salicifolia*

Species	Month	Part	CP (%)	EE (%)	Ash (%)	Cu (ppm)	Zn (ppm)	Fe (ppm)
H. rham.	May	Leaves	28.4	2.8	11.8	0.20	39.6	684
	July	Leaves	17.3	4.3	6.3	5.6	27.4	642
	Sept.	Leaves	20.3	4.7	6.5	1.4	31.8	561
	April	Bark	17.4	2.3	4.7	1.4	6.4	183
	April	Branch	13.7	1.9	6.8	1.8	15.2	177
H. Sali.	July	Leaves	21.6	4.6	7.5	3.8	24.8	828
CD			4.9	1.2	2.3	1.9	11.6	265

Table 10.3: Comparison of Fodder Values Seabuckthorn of with some other Fodder Species

Plant species	Age (yr)	CP	EE	CF	NFE	P	Ca	Ash
H. rhamnoides	9	24.2	4.2	17.4	45.1	0.26	1.1	4.9
Caragona korshinskii	6	23.9	2.9	23.1	38.1	0.27	2.1	7.4
Sophora viciifolia	9	28.6	2.8	17.9	39.9	0.34	1.6	6.7
Campylotropis macrocarpa	8	30.3	4.0	15.5	30.1	0.47	1.5	7.6

Table 10.4: Mineral Content of Leaves of Seabuckthorn.

Mineral	(g/kg) content	Mineral	(mg/kg) content
Ca	7.77-9.54	Fe	381.6-410.9
P	1.66-1.78	Mn	46.8-50.2
S	1.00-1.22	Cu	4.8-5.3
Cl	3.55-6.07	Zn	33.2-35.8
K	12.03-12.73	Co	0.35-0.37
Na	1.17-1.19	Mg	1.65-1.81

Table 10.5: *In- sacco* degradability of Seabuckthorn Leaves.

Parameters	Hr	Ac	As	Al	Bv	Dh	Fr	Ll	Ma
Degradability at 48 h (%)									
DM	71.5	47.2	30.4	85.3	42.7	45.2	88.8	72.3	76.2
CP	68.1	34.5	25.5	92.9	64.4	57.7	86.4	72.3	87.1
NDF	22.6	28.4	25.0	75.0	20.2	39.2	84.0	58.5	70.4
NDF-N	30.3	11.7	0.0	90.3	27.8	51.8	87.9	56.5	79.3
DM (%)									
PD	75.6	55.2	32.4	85.1	42.3	56.5	93.1	75.6	85.0
ED	61.4	33.6	24.4	57.3	27.1	26.5	56.2	51.9	52.5
c/h	0.614	0.039	0.095	0.095	0.049	0.035	0.065	0.069	0.047
CP (%)									
PD	77.4	42.4	22.5	93.1	53.5	75.2	95.8	78.6	95.2
ED	51.9	14.8	21.9	64.4	51.8	35.9	60.1	43.7	57.6
c/h	0.048	0.030	0.091	0.069	1.223	0.028	0.054	0.043	0.067
N-fraction (%)									
RDN	14.1	4.8	7.3	19.7	16.9	11.1	13.2	14.6	16.3
UDN	13.6	28.7	28.4	11.4	16.1	20.6	8.9	19.1	12.2
DUN	7.9	16.0	11.6	4.6	9.4	11.2	2.6	5.8	5.8

Hr-*Hippophae rhamnoides*; Ac-*Acacia catechu*; As-*Albizzia stipulate*; Al-*Artocarpus lakoocha*; Bv-*Bauhinia variegate*; Dh-*Dendrocalamus hamiltonii*; Fr-*ficus roxburghii*; Ll-*Leucaena leucocephala*; Ma-*Morus alba*; PD-Potential degradability; ED-Effective degradability; c-Degradation rate; RDN-Rumen degradable nitrogen; UDN-Rumen undegradable nitrogen; DUN-Rumen undegradable nitrogen but available post ruminally.

Table 10.6: Potential and rate of Gas Production (ml/200mg), Digestible Organic Matter, Metabolizable energy and intake potential of various Tree Forages.

Parameters	Hr	Ac	As	Al	Bv	Dh	Fr	Ll	Ma	Mo*
PGP	31.1	24.4	17.4	51.1	23.1	33.8	42.6	37.2	35.4	49.5
RGP	0.065	0.046	0.059	0.038	0.031	0.026	0.046	0.058	0.062	0.085
OMD, %	45.5	45.6	44.8	63.2	42.1	40.9	85.5	69.7	64.2	74.1
ME, MJ/kg	6.6	10.5	6.8	9.1	6.7	6.7	9.2	13.1	11.3	95
Intake, 1/kg	72.7	58.2	65.8	58.8	46.5	46.5	62.0	69.6	61.7	-

Hr-*Hippophae rhamnoides*; Ac-*Acacia catechu*; As-*Albizzia stipulate*; Al-*Artocarpus lakoocha*; Bv-*Bauhinia variegate*: Dh-*Dendrocalamus hamiltonii*; Fr-*Ficus roxburghii*; Ll-*Leucaena leucocephala*; Ma-*Morus alba*; Mo-*Moringa oleifera*; PGP-Potential gas production (ml/200mg). RGP-Rate of gas production; OMD-OM digestibility; *from Makkar and Becker (1996).

Table 10.7: Nutrients of Berries about five individual Plants of Chinese Seabuckthorn with good Eco-economical Characters.

Variety	Vc content (mg/100g)	Total content of organic acid (%)	Content of crude fat (%)	Content of crude protein (%)	Content of soluble total sugar (%)
GW1-34	1377.8	12.30	10.22	1.65	10.42
CK	1295.2	13.20	9.86	0.98	9.86
95-B	345.90	8.45	6.50	1.46	5.50
95-C	388.90	7.56	7.45	1.84	6.80
95-D	365.70	7.83	8.79	1.75	7.46

Table 10.8: Nutrients of Stems and Leaves of five individual plants of Chinese Seabuckthorn with good Eco-economical Characters.

Variety	Part name	Content of crude protein (%)	Content of crude fat (%)	Content of crude fiber (%)	Content of mineral ash (%)	Content of total phosphorus (%)	Content of calcium (%)	Content of water (%)
GW1-34	Stem	9.888	1.81	37.17	3.02	0.146	0.31	2.921
	Leaf	5.200	8.56	13.57	8.45	0.154	1.95	2.60
CK	Stem	7.569	1.08	49.18	1.77	0.149	0.18	3.05
	Leaf	16.025	11.26	12.41	9.81	0.175	0.37	2.77

References

Agrwal, Aruna. 2001. Role of Seabuckthorn (*Hippophae rhamnoides*) in the management of age related memory disorder, In: *Int. Workshop on Seabuckthorn*, New Delhi, India. pp.68.

Angel, Prooroan. 2005. The utilization of the Seabuckthorn in Romania, past, present and future. In: *Second International Seabuckthorn Association conference*-Organised by ISA and CACSD, Beijing (China) from Aug., 26th to 29th, 2005.: p:13.

Chauhan, A.S., Rekha M.N., Ramteke R. S. & Eipeson W. E. 2001. Seabuckthorn (*Hippophae rhamnoides L*) berries: Harnessing its potential for processing. In: *Int. Workshop on Seabuckthorn*, New Delhi, India. pp. 60.

Garonovich, I. M. 1980. Ascorbic acid contents of leaves and fruit of Seabuckthorn. In: *Vestsi-Alademoo Navuk-BSSR* 5, 10-14.

Lal Brij, Ahuja Initials & Gupta A.K. 2001. Application of Seabuckthorn in amchi system of medicine, In: *Int. Workshop on Seabuckthorn*, New Delhi, India. pp.66.

Li., Daigcong, Hhang, Jin., Jiany, Jan. & Gangshuan, Bai. 2005. Study on introduction and cultivation of Seabuckthorn variety in the semiarid region of loess plateau. *Journal of the Global Seabuckthorn research and development* 3 (2) , 46-52.

Patro, B.S. Simuel R., Sreekanth A., Dwivedi S.K., Chaurasia O.P., Chttopadhyay S., Singh Brahma & Banerji A. 2001. Anti-oxygenic activity of India Seabuckthorn (*Hippophae Sps.*) In: *Int. Workshop on Seabuckthorn*, New Delhi, India :pp.70.

Rongsen, L. 1992. Seabuckthorn-A multipurpose plant specie for fragile mountains. ICIMOD Occasional paper, No. 20, Kathmandu, Nepal.

Rongsen, L. 1994. New achievements in seabuckthorn development in China ICIMOD. Kathamandu, Nepal. pp. 62.

Sharma, V.K., Chauhan, J.S. Virendra Singh & Desy Rani. 2005. Project entitled,"Nutritional evaluation of Seabuckthorn (*Hippophae L.*) for productive performance in cattle", granted by Indian council of Agriculture Research.

Singh, V. 1994. A report on arboredal Seabuckthorn (*Hippophae salicefolia*) from dry temperate Himalayas. *J. Tree Sciences* 3, 67-68.

Singh, V. 1995. Rooting rates of hardwood cuttings of Seabuckthorn (*H. rhamnoides*). *J. Tree Sciences* 14, 87-88.

Singh, V., Singh B. & Awasthi, C.P. (1995) Distribution, taxonomy and nutritional value of Seabuckthorn (*Hippophae.L*) in dry temperate Himalays. In: *Proc. The workshop of Seabuckthorn*, Beijing, China. pp. 52-59.

Singh, V. & Dogra K.K. 1996. Characteristics, utilization, regeneration, biomass yield and nutritional value of Seabuckthorn. *Indian Forester* 122, 486-491.

Singh, V. 1998. Goegraphical adaptation of Seabuckthorn (*Hippophae L.*) in dry temperate Himalayas. In: *Proc. Int. Symp. Seabuckthorn*, Ulan-ude, Russia.

Singh, V., Singh R.K. & Singh B. 1999. Fodder values of foliage of seabuckthorn growing in Lahual valley, dry temperate Himalayas. In: *Proc. Int. Workshop on Seabuckthorn*, Beijing, China.

Singh, V. 2001. Seabuckthorn (*Hippophae* L.) - A potential fodder for livestock. In: Int.Workshop on Seabuckthorn, New Delhi, Indian. pp. 32, 48, 53, 63, 64.

Singh, B., Bhat, T.K. & Singh Virendra. 2005. Seabuckthorn leaves as cattle feed- A comparative evaluation. *Indian J. Anim. Nutir.* pp. 22, 16.

Soloennko, L.P., Koshelev J.A., Ageeva L.D. & Skoukovsky B.A. 1993. Solvent of cake of seabuckhtorn leaves as a high valuable additive to fodder. In: *Proceedings of International Symposium on seabuckhtorn*, pp.423-425.

Tiwari, S. C. & Singh, S. S. 2001. A report on Seabuckthorn from Arunchal Pradesh, North-Eastem India, In: *Int. Workshop on Seabuckthorn*, New Delhi, India: pp.42.

Uniyal, M.R. & Uniyal, R.C. 2001. New source of Ayurvedic drug amlavetas (*Hippophae sps.*), Commonly found in Himalayan region, In: *Int. Workshop on Seabuckthorn*, New Delhi, India. pp.66.

Zhanmong, W. 1992. The fodder values of seabucthorn and tiger bushes. *Hippophae* 3:18-21.

□□□

11

Phytochemical Investigation on Indian Seabuckthorn

K.P. Suja, K.S. Pradeep Kumar and A. Banerji

CONTENTS

11.1 INTRODUCTION

Seabuckthorn (*Hippophae* spp.), a hardy deciduous shrub of Elaeagnaceae family, grows widely in Europe, Russia, Central Asia, temperate regions of South Asia and China. In India, it abundantly grows in Ladakh region of Jammu & Kashmir, cold deserts of Himachal Pradesh, dry temperate areas of Kumaon and Garhwal, Uttarakhand and Sikkim (Bhatt *et al.*, 1993). All the parts of seabuckthorn find applications in medicine, as nutraceuticals or in cosmetics or agriculture. The diversity of the biochemical profile explains the varied bioactivities exhibited by seabuckthorn and its products. Presence of a cocktail of vitamins, essential fatty acids and phenolics contribute to the nutritional qualities of seabuckthorn. It is a source of high value oil containing several bioactive substances like essential fatty acids, carotenoids, phytosteroids, vitamin E etc. (Beveridge *et al.*, 1999). It provides one of the richest sources of pro-vitamin A (carotenoids), vitamins E and C, flavonoids and lipids that make it highly valuable for pharmaceutical and cosmeceutical purposes (Beveridge *et al.*, 1999). Leaves are also an important source of polyphenols, vitamins, tannins, free and esterified sterols, triterpenoids, amino acids and proteins. Seabuckthorn and its products have been in use in folk medicines of China, Russia, Mongolia, India and Tibet. Seabuckthorn based preparations feature in several pharmacopoeia (Li and Wang, 1998). In traditional Chinese medicine, seabuckthorn products have been used as health drinks, for cardiovascular and digestive diseases and ailments related to lungs and skin. Chinese Olympic athletes were served with seabuckthorn based health drinks to improve their physical performance and endurance. The systematic clinical trials to validate the bioactivities of sea buckthorn based products were initiated in Russia only after 1950's. In the early stages of Soviet space programme, the astronauts were given seabuckthorn based health foods as vitamin source and medicine for protection against radiation and improving the body immunity during oxygen deficiency in their space flights. However, it may be noted that the yields of the phytochemicals from seabuckthorn show enormous variations depending on geographical location, genetical make up of the source, ecology, stage of collection etc. In addition to inherent reasons, the apparent variability are also due to different methods used for extraction and estimations of phytochemicals. In India, Phytochemical investigations on seabuckthorn (SB) have been initiated only recently (Singh and Khosala, 2001). In the present review a biochemical profile of seabuckthorn (*Hippophae* rhmnoides ssp. turkestanika)is presented.

The plant material for the study was collected from the Ladakh (India) by Field Research Laboratory (FRL), Defence Research and Development Organisation (DRDO), Leh from *Hippophae rhmnoides* ssp. turkesta-nika.

Sequential extraction of plant material was carried out in a Soxhlet extractor using solvents of varied polarities. In an alternative approach, the plant material was refluxed with appropriate solvents. The solvent from the extract was removed under reduced pressure and the concentrate thus obtained was separated into different fractions by partitioning with a variety of solvents such as petroleum ether (PE), diethyl ether, chloroform, ethyl acetate (EtOAc), methyl ethyl ketone (MEK), n-butanol. The solvents were removed using a rotary evaporator (Heidolph, Laborota 4001) at reduced pressure. The separation of individual components was carried out by column chromatography, preparative thin layer chromatography (TLC), centrifugal chromatography (Chromatotron MODEL 79247) and gel permeation chromatography. The separation of compounds/extracts was monitored by TLC, circular paper chromatography, high performance liquid chromatography (HPLC). For the molecular characterization, optical spectroscopy (IR, UV), ^{1}H NMR, ^{13}C NMR (Bruker ADVANCE DPX_{300} series) and Mass spectroscopy (JEOL JMS 600, Shimazdu GC-MS 5050 series) were used. Different parts of the plant were studied individually.

11.2 EXTRACTIVES FROM SEABUCKTHORN

Fruits were separated into pulp and seed; extraction was carried out sequentially using Soxhlet extractors. The results are presented in the Table 8.1. As can be seen the pulp is the richest in both polar as well as non polar extractives. The water extractives of seabuckthorn leaves, obtained after sequential extraction using pet ether, chloroform and ethanol was 4.55 per cent.

Table 11.1: Extractives from Seabuckthorn

Sample	PE	$CHCl_3$	MeOH
Pulp	26.38%	11.26%	47.38%
Seed	3.13%	2.87%	15.73%
Leaves	4.67%	2.24%	24.575

11.2.1 Carotenoids

A red oil was obtained on removal of Pe from the extract of seabuckthorn fruit (SBF). It was fractionated by chromatography on silica gel column using mixture of Pe with increasing proportions of ETOAc. Several fractions were collected and analyzed by TLC using appropriate solvent systems. (Pe-ETOAc; 7: 3). Four prominent coloured spots (2 red and 2 yellow) could be seen. They turned blue when sprayed with an alcoholic solution of antimony trichloride indicating their carotenoid nature. Chromatotron (centrifugal chromatography) was very convenient for the preparative scale separation of compounds. Preparative TLC (silica gel, Pe : ETOAc, 7:3) was also used successfully for separation of the individual compounds. The identities and nature of carotenoids were established spectroscopically. β-carotene was the main fraction and was characterized by its absorption at l max (436,450,480) nm. The other carotenoids were present in small quantities. They were, lycopene (449, 469, 501 nm), β-cryptoxanthin (425,449,476), zeaxanthin (425, 450, 474), rubiaxanthin (440, 464, 494) and gazaniaxanthin (440, 464, 494). Quantitation was carried out by spectroscopic techniques. Carotenoids were also present in the seed oil. The amount of β-carotene in the pulp (1007 mg/100g of pulp) was much higher than that of seed (949 mg/100g). The non-polar fractions of leaves contain only very small amount of carotenoids. Pe extract (1.9 %) of SBL was dark green in colour. The chlorophyllous materials were removed by partitioning with 70% aq. MeOH and hexane. It was filtered and the filtrate partitioned with hexane. The yellow hexane extract was subjected to column chromatography (silica column) using increasing amount of ETOAc in Pe. The amount of carotenoid obtained was very low.

11.2.2 Steroid, Triterpenoids, Fatty Substances

The chloroform extract of sbf was also obtained as red oil. in addition to residual β-carotene, the oil shows the presence of fatty acids/esters, tocopherols, steroids and triterpenoids. Column chromatography of the $CHCl_3$ extract yielded several fractions. Sitosterol, b-amyrin, uvaol, ursolic acid, oleanolic acid, betulinic acid were among the compounds identified by ^1h and ^{13}c NMR spectroscopy. Ursolic acid was further purified by chromatotron. It is important to note that ursolic acid is known for many medicinal properties such as anti-inflammatory, anti-tumour (induction of apoptosis), anti-microbial, anti-viral and anti-hyperlipidemic activities (Liu, 1995). It is under clinical trial for possible use in skin melanoma (Harmand *et al.*, 2005). Sitosterol and related phytosterols are reported to have anti-ulcerogenic activity (Galati *et al.*, 2003). Sitosterol and ursolic acid are the major component in the chloroform extract of SBF in addition to steroids and triterpenoids, unsaturated fatty acids were also obtained (GLC) by column chromatography. Presence of tocopherols was also inferred.

For the fatty acid analysis of the oil, the methyl esters of the fatty acids were prepared ($MEOH-H_2SO_4$) directly and analysed by GC-MS. Following fatty acids were identified: palmitic (23 %); palmitoleic (42 %), stearic (0.57 %); oleic (29 %); linoleic (4 %) and linolenic (0.5 %) acids. Presence of palmotoleic and linolenic acids is noteworthy due to their possible actions on many physiological processes such as cholesterol lowering, stroke reducing effects etc. (Yamori *et al.*, 1986). For the study of non-fatty constituents, the oil was saponified using ethanolic NaOH. the unsaponifiable portion was fractionated by elaborate column chromatography. The presence of steroids and triterpenes referred above were also isolated. The chloroform extract of seeds and leaves were also studied by column chromatography. Seed extract yielded mainly sitosterol and tocopherols. The main constituent of chloroform extract of SBL was ursolic acid (TLC; Pe/ETOAc, 7:3) and its amount was higher than that in fruit or seed. In addition to steroids, triterpenes and fatty acids, a white neutral substance was also isolated from the Pe eluate from the column chromatographic fraction of Pe extract SBL. The product did not answer to colour tests for phenolics or steroids. The IR spectrum showed absorptions (474 and 722 $cm^{-1)}$ for alkanes and for O-H group (3406 cm^{-1}). The carbonyl absorption was absent. The highest peak in the MS (FAB) was recorded at 256 mu. In ^{1}h NMR spectrum, signals were observed at 4.7 (singlet), 1.18 (singlet) and 0.7 (triplet) ppm with signal intensity in the ratio of 3:8:1. A small peak was also observed at 4.8 with a signal intensity of 1 along with the signal at 4.7 ppm.^{13}c NMR spectra furnished signals at 14.09, 22.70, 29.39, 29.73 and 31.96ppm. Further investigation is needed for full characterization of the compound.

11.2.3 Ascorbic Acid

Ascorbic acid plays important role in imparting nutritional quality to seabuckthorn oil. The content of ascorbic acid varies considerably with season, location and age. Ascorbic acid content of SBF and SBL were estimated spectrophotometrically. The estimation is based on the reduction of 2,6-dichloro phenol indophenol by ascorbic acid (Annon., 1966). Ascorbic acid content of SBF were compared with that of SBL, lemon and amla. From experiments it was found that the fruit contains approximately 66 mg/100g and the leaves contains 126 mg/100g of ascorbic acid while, lemon and amla contain 14.33 mg/100g and 310 mg/100g, respectively.

11.2.4 Phenolics, Flavonoids

The fresh leaves (SBL) were collected from the campus of FRL, Leh (June 2005). SBL was extracted at Phytochemical Research Laboratory at FRL, Leh using solvents like chloroform and ethanol. The concentrates were obtained after removal of solvents at reduced pressure. Fractionation and characterization of the extracts were carried out at the Regional Research Laboratory, Trivandrum (RRL-T) for detailed phytochemical investigations. Different extracts from SBL were studied individually. The ethanol extract, on removal of solvent gave a dark brown extractive, which was refluxed with Pe followed by $CHCl_3$ to remove less polar constituents. The resulting extract gave positive ferric reaction (blue black) reminescent of gallic acid derivatives. The Shinoda's test was positive (though there was delay in colour development) suggesting the presence of flavonoids. A TLC analysis of the crude extract shows the absence of free flavonoids or phenolics. The flavonoids/ phenolics must be present mainly in the bound state such as glycosides. The extract is an intricate mixture of glycosides. Attempts to isolate individual compounds from the crude extract by gel permeation chromatography or adsorption chromatography were not successful. Purification of flavonoids/ phenolics through complexation of heavy metal salt (lead acetate) was also of not much utility. Extraction with MEK was useful in obtaining phenolics/ flavonoids rich fraction. To resolve the complexity of the polar extracts, enzymatic and acid hydrolysis were tried. No significant change was observed by using various enzymes such as b-glucosidase, pectinase and diastase. The crude extract was therefore, hydrolyzed by sulfuric (7 % aqueous) or hydrochloric (10 %) acid at steam bath for two hours. The reaction mixture was extracted with ETOAC. The concentrate obtained from the extract was put for extensive adsorption chromatography followed by gel permeation chromatography. Three compounds, a, b, c were isolated in small quantities. Compounds a and b gave yellow spots in TLC (toluene : ethyl acetate: acetic acid; 5:4:1) while c appeared as a cream coloured spot which gave a blue black with $FECL_3$ spray. Contrarily to reports on occurrence of large quantity of flavonoids in several species of *Hippophae*, only a small amount could be isolated from the plant collected from Ladakh.

Compound a: The spot for compound a corresponded to isorhamnetin in TLC (toluene: ethyl acetate: acetic acid; 5:4:1). However, other properties of a were different from isorhamnetin. It gave a green colour with alcoholic $FECL_3$ and dissolved in aqueous NAOH giving a yellow colouration. Shinoda test gave pink colour. A yellow precipitate was obtained with ethanolic lead acetate solution. UV spectrum showed a prominent absorption at 278 nm along with minor peaks at 225 and 370 nm. In the ^{1}h NMR spectrum (Cd_3OD), a signal at 4.5 ppm (singlet, 3h) was observed. Signals for aromatic protons (7.2 and 6.5 ppm and a high field proton at 9.4ppm) were also observed. On acetylation, an acetate was obtained. The ^{1}h NMR spectra (in $CDCL_3$) of the acetate did not bring about much change in the chemical shifts except that the signal at 4.5 shifted to 5.1 ppm.

Compound b: In TLC, compound b apparently corresponded to quercetin in colour and RF (TLC ; toluene: ethyl acetate: acetic acid; 5:4:1) value. It gave a typical ferric reaction (green) and pink colour was obtained in Shinoda's test. It dissolves in aqueous. Alkali giving yellow colour. Yellow precipitate was obtained with lead acetate solution. Its UV spectrum (256, 268 sh, 371 nm) was also similar to that of quercetin and related compounds. In ^{1}h NMR spectrum, prominent multiplets were obtained at 2.6 and 2.4 ppm. Clusters of ill-resolved multiplets appeared between 6 to 8 ppm. On treatment with AC_2O-Pyridine, acetate was obtained. But acetylation did not resolve the spectrum.

Compound c: It is a white amorphous powder, soluble in ethanol/ methanol. It gave prominent colour with $FeCl_3$ (bluish black colour). It dissolves in alkali (NAOH) solution giving pink colour. It did not answer Shinoda test. From the proton and carbon NMR spectra, compound c appeared to be a small molecule. It gave acetate suggesting that c has many hydroxyls. It gave a prominent UV absorption at 274 nm in methanol. These properties suggest that compound c could be gallic acid. This contention was supported by comparing with NMR, TLC and paper chromatography with an authentic sample of gallic acid.

The concentrate from the methanolic extract of SBF (fruits) gave a brown mass. It answered tests for flavonoids but separations of individual components was not successful. SBF- MEOH concentrate therefore was partitioned with MEK and water. The concentrate from MEK extract gave a dark-green colour with $FeCl_3$ indicating the presence of phenolics. Red colour was obtained in Shinoda's test suggesting the presence of flavonoids. Presence of glycosides/ carbohydrates was inferred from a positive Molisch's (b-naphthol and H_2SO_4) test. The UV spectra using shift reagents indicated the presence of flavonoids. To further resolve the fraction, acid hydrolysis was carried out. UV spectral studies using shift reagents like $AlCl_3$, $AlCl_3$ + HCl, sodium methoxide, sodium acetate and sodium acetate + boric acid showed that the major compound was isorhamnetin. Spectral studies of minor fractions showed the presence of quercetin and kaempferol. The presence of gallic acid suggest the possible presence of gallo-tannins in the mixture before hydrolysis. In addition to phenolics and flavonoids, a colourless crystalline carbohydrate derivative, quebrachitol (inositol -2 -methyl ether) was also isolated.

A sample of commercial seabuckthorn flavones extract of chinese origin (SBFC) was obtained through the courtesy of DIPAS (DRDO) Delhi. Majority of SBFC was insoluble in the common solvents including polar solvents such as ethyl acetate, ethanol, methanol and water. Crude SBFC gave only weak response to Shinoda test. The flavonoids from SBFC could be isolated by repeated solvent extraction, column chromatography and preparative TLC. On TLC, two yellow spots were obtained. The TLC pattern of d and e (toluene: ethyl acetate: acetic acid; 5:4:1) was similar to that of compound a and b from SBL extract. However, further studies revealed that they were different. Individual compounds could be studied by prep TLC. The spectral data on isolated compounds were collected. Compound d showed l max (254, 300 sh, 370) and compound e (255, 296 sh, 370). An acetate of mixture containing containing d and e was prepared (acetic anhydride/pyridine) and analysed by NMR. Complex NMR signals were obtained suggesting it to be a complex mixture. The acetate of flavone was also prepared and studied by NMR. The mass spectrum (FAB) of the mixture showed the highest peak at 414 mu. Tannins and glycosides were not detected in the sample. The bioevaluation of the purified sample was carried out in collaboration with Defence Institute of Physiology an Allied Sciences, (DIPAS) Delhi. The cytoprotective and free radical scavenging activity (unpublished work) of the sample was found to be very high. Therefore, experiments are being carried out to procure the purified sample in sufficient quantities for future work.

11.3 ANTIOXIDANT ACTIVITY OF SEABUCKTHORN EXTRACTS

Antioxidants delay or prevent the undesirable oxidation of the important biomolecules (Klein, 2001). They may provide protection against damage caused by reactive oxygen species (ROS) (Singh, 2005). The ROS are believed to be the causative factor for various degenerative diseases including aging. Natural antioxidants include vitamins, enzymes, phenolics, flavonoids etc. Seabuckthorn is a rich source of such compounds. The antioxygenic, antiatherosclerotic, anti-aging properties of SB were reported earlier. In the present study the antioxidant activity of SB extracts were studied for their radical scavenging activity using DPPH model system. The free radical scavenging power of Pe, $CHCl_3$, MeOH and Mek extracts of SBF were studied. The radical scavenging power of these extracts were very low. Very significant activity was observed for the extracts from the leaves of seabuckthorn (SBL). The polar fractions such as MeOH and Mek extracts showed very high radical scavenging power. This may be due to the presence of phenolic acids and flavonoids. The activity was measured at 500 and 1000 ppm levels of the extracts. SBL-MeOH at 1000 ppm level showed 92 % radical scavenging power and at 500ppm, 77 %. SBL-MEK extract showed 42 % activity at 200 ppm. The Ec_{50} values of selected samples were studied by the above method. The antioxidant activity of seabuckthorn tea leaves (from FRL, Leh) showed very high antioxidant activity (Ec_{50} 1,200) compared to SBFC (Ec_{50} 5,700), *rhodiola* extract (Ec_{50} 5,900), DARL tea (Ec_{50} 6,400) and ginseng powder (Ec_{50} 18,000) (Table 8.2). Ec_{50} value of quercetin is very high compared to isorhamnetin.

Both compounds are reported to be present in seabuckthorn. Antioxidant activities of a-tocopherol, ellagic acid, ursolic acid etc. were also studied by the above method. At the concentration level of 200 ppm, ellagic acid showed high radical scavenging power (88 %) compared to BHA (70 %), a-tocopherol (51 %) and ursolic acid (8 %).

Table 11.2: Ec_{50} (DPPH) values for compounds/Extracts

S. No.	Sample	EC50
1	SB tea extract (FRL)	1,200
2	SB Flavone (Chinese)	5,700
3	*Rhodiola* extract	5,900
4	Nilgiri tea	100
5	Quercetin	35
6	Isorhamnetin	620
7	BHA	115
8	Ginseng powder	18,000
9	DARL tea	6,400

11.4 SEABUCKTHORN AS DYES

The great chemodiversity of seabuckthorn constituents (for example carotenoids, tocopherols, sterols, polyphenolics, flavonoids, tannins, etc.) suggests their possible use as a food as well as textile dyes. A detailed investigation has been carried out to evaluate the dyeing properties of different extracts. The non-polar extracts, which are rich in carotenoids, impart yellow color to food materials such as vegetable oil, butter, milk and curd. The colours are stable and have high tinctorial properties. The use of seabuckthorn extracts for food materials has added advantage since they enrich them with important vitamins. The polar extracts of SB fruit also gave similar colors but with low tinctorial values and less stability. Leaves of SB are rich in polyphenolics and tannins. Dyeing properties of the polar extracts were investigated on cotton, silk, nylon and wool and detailed dyeing characteristics were determined. Camel brown color was obtained in all these fabrics. The range of color and tinctorial values could be enhanced substantially by the judicious use of mordants. Thus, the use of seabuckthorn as a source of natural dyes will contribute to its value-addition and help in providing additional economic benefit to the farmers.

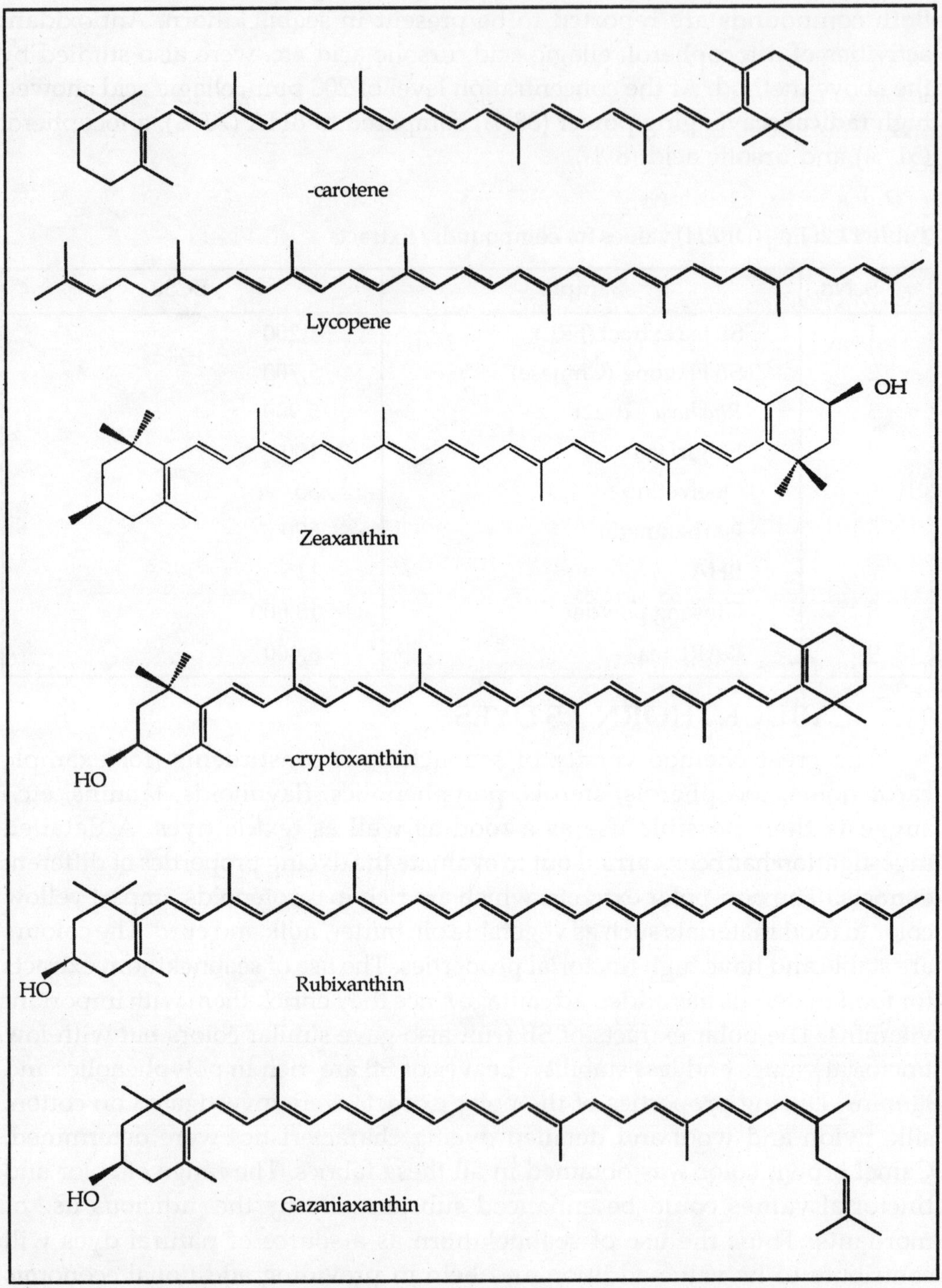

-carotene

Lycopene

OH

Zeaxanthin

HO

-cryptoxanthin

HO

Rubixanthin

HO

Gazaniaxanthin

Fig. 11.1: Structures of carotenoids present in seabuckthorn

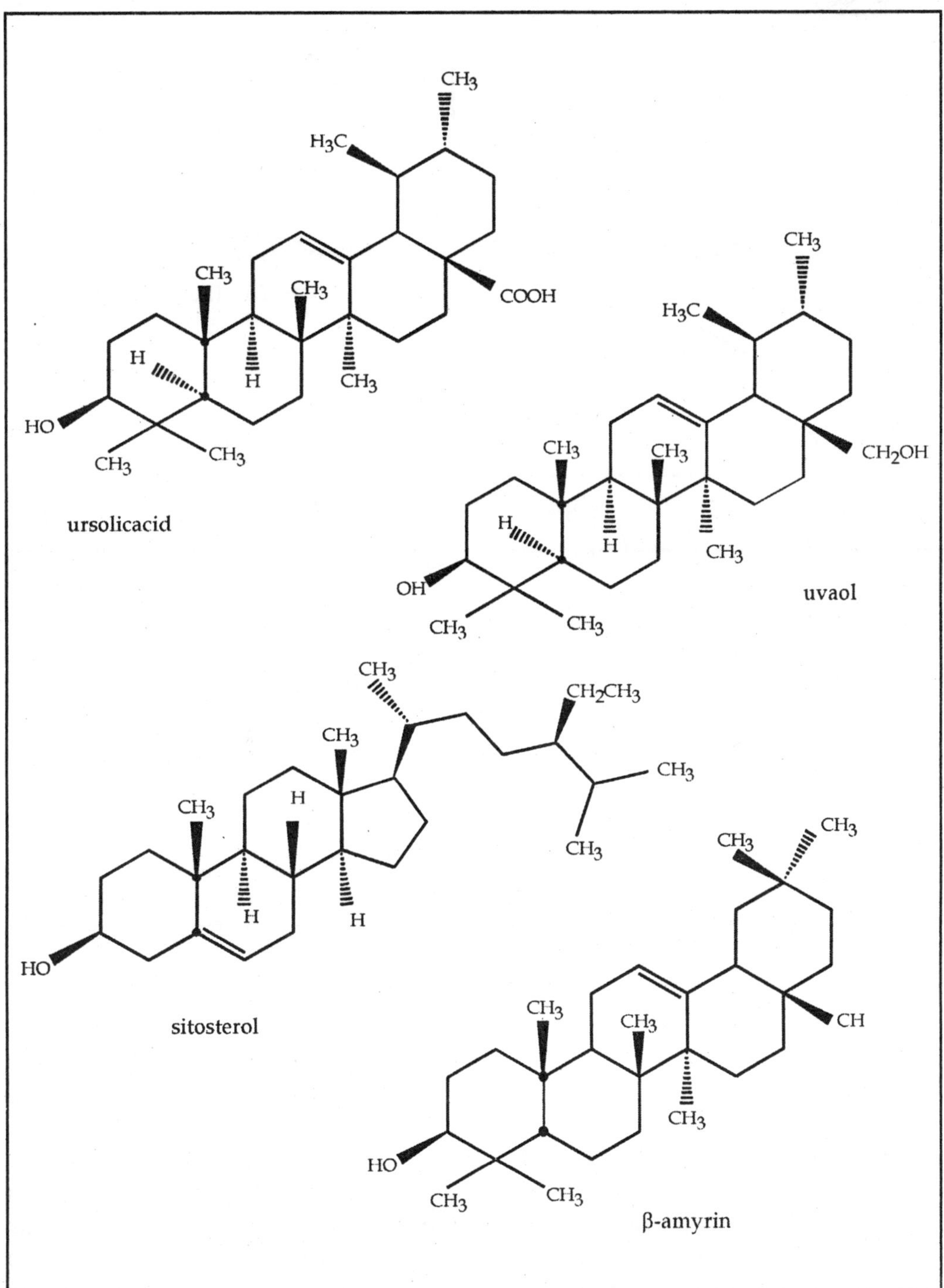

Fig. 11.2: Structures of trepenoids/steroids of seabuckthorn

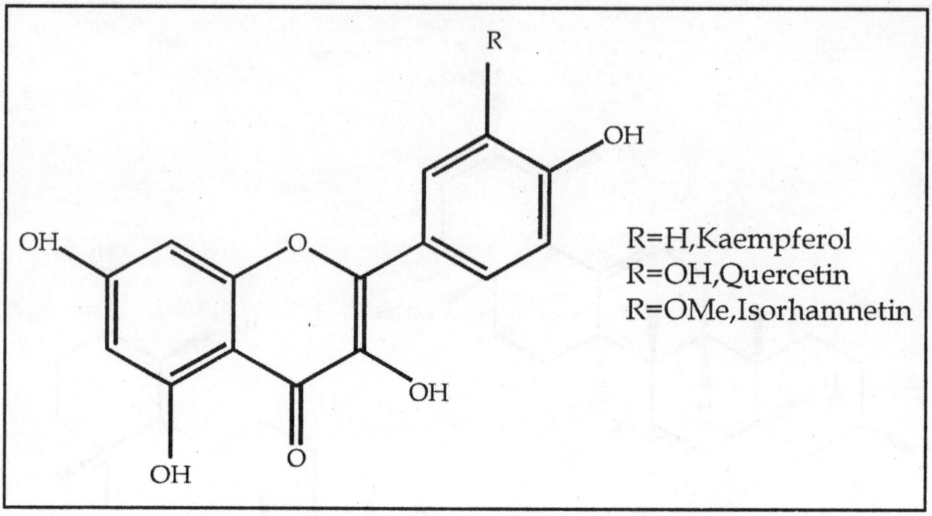

Fig. 11.3. Structures of flavonoids in seabuckthorn

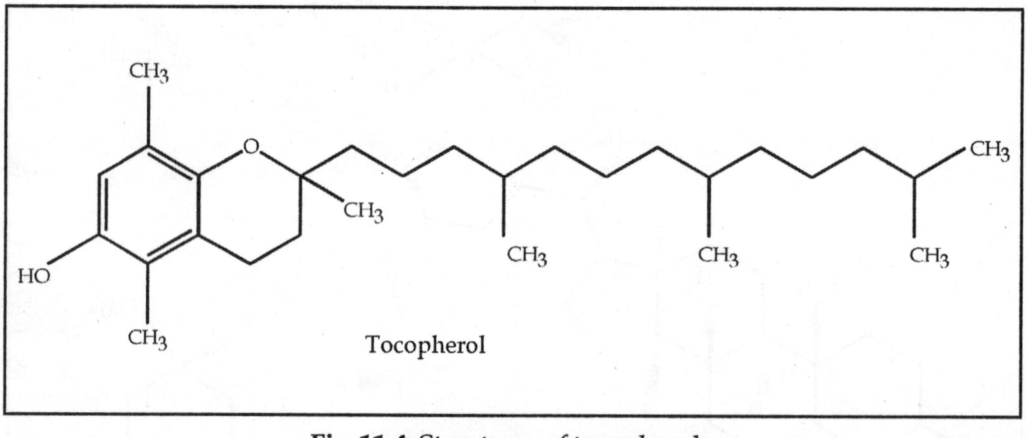

Fig. 11.4: Structures of tocopherols

11.5 CONCLUSIONS

Seabuckthorn is being growth in many countries and has been in regular use as medicine and food supplement in Mongolia, Tibet, Russia and China for several centuries. But during the last decade, it has attracted the global attention because of the awareness of its multifarious utility. In India also there is realization of possible utility of seabuckthorn. Seabuckthorn products have been found to be useful for skin care, sunscreen, and cardiovascular diseases, wound healing etc. It contains a cocktail of nutraceuticals like vitamins (carotenoids, tocopherols), essential fatty acids, phenolics (flavonoids, tannins etc). The non-polar fraction of seabuckthorn includes β-carotene, sitosterol, ursolic acid, fatty acids etc. The polar fraction mainly contains phenolic acids and flavonoids. The radical scavenging power of the polar extracts are found to be high in a DPPH· model antioxidant system compared to that of non-polar extracts. The yield of these phytochemicals shows considerable variations due to geographical, climatic, ecological reasons. In addition, the reported variations could also be due to errors incurred in the methods of extraction and estimation. Most of the work on seabuckthorn has been confined to *H. rhamnoides*. Studies on other species of *Hippophae* would enrich our bioresource biochemically as well as genetically.

References

Alberto, D., Carmen Gomez-Cordoves. & Begona Bartolome. 2003. Commercial Dietary Antioxidant Supplements Assayed for Their Antioxidant Activity by Different Methodologies, *J.Agric.Food Chem.* 51, 2512-2519.

Annon. 1966. Methods of Vitamin Assay, The Association of Vitamin Chemists, Interscience Publishers, New York, 3rd edn., p.287.

Beveridge, T., Thomas, S.C.Li., Oomah, B.D. & Smith, A. 1999. Seabuckthorn Products: Manufacture and Composition, *J. Agric. Food Chem.* 47, 3480 – 3488.

Bhatt, A.K., Agrawal, H.O. & Pratap, T. 1993. Seabuckthorn (*Hippophae* L.) and sustainable mountain development. In: "*Himalayan Biodiversity conservation strategies*" (Ed. U.Dhar), p 439-450, Gyanodaya Prakashan, Nainital.

Galati, E.M., Mondello, M.R., Giuffrida, D., Dugo, G., Miceli, N., Pergolizzi. and Taviano, M.F. 2003. Chemical characterization and biological effects of Sicilian *Opuntia ficus-indica* (L.) fruit juice: antioxidant and antiulcerogenic activity, *J.Agric.Food.Chem.* 51, 4903-4908.

Harmand, P O., Duval, R., Delage, C. & Simon A. 2005. Ursolic acid induces apoptosis through mitochondrial intrinsic pathway and caspase-3 activation in M4 Bell melanoma cells, *Int. J Cancer* 114, 1-11.

Klein, E. & Weber, N. 2001. *In Vitro* test for the effectiveness of Antioxidants as Inhibitors of Thiyl Radical induced Reactions with Unsaturated Fatty Acids, *J. Agric. Food Chem.* 49, 1224-1227.

Li, T.S.C. & Wang L.C.H. 1998. Physiological components and health effects of ginseng, Echinacea and seabuckthorn, In, "*Functional foods, Biochemical processing aspects*" (Ed. G. Mazza), p 329-356, Technomic, Lancaster.

Liu, J. 1995. Pharmacology of oleanolic acid and ursolic acid; *J. Ethnopharmacol* 49, 57-68.

Singh, V. & Khosla, P.K. 2001. Seabuckthorn- A resource for health and environment in twenty first century, *Proceedings of international workshop on sea buckthorn,* New Delhi, Feb. 18-21.

Singh, V. 2005. Seabuckthorn (*Hippophae* L.): A multipurpose Wonder Plant (Ed. V.Singh), p .3-69, Daya Publishing House, New Delhi, India 13.

Yamori, Y., Nara, Y., Tsubouchi, T., Sogava, Y., Ikeda, K. & Horie, R. 1986. Dietary prevention of stroke and its mechanism in stroke prone spontaneously hypertensive rats-preventive effect of dietary fibre and palmitoleic acid. *J. Hypertension* 4, 449-452.

□□□

12

Process for Extraction Oil and Juice from Seabuckthorn

V.V. Venugopalan, A. Ranjith, K. Sarinkumar, C. Arumughan

CONTENTS

12.1 INTRODUCTION

Seabuckthorn a fascinating plant of Elaeagnaceae family with very high nutraceutical and therapeutic values, is grown in Asia, Europe, and Northern America. The most important part of the plant is berries, with long history of applications in Tibetan and Mangolian medicines. The berries are rich in vitamins, polyphenols organic acids and bioactive lipids. Presently, a wide spectrum of physiological effects of SB berries and berry products have been reported which includes inhibition of LDL cholesterol oxidation and platelet aggregation, reduction of atopic dermatitis, immunomodulation, cytoprotective effects and protection from gastric ulcer (Baoru and Kallio 2002; Clair *et al.*, 2002; Anu *et al.*, 2000; Katajisto *et al.*, 1999 and Geetha *et al.*, 1999). Being a good source of bioactive phytochemicals, hundreds of industries in China and Russia are engaged in the production of nutraceutical and cosmaceutical products from seabuckthorn. In India, seabuckthorn is growing in cold deserts of Trans-Himalayan regions like Ladakh, Lahaul and Spiti. India, with second largest area under sea buckthorn in the world, effective exploitation of this valuable wealth for commercial and health care applications has not been given attention. Reports describing the processing of SB berries are rather limited. Juice, pulp oil, seed oil, cream and pigments are the main commercial products from SB berries. In a recent review Beaveridge *et al.*, (1999) have discussed the available processes for oil and juice (Jianfeng *et al.*, 2002). Mechanical pressing, vegetable oil extraction and solvent extraction are the major methods suggested for oil extraction. In mechanical pressing, the pulp oil recovery is partial and extraction efficiency depends to the selection and design of pressing unit. Possible destruction of heat sensitive bioactive constituents during desolventization and presence of residual solvents in oil restricts the application of solvent extraction technique. Supercritical CO_2 extraction techniques yields superior quality solvent free oil enriched with active constituents, but being a dry process, production of juice and nutritive pulp cannot be achieved by this. Heitscher *et al.*, (1996) suggested a wet enzymatic processing method for pulp oil and juice (Beveridge *et al.*, 1999). The juice obtained by the conventional processing technique is reported to be turbid with high content of suspended solids and pulp oil and found to form undesirable oily layer on the top on storage. Centrifugation of unheated juice causes rapid separation into a floating cream phase with oil, opalescent clear juice in middle and sediment. Zhang *et al.*, (1989) suggested the use of stalk centrifuge or cream separator for the separation of this cream from the juice (Zhou and Chen, 1989). Liu and Liu (1992) reported the application of pectin methyl esterase to break down the pectins in pulp to obtain clear juice (Zhang *et al.*, 1989). Heilscher and Lorber (1996) reported the use of crystalline sugar for the sedimentation and subsequent centrifugation for clear juice (Stastova *et al.*, 1996). In this, development of an integrated wet process for various products from fresh seabuckthorn berries, employing a specially designed mechanical dewatering unit, where maximum recovery of juice and pulp oil is obtained is reported. This process developed at pilot scale, enabled simultaneous production of superior quality pulp oil, and oil free clear juice with maximum retention of bioactive phytochemicals from berries. This green process which recovers all the major seabuckthorn products simultaneously without using solvents or chemicals, can be easily adopted for the commercial processing of fruits in ecologically sensitive high altitude areas where berries are being cultivated. So far no attempt has been made to utilize the SBT berries grown in Trans-Himalayan regions of India, that hosts unique geo-climatic conditions. The present study reports the processing of SBT berries, detailed chemical evaluation of process streams and products for the first time.

12.2 PROCESS DEVELOPMENT

Fig. 12.1 represents the schematic flow diagram of the present process. Fresh Sea buckthorn berries with moisture content of 70-75% were first cleaned by spraying with water for removal of impurities followed by blanching with hot water to destroy enzymes and to facilitate extraction of juice and oil from fruits. The berries were then subjected to a mechanical dewatering, which is the key step involved in our process, where 80% of the water and pulp oil and 60 % of the soft solid parts (pomace) of the berry were separated as a liquid phase, leaving a solid cake comprised of residual pulp and unbroken seed. This effective dewatering is achieved by multi stage pressing using a specially designed high-pressure screw press with compression ratio is about 10. The liquid phase is then subjected to clarification about 70-80^{0}C for 30 minutes with continuous stirring in a steam heated jacketed vessel, followed by centrifugation in a continuous 3 phase centrifuge, through which an efficient separation of pulp oil, clear oil free juice and a solid sludge is achieved. A series of experiments with a batch size of 30-100 Kg were conducted in pilot scale level, for standardization of the process parameters and for achieving maximum process efficiency.

12.2.1 Super Critical CO_2 Extraction of Seed Oil

SB seeds separated from the dried residue was grounded and sieved through a 42 mesh. A series of experiments (1 kg/batch) were carried out for extraction of seed oil using pilot model super critical extraction unit (SFE-2L,

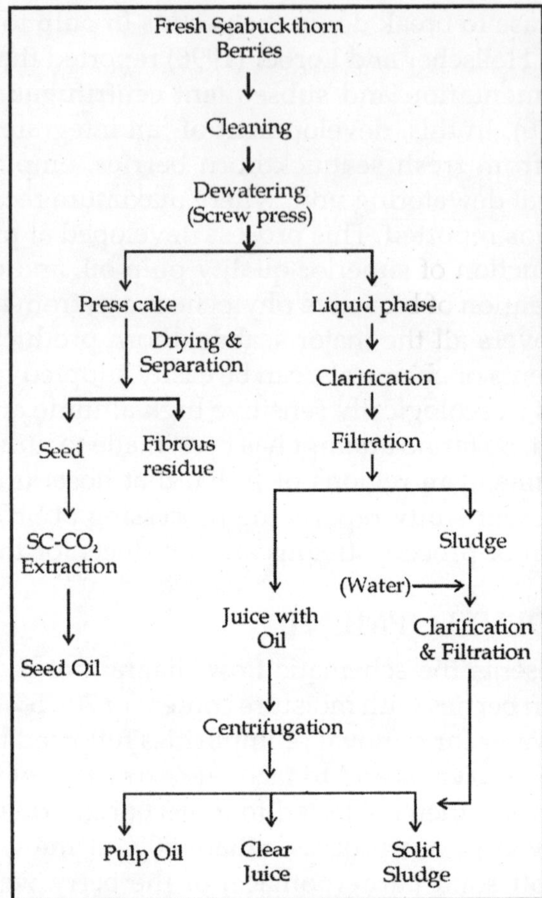

Fig. 12.1: Integrated processing of fresh seabuckthorn berries for pulp oil and juice

Thar Designs, Inc., Pittsburgh, USA) at 60 °C and at 450-bar pressure with a gas flow rate of 60 g/min for a process duration of 3 hours. The seed oil was collected in the cyclone separator.

12.3 RESULTS AND DISCUSSION

Some of the unit operations involved in the present process are similar to available reports, but there are variations in terms of process efficiency and product quality. Our process is standardized through pilot scale operations, material balance study, quality evaluation of process streams and end products, commercial feasibility etc. and ensures superior quality oil and juice with higher yield. Also, this is a first attempt made for commercial level processing and chemical evaluation of Indian seabuckthorn berries. Higher yield of pulp oil and juice in present process is achieved by employing a combination of specially designed high-pressure screw press and a continuos high-speed centrifuge. The process is eco-friendly and of zero wastage since the co-products like residual pulp, and solid pulp in this process can be utilized for residual oil recovery or usage as animal feed.

Results of the compositional analysis of process streams from fresh berry processing are shown in Table 12.1. Pulp oil yield was 2.6-3.0 % of fresh weight of Indian berries with recovery of 65-70 % of oil present in berries. The yield of the clear juice is 70-80 % on berry weight, with 10-13 % soluble solids

Table 12.1: Compositional Analysis of Process Streams (Mean ± SD)

Trial No.	Process streams	Total solids (g kg^{-1})	Total fatty matter (g kg^{-1})	Total protein (g kg^{-1})	Ash content (g kg^{-1})	Total soluble sugars (g kg^{-1})	Crude Fiber (g kg^{-1})
I	Berries (Soft parts)	279±2	46±1	18.1±3.5	10.2±1.1	95±6	24.4±0.9
	Juice	132±7	0.8±0.0	5.2±1.2	7.7±0.6	58±3	-
	Fibrous Residue[a]	362±9	30±3	59±4.3	7.2±1.3	118±5	98±7.3
	Sludge	202±3	68±7	24±2.2	8.2±1.9	45±7	10.5±0.6
II	Berries (Soft parts)	262±2	44±1	21.2±3.1	10.2±0.8	92±8	23.5±2.1
	Juice	107±6	0.3±0.1	4.1±0.4	6.1±1.2	49±2	-
	Fibrous Residue[a]	387±1	32±2	64±9.4	6.3±2.1	118±9	121±8.2
	Sludge	191±2	67±4	20.4±7.1	9.6±1.3	48±4	8±1
III	Berries (Soft parts)	269±1	44±1	21.6±3.4	9.9±1.2	98±6	24.4±4.2
	Juice	109±3	0.4±0.1	4.5±2.9	5.5±0.9	49±1	-
	Fibrous Residue[a]	349±0	29±3	61.2±7.4	8.1±1.9	113±7	101±9.1
	Sludge	195±2	64±2	21.2±2.9	10.1±0.8	47±6	5.5±1.3

[a] analyzed and expressed without seed.

and was almost free from oil (<0.1 %). The by-products of the process were fibrous residue and sludge with about 3% and 6.5% total fat, respectively, that together comprised about 25-30% of the oil present in the soft parts of the berry (pulp). Solvent or supercritical fluid could be employed to recover this residual oil that was not attempted here. The seed in the fibrous residue was easily separated after drying followed by air classification. The seed was dried and grounded followed by extraction of oil using supercritical CO_2 as described before. Chemical evaluation of the products was done in terms of bioactive molecules and compared with that of raw berries. Pulp oil obtained by this process was found to be extremely rich in carotenoids, tocopherols and sterols along with 16:1 as the major fatty acid, which is unique among oils of plant origin (Tables 12.2, 12.3). By the present process about 70% of the total pulp oil was recovered with maximum retention of bioactive was achieved with out the usage of organic solvents.

Table 12.2: Fatty Acid Composition (wt. %) of Pulp Oil and Hexane and Supercritical CO_2 (SC- CO_2) Extracted Seed Oils (Mean ± SD)

		Fatty acids					
		16:0	**16:1**	**18:0**	**18:1**	**18:2**	**18:3**
Pulp oil	I	33.8±1.5	46.4±2.0	1.0±0.5	13.4±0.9	5.4±1.3	tr
	II	31.2±1.4	49.1±3.1	0.8±0.1	13.9±3.5	4.6±0.6	tr
	III	30.9±1.3	45.6±2.6	0.9±0.4	16.8±1.7	5.8±0.4	tr
Seed Oil	Hexane	17.9±2.0	2.1±0.5	4.1±1.1	29.8±1.8	26.1±1.4	18.8±2.1
	SC- CO_2	17.2±2.0	3.6±0.4	3.8±0.7	30.0±2.0	27.7±1.4	16.7±1.3

Table 12.3: Carotenoid, Tocopherol and Sterol content in the Soft parts of the Berries, Pulp oil and Hexane and Supercritical CO_2 (SC- CO_2) Extracted Seed Oils (Mean ± SD)

Composition	Soft parts of berries	Pulp oil			Seed oil	
		I	**II**	**III**	**Hexane**	**SC-CO_2**
Carotenoids (mg kg^{-1})	128±4	2810±45	2636±34	2450±24	387±10	40316

Contd...

Table 12.3 Contd...

Tocopherols/tocotrienols (mg kg^{-1})						
α-T[a]	47.2±4.0	987±38	1141±19	1067±61	520±24	587±21
α-T3[b]	6.6±1.5	182±24	131±13	94±11	11±1	35±6
β-T[a]	1.1±0.2	28±6	22±4	21±4	19±3	21±4
γ-T[a]	4.5±0.5	91±4	134±18	109±20	306±21	315±25
γ-T3[b] +δT[a]	6.1±1.2	107±14	154±19	134±11	46±6	39±7
δt3[b]	0.90±0.1	14±2	17±1	15±0.8	10±1	15±2
Total	66.4±4.1	1409±48	1323±24	1439±84	912±34	1012±45
Sterols (mg kg^{-1})						
β-sitosterol	180±14	3200±85	3221±63	3034±84	13305±132	13774±194
Stigmasterol	80±13	1203±42	1058±61	1062±67	1125±111	1235±99
Campesterol	tr	tr	tr		1770±50	1878±99
Total	260±26	4403±61	4279±52	4096±98	16200±260	16888±150

[a] T : Tocopherol; [b] T3 : Tocotrienol

As a part of the integrated approach to processing of SB berries, a comparative evaluation of seed oil obtained by the hexane and super critical (SC-CO_2) extraction was performed with the seeds obtained separated from the residue. Yield of hexane extracted oil was slightly more (6.1% by wt.. on seed) as compared to that of SC-CO_2 extracted oil (5.2% by wt. on seed). All the phytochemical constituents (carotenoids, tocopherols/tocotrienols and sterols) were higher in the oil extracted with SC-CO_2 than that in hexane (Table 12.3). Lower content of phytochemicals in the hexane extracted oil could be due to non-oil components such as wax and gum extracted with hexane resulting higher extract yield with lower concentration of phytochemicals. The seed oil differed significantly in its chemical characteristics from that of pulp oil. 16: 1 fatty acid was the most predominant fatty acid in pulp oil, its content in seed oil was as low as 2.1% (Table 12.2). Tocopherol and carotenoid content in the seed oil was lower than that of pulp oil

(Table 12.3). However, seed oil had about 4 times higher content of sterols in comparison with that of pulp oil. The juice obtained retained more than 60% of organic acids, 70% of vitamin-C and 40% Polyphenol and 50% of flavonoids present in soft parts of fresh berry (Table 12.4). Also after addition of water to facilitate dewatering is reported processing and consequent dilution of juice, concentration of total solids and bioactive remain fairly high in fresh juice that could be attributed to extraction efficiency. The final juice obtained can further, diluted in the ratio 1:2 to formulate ready to drink health beverage, rich in bioactive phytochemicals.

Table 12.4: Vitamin C, acids, polyphenols and flavonoids in berries & juice (Mean ± SD)

Composition	Soft parts of berries	Juice		
		I	II	III
Vitamin C (mg kg^{-1})	2232±72	1840±18	1683±21	1752±29
Total titrable acids (g kg^{-1})[a]	-	35.6±0.3	36.9±0.6	36.4±0.4
pH	-	3.2	3.3	3.1
Total polyphenols (mg kg^{-1})	5646±87	2821±11	2392±26	2757±39
Flavonoids (mg kg^{-1})	575±29	400±18	340±33	401±26

[a] Expressed in terms of malic acid

12.4 CONCLUSIONS

An efficient process at pilot scale through process and products integration was developed to produce biologically active phytochemicals rich pulp oil and juice for nutraceutical applications from fresh seabuckthorn berries. Fresh berries were subjected to high pressure dewatering using continuous screw press. The separated liquid phase containing pulp oil was clarified at 80°C followed by high-speed continuous centrifugation to obtain pulp oil, clear juice and sludge. The pulp oil yield was 2.6-3% of fresh berries with 65-70% extraction efficiency with remaining 20-30% of oil distributed between residue and sludge. Pulp oil was very rich in carotenoids (2,450-2,810mg/kg) tocopherols (1,177-1,439 mg/kg) and sterols (4,096-4,403 mg/kg) with a characteristic fresh berry flavor and 16:1 as the major fatty acid (45.6-49.1%). The clear juice obtained was free from oil and with high amount of bioactive phytochemicals like vitamin C (1,683-1,840 mg/kg) polyphenols (2,392-2,821 mg/kg) and flavonoids (340-400 mg/kg). The juice was very acidic (pH 3) with high amount of organic acids (3.56-3.69%). The seeds in the press cake were separated and extracted for oil using supercritical CO_2. Seed oil contained carotenoids (403 mg/kg), tocopherols (1,012 mg/kg) and sterols (16,888 mg/kg).

References

Anu. K. Johansson., Helena, Korte., Baoru, Yang., John, C. Stanley., & Heikki P Kallio. 2000. Sea buckthorn berry oil inhibits platelet aggregation. *J. Nutr. Biochem.* 11,October, 491-495.

Baoru Yang., & Heikki, Kallio. 2002. Composition and physiological effects of sea buckthorn (*Hippophae*) lipids. *Trends in Food Science and Technology* 13, 160-167.

Beveridge, T., Li, T. S. C., Oomah, B. D., & Smith, A. 1999. Seabuckthorn Products: Manufacture and Composition. *J. Agric. Food Chem.* 47(9), 3480-3488.

Clair, Eccleston., Yang, Baoru., Raija, Tahvonen., Heikki, Kallio., Gerald, H. Rimbach., & Anne, M. Minihane. 2002. Effects of antioxidants rich juice (Seabuckthorn) on the risk factors of Coronary heart diseases in human *J. Nutr. Biochem.* 13, 346-354.

Geetha, S., Sai Ram, M., Singh, V., Ilavazhagan, G., & Sawhney, R.C. 2002. Anti-oxidant and immunomodulatory properties of sea buckthorn (*Hippophae rhamnoides*)- an in vitro study. *J Ethano Pharmacology* 79, 373-378.

Heilscher, K., and Lorber, S. 1996. Cold working process for obtaining clear juice, sediment and oil from sea buckthorn berries and their juice GFR Patent DE 44 31 C1.

Jianfeng, Xing, Baoru, Yang, Yaling, Dong, Bingwen, Wang, Junxian, Wang, & Heikki, Kallio. 2002. Effects of seabuckthorn seed and pulp oils on experimental models of gastric ulcer in rats. *Fitoterpia* 73, 644-650.

Katajisto., Olli, J Peltola., & Heikki, P. Kallio. 1999. Effects of dietary supplementation with sea buckthorn (*Hippophae rhamnoides*) seed and pulp oils on atopic dermatitis. *J.Nutr.Biochem* 10, November, 622-630.

Liu & Liu 1992. Production and use of clear juice and oil Sea Buckthorn berries, Heitscher, Karl, Lorber, Sietfried,(Germany) Ger DFE 4331394,(1, 15 Feb 1996).

Stastova, J., Jez, J., Bartlova, M., & Sovova, H. 1996. Rate of the vegetable oil extraction with super critical CO2. III. Extraction from sea buckthorn. *Chem. Eng. Sci.* 51 (18), 4347-4352.

Zhou, X., and Chen, C. 1989. Research on the process technology of turbid type sea buckthorn beverage. *Proceedings of the international symposium on sea buckthorn (Hippophae rhamnoides. L.)*, Xian, China, October, 19-23, pp. 310-313.

Zhang, Y., Xu, J., & Yu, N. 1989. Research on the process technology sea buckthorn and oil. *Proceedings of International Symposium on Sea Buckthorn (H. rhamnoides* L.*)* Xian, China, Oct-19-23, pp. 307-310.

□□□

13

Bioactive Substances and Therapeutic Uses of Seabuckthorn

T. Parimelazhagan, O.P. Chaurasia and Zakwan Ahmed

CONTENTS

13.1 INTRODUCTION

Increasing health concerns today necessitate reliance on alternate herbal remedies for sustaining good health. Seabuckthorn is a plant, which has rich medicinal properties, and its fruits contain high amount of bioactive substances. It is one of the richest natural source of vitamin C, besides having many other important human health related molecules. Seabuckthorn oil is one of the most versatile natural oil and has been used in Asia for treating various skin conditions. It is packed with carotenoids, tocopherols, rare fatty acids and phytosterols, giving it synergistic power to protect cell membranes and enhance cell regeneration (Derevich and Shindyapkin, 2004; Alam, 2004). Seabuckthorn is traditionally used in the treatment of gastric ulcers, and laboratory studies confirm the efficacy of the seed oil for this application (Xing, 2002). Seabuckthorn oil is one of the imperative products obtained from both berries and seeds and it contains different kinds of nutrients and bioactive substances including vitamins, fatty acids, free amino acids and elemental components. One of the many special features of seabuckthorn is the exceptionally high oil content. Seabuckthorn oil is also rich in bioactive non-nutrient phytochemicals which include flavonoids, glycosides, phytosterols and their esters, organic acids etc. This marvelous oil has copious amounts of vitamin E, β-carotene, antioxidants and carotenoids which helps to substantiate its relatively high success rate as a skin repairing and conditioning oil. The presence of valuable chemical and nutritionally important constituents in seabuckthorn oil, and from the scientific knowledge of their importance, it is clear that seabuckthorn oil is one of the most important source of these materials. The excellent combination of the high content of Carotenoids and vitamin E makes valuable oil in the commercial market. This review provides an overview of various bioactive substances present in the seabuckthorn oil and their potential uses for human welfare.

Since the ancient Tibetans started using seabuckthorn more than one thousand years ago, hundreds of Asian traditional recipes have been developed and carried on through generations, utilizing the nutritional and medical properties of the berry. Through the long history of application as food and medicinal ingredients, seabuckthorn berry is known to be effective in treating wounds, inflammation, mucous-membrane related disorders and diseases such as cough, sputum, gastric and gynecological problems (Risto and Baoru, 2003). The clinical trials and scientific studies conducted during the 20th century in several countries confirm medicinal and nutritional value of seabuckthorn oil. Herbal remedies made of seabuckthorn oil are most frequently used for the treatment of diseases of skin and digestive system.

Traditional use of seabuckthorn oil to promote the recuperation of skin injuries and support the healing of skin diseases, well agrees with the data of modern clinical studies (Alam, 2004a). Berries from seabuckthorn, a shrub still virtually unknown in North America are currently being incorporated as a functional food and into nutraceutical products in Europe and Asia (Li and Schroeder, 1999). Traditional products from the berries include juices, liquors, wine, jams, candy and ice-cream. However, the berry's unique chemical and nutritional composition has offered economic potential as a health food in North America (Oomah and Mazza, 1999).

The presence of valuable chemical and nutritionally important constituents in seabuckthorn, and from the scientific knowledge of their importance, it is clear that seabuckthorn can be used as alternative nutritional source. The present study describes comprehensive source of overall information on seabuckthorn oil, so as to obtain a clear perception over the chemical and medicinal aspects of seabuckthorn oil for the future studies.

13.2. BIOACTIVE SUBSTANCES

During the past few decades, inspired by the long traditional use and the increasing public recognition of the benefits of natural products, scientists have carried out extensive research on seabuckthorn. This has resulted in an improved understanding of the health effects and the chemical composition of the berry. It is known that the berry is rich in multiple lipophilic and hydrophilic bioactive compounds such as vitamin C and E, carotenoids, flavonoids, fatty acids, plant sterols, lignans and minerals (Risto and Baoru, 2003). Seabuckthorn is a kind of food that can be used as medicine containing multiple vitamins, trace elements, amino acids and other bioactive substances. According to Russia's research, more than 190 bioactive substances had been found in seabuckthorn pulps, 103 kinds in seabuckthorn oil including 6 kinds of vitamins, 22 kinds of fatty acids, 42 kinds of fats and 33 kinds of flavone and phenolic compounds. The oil of the seabuckthorn berry contains fatty acids of virtually all-important series, omega 3, 6, 7 and 9 and is exceptionally rich in palmitoleic acid (omega 7), rarely found in the plant kingdom. The oil is also rich in natural antioxidants, tocopherols, tocotrienols, carotenoids and plant sterols. The fruit residue, which includes the outer peel, is rich in the colorful carotenoids and vitamin E; the seed has the highest level of the unsaturated fatty acids and sterols.

The vitamin C concentration in berries varies from 360 mg/100 g of berries for the European sub-species *rhamnoides* (Rousi and Aulin, 1977; Yao *et al.*,

1992) to 250 mg/100 g of berries for the Chinese sub-species *sinensis* (Yang *et al.*, 1988, Wolf and Wegert 1993). Vitamin E concentration can be up to 160 mg/100 g of berries (Zhang, 1989; Eliseev, 1989; Wahlberg and Jeppsson, 1992). Seabuckthorn is also rich in flavonoid and contains appreciable amount of water-soluble and fat-soluble vitamins (Zhang, 1989; Solonenko and Shishkina, 1989; Schapiro, 1989). The juice provides a nutritious beverage, high in suspended solids, and very high in vitamins especially vitamin C and carotenoids (Alam, 2004). The chemical composition of seabuckthorn berries varies with the origin, climate and method of extraction. Indian seabuckthorn juice gave an average pH of 3.13 and an average titratable acidity of 1.97% calculated as malic acid (Beveridge *et al.*, 2002).

The high vitamin concentration makes seabuckthorn fruit highly suitable for the production of nutritious soft drinks. The seabuckthorn juice is yellow in colour due to high amount of carotene. There are many mineral elements present in berries/ juice of seabuckthorn. Potassium is the most abundant of all the elements investigated in berries or juice (Kallio *et al.*, 1999; Tong *et al.*, 1989). Large variation in concentration of acids has been also reported among different origins. Of all organic acid components in seabuckthorn juice, malic and quinic acids were major acids together constituting around 90% of all the fruit acids in different origins (Kallio *et al.*, 1999). Seabuckthorn berries (Juice) is rich in various free amino acids. Eighteen kinds of free amino acids were detected from seabuckthorn (Chen, 1988). It contains more aspartic acid (426.6 mg/100g) (Zhang *et al.*, 1988). The profile of the volatile compounds of seabuckthorn juice is very typical and distinctive from that of other common berries. The compounds were mainly esters of short chain, branched or n-fatty acids and alcohols. The profile of the volatiles is clearly dependent on the time of harvesting of the berries (Yang, 2001). Seabuckthorn fruits are rich in pigments and lipoproteins are located in membranes and the fleshy mesocarp (Pintea *et al.*, 2001). Seabuckthorn berries contain up to 13% soluble sugars, mainly glucose, fructose and xylose and 3.9% organic acids, mainly malic and succinic acid (Ma and Cui, 1989). Seabuckthorn is rich in proteins and free amino acids. A total of 18 amino acids have been found in seabuckthorn fruit (Zhang, 1989; Mironov, 1989). There are at least 24 chemical elements present in seabuckthorn juice viz. nitrogen, phosphorous, iron, manganese, boron, calcium, aluminium, silicon and others (Wolf and Wegert, 1993; Zhang, 1989; Tong *et al.*, 1989). The leaves and young sprouts make a wonderful vitamin rich and well-ingested fodder. At the same time, in fruit processing plants, the leaves of seabuckthorn make tons of raw materials. In the course of extraction of oil from the seabuckthorn leaves, which is used in perfumery and a significant portion of waste products (leaf cake after oil extract) used as additives to fodder (Solovenko *et al.*, 2003). The leaves contain b-amyrinoleyl alcohol acid. It can dilate the cardiac and cerebral vessels, facilitate blood circulation and slightly lower the blood pressure (Ge Xiaoyan *et al.*, 1986).

13.3 SEABUCKTHORN OIL

Seabuckthorn oil is one of the most versatile natural oils. It is packed with carotenoids, tocopherols, rare fatty acids and phytosterols, giving it synergistic power to protect cell membranes and enhance cell membranes and enhance cell regeneration. Seabuckthorn oil appears to be unsurpassed as a natural source of vitamin A. Seabuckthorn oil extraction can be done using several methods. The nutritional qualities of seed and pulp oils from seabuckthorn depends upon the method of extraction.

13.3.1 Oil Extraction

Seabuckthorn oil extraction from pulp and seed were practiced by the following methods, i.e. pressing (screw press), solvent extraction using *n*-Hexane, supercritical fluid extraction using CO_2 and aqueous extraction. Oil pressing was carried out by using screw press. Prior to pressing, the screw was cleaned properly by water and allowed to heat for 10 min through electrical connections to raise its temperature. After heating, the seeds and pulps were pressed for 5 min to achieve steady flow of oil. Extracted oil was collected in a container and heated up to 50-60 ^{0}C, then cool it for 20 min. After cooling, the oil will be drained out which were wrapped with aluminum foil for protection from light and were stored at -20^{0}C for further use. The extraction by pressing is without residue, but the extraction rate was slow and colour looks like dark. Solvent extraction using *n*-Hexane (boiling range 45-60 ^{0}C) was done by soxhlet apparatus following a standard method (AACC, 2000). Prior to extraction, seeds were prepared by grinding them in a Mixer. During *n*-Hexane extraction, the oil temperature was maintained at 40 ± 1^{0}C. The extraction rate is high by solvent extraction, but it has solvent residues in the oil affecting the quality of oil. Supercritical CO_2 fluid extraction (SCFE) is one of new technology developed in recent years, which control the CO_2 under the state of supercritical between its liquid and gas (Cenkowski *et al.*, 2006). At this point, the solubility is most and easy to extract the all compounds from the seabuckthorn seeds. Supercritical CO_2 extraction yields extracts, which are free of microbes or their spores. It is also totally solvent free. The compounds of seabuckthorn seeds and fruits will not be damaged during its processing, not only the quality of product can be insured, but also there is environment pollution. Because the oil is out with pressure, the gas of CO_2 contained within oil will be pressed out and play the role of protecting by separating the air an oil. In aqueous extraction, 1 kg of fresh fruits was added to water for 30 min at 60 °C in a water bath stirred continuously. After heating, cool it for 10 min and take out the fruits in a muslin cloth and squeeze it three times to get complete oil extract. The extract was heated for 30 min at 50 °C then the mixture was centrifuged for 15 min at 10,000 rpm at 20 °C. The liquid fraction was decanted and centrifuged again at room temperature for 10 min at 10,000 rpm to recover pulp oil (Parimelazhagan *et al.*, 2005).

13.3.2. Seabuckthorn Pulp Oil

Seabuckthorn pulp oil is pure natural oil, its purity allowing it to be effective in lower dosages. It has long been used for treating various skin conditions. Seabuckthorn berry oil contains a high level of palmitoleic acid (Jablczynska *et al.*, 1994; Ul'chenko *et al.*, 1995; Moracova *et al.*, 1995; Zadernowski *et al.*, 1997). The oil is attracting more and more attention because of the increasing interest in the physiological role of the monounsaturated fatty acids (Colquhoun *et al.*, 1996; Gomez-Tubio *et al.*, 1995 and Perez-Jimenez *et al.*, 1999). The high content of carotenoids (up to 7 g/kg), tocopherols (up to 7 g/kg), and phytosterols (up to 20 g/kg) (Pintea *et al.*, 2001; Moracova *et al.*, 1995; Zadernowski *et al.*, 1997) are special characteristics of the oil from pulp of the berries. The existing knowledge shows that the composition of seabuckthorn berries varies greatly according to their origin, and climatic and geological conditions of the growth areas (Chen *et al.*, 1990; Jablczynska *et al.*, 1994; Moracova *et al.*, 1995). Among pigments, β-carotene (important antioxidant in the food industry) is the most abundant in the pulp oil and constitutes 15 to 55% of total amount of carotenoids (Yang and Kallio, 2002). Important sterols include cholesterol, phytosterol and steroid hormones, and their contents in seed, pulp and fresh whole berries (spp. *rhamnoids* and *sinensis*) are 1,200-1,800, 240-400, and 340-520 mg/kg of berries, respectively (Yang *et al.*, 2001). Its effectiveness is based on a combination of lipophilic compounds, working synergistically to protect and regenerate stressed skin cells and other protecting organs. Seabuckthorn oil has a tremendous effect in skin regeneration. This was found especially useful in treating various conditions requiring enhanced cell regeneration, for example, burns, scalds or eczema-like damage. Its yellow colour is due to its high content of carotenoids (Parimelazhagan *et al.*, 2004). The wide chemical and phytochemical composition of seabuckthorn oil has recently reviewed (Alam, 2004a) and same has presented in Table 13.1. It was found to vary with the origin, climate and method of extraction (Xurong and Tigerstedt, 2001).

Table 13.1: Main bio-constituents of Seabuckthorn oils from seed, fruit pulp (juice) and fruit residue after removing juice

Constituents	Concentration (mg/100g)		
	Seed oil	Pulp oil	Fruit residue oil
Vitamin E	207	171	300-600
Vitamin K	110-230	54-59	–
Carotenoids	30-250	300-870	1,280-1,860
Total acids	11	38	–
Total flavonoids	–	–	550
Total sterols	1,094	721	–
Unsaturated fatty acids	87%	67%	70%
Saturated fatty acids	13%	33%	30%

Seabuckthorn pulp oil is especially rich in Omega-7- series of fatty acid (palmitoleic acid). This rare fatty acid is a key component in pulp oil and it can act as a replacement for other fatty acids in cell membranes. Palmitoleic acid has fluidity similar to that of polyunsaturated fatty acids but low susceptibility to lipid peroxidation. An increase in the level of dietary palmitoleic acid has been suggested to improve the metabolism of vascular smooth muscle cells (Yamori *et al.*, 1986). Oil from the soft parts of seabuckthorn berries is known to contain a high level of palmitoleic acid (Quirin *et al.*, 1993; Jablczynska *et al.*, 1994; Zadernowski *et al.*, 1997). Both the pulp and seed oils rich in tocopherols, tocotrienols and plant sterols. In addition, the pulp oil contains especially high level of carotenoids (Table 13.3). Predominant tocopherols in the pulp oil were a-tocopherol and β-tocopherol, with α-tocopherol constituting 85% of the total tocopherols and tocotrienols identified (Cenkowski *et al.*, 2006). Processing of pulp, which included, thawing berries, macerating, extracting juice, pressing cake, drying, threshing and sieving had a deleterious effect, decreasing a-tocopherol concentrations in pulp-flake oil. Concentration of total carotenoids increased with extraction time (3 to 6 h). Screw pressed oil contained higher amounts of total carotenoids compared to the super critical CO2 fluid extraction technique (Cenkowski *et al.*, 2006). Total carotenoids in chloroform/methanol-extracted oil from pulp were 347.1 mg/100g oil (Table 13.3). Beveridge (2003) reported a wide range of total carotenoids from 330 to 1,000 mg/100g oil, depending on plant subspecies or cultivar.

13.3.3. Seed Oil

Seabuckthorn seed oil is an excellent source of essential fatty acids (Table 13.2). It contains both linoleic and alpha linolenic acid in a well-balanced

Table 13.2: Fatty acid composition of oils from seed & pulp of seabuckthorn (*H. rhamnoids*)

Oil	Fatty acid %						Reference
	Palmitic	Palmitoleic	Stearic	Oleic	Linoleic	Linolenic	
Seed	6-10 6-7.5	- -	2-4 2-3	15-20 13-14	35-40 35-36	20-35 35-40	Yang, 2001; Cenkowski *et al.*, 2006
Pulp	15-40 20-36	15-50 20-40	1-2 1-2	10-20 3-4	5-15 12-15	5-10 2-5	Yang, 2001; Cenkowski *et al.*, 2006

Table 13.3: Carotenoids and tocopherol concentrations (mg/100g oil) in seabuckthorn seed and pulp oil

Oil	Carotenoids	Tocopherol	Species	Reference
Seed	10-50	100-200	*rhamnoids*	Risto and Baoru, 2003
	22.5	273.6	*rhamnoids*	Cenkowski *et al.*, 2006
	67.5	92.7	*caucaus*	Mironov, 1989
	-	100-300	*sinensis*	Kallio *et al.*, 2002
Pulp	100-40	100-400	*rhamnoids*	Risto and Baoru, 2003
	347.1	261.2	*rhamnoids*	Cenkowski *et al.*, 2006
	350.0	216.0	*rhamnoids*	Zhang *et al.*, 1989
	-	400-700	*sinensis*	Kallio *et al.*, 2002

1:1 ratio. Seabuckthorn seed oil contains 150 to 220 mg natural tocopherols per 100 g of oil. Alpha-tocopherol is the most active form of vitamin E in humans and is a powerful biological antioxidant. An extremely high level of alpha-tocopherol (1,046 mg/100 g pulp oil) was reported for a cultivar of spp. *mongolica* from Altai (Jablczynska *et al.*, 1994). The total content of tocopherols and tocotrienols in seeds varied from 40 to 220 mg/kg in sp. *sinensis*, from 280 to 300 mg/kg in sp. *rhamnoides* and was about 250 mg/kg in sp. *mongolica*. Seeds of the latter two subspecies were clearly richer in tocopherols and tocotrienols than those sp. *sinensis* (287.4 $\pm$ 11.7 mg/kg in sp. *rhamnoides* and 249.8 $\pm$ 5.1 mg/kg in sp. *mongolica* vs. 126.7 $\pm$ 59.8 mg/kg in *sinensis* (Kallio *et al.*, 2002).

Changes in vitamin E content in seeds of seabuckthorn (*H. tibetana* and *H. rhamnoides* L. sp. *sinensis*) growing naturally in Gansu Province (altitude 2,865-2,895 m, China) from the 30th to the 150th day after flowering (DAF) (corresponding to the period from mid May to mid October) has been reported (Lieu *et al.*, 1990). The total content of vitamin E in seeds of sp. *sinensis* increased sharply after the 107th DAF, reached a maximum level (800 mg/kg fresh weight) on the 113rd DAF (early to mid September) and decreased during the following period. For the seeds of *H. tibetana* the total vitamin E content increased steadily from the 107th to the 139th (over 800 mg/kg) DAF (early October) and decreased during the following period. According to the above results and previous investigations (Lieu *et al.*, 1990), the highest level of total tocopherols and tocotrienols could only be obtained in seeds from the berries normally considered as being over-ripe, the exact optimal harvesting date depending on origins and growth conditions. Phytosterols are also present at around 1%, consisting mainly of β- sitosterol. Also present are minor amounts of tocotrienols, a fascinating new group of active phytocompounds (Franke and Muller, 1983). Our experimental results showed that the solvent extraction (*n*-Hexane) of fruit pulp (juice), fruit residue after removing juice and seed contain 20 to 25%, 18 to 20% and 10 to 15% oil, respectively (Parimelazhagan *et al.*, 2004 and 2005). The seed oil, which is unsaturated oil, shows promise because of its light absorption and emollient properties as an ingredient in cosmetics, phytopharmaceutical and UV skin protectant preparations. It absorbs strongly in the UV-B range (290-320 nm) and may therefore, be used as a natural sun-screen absorber (Beveridge *et al.*, 1999).

Seabuckthorn seed and pulp oils were analysed for different physico-chemical characteristics and are shown in Table 2. The results indicate that the oils from the seed and pulp have identical quality except in carotenoids and vitamin E (tocopherol) content. In the present study, it was observed that among the seed and pulp oil assayed for their vitamin E content, the seed oil exhibited high content than the pulp oil (Parimelazhagan and Ahmed, 2005). The most important phytonutrients of health significance are tocopherols, present in the seabuckthorn oil in exceptionally high concentration, that also are subject to significant variation due to geo-climatic conditions. Refractive Index and Optical Rotation are very stable parameters and were used for checking the identity of oils. In this experiment, acid and peroxidase values of both the oils were estimated (Table 13.2) and results revealed that values are reasonable according to quality parameters (Parimelazhagan and Ahmed, 2005). These two parameters can be helpful to ensure quality of products. Our results agree with data from other sources of literature (Lu, 1993; Mironov, 1989; Zhang, 1989; Bernath and Foldesi, 1992).

Table 13.4: Physico-chemical Properties of Seabuckthorn oil

Parameters	Pulp oil	Seed oil
Refractive Index	1.46 ± 0.002	1.41 ± 0.017
Optical rotation	2.10 ± 0.05	2.14 ± 0.065
Acid value	8.80 ± 0.14	10.0 ± 0.235
Peroxidase value (meq O_2 / kg)	1.36 ± 0.02	1.42 ± 0.038
Carotenoids (mg/ 100 g)	228.1 ± 2.5	47.15 ± 2.38
Vitamin E (mg/ 100 g)	146.2 ± 1.5	185.5 ± 1.21

All the values are mean ± SE

Differences between pulp and seed oil may depend on samples or on biochemical factors. The general concentration in the pulp oil and in the seed oil are in the same range but the degree of unsaturation of the seed oil is 2.5 fold higher than in the pulp. This causes the instability of the seed oil. Seed oil was comprised primarily of b-sitosterol (97%) with trace amounts of campesterol (Cenkowski *et al.*, 2006). Yang *et al.* (2001) reported that b-sitosterol is the major sterol in seeds of ssp. *sinensis*. Chloesterol and stigmasterol were not detected in seed oil samples. The concentration of b-sitosterol in seed oil changed with extraction method, namely, solvent extraction (746.3 mg/100 g oil), SCFE CO_2 (667.8 to 748.1 mg/100 g oil), and screw pressing (635.0 mg/100 g oil). The amount of b-sitosterol and campesterol increased by more than 10% when extraction time was increased from 3 to 6 h (Cenkowski *et al.*, 2006).

13.3.4. Products

Since the discovery of the nutritional value of seabuckthorn, hundreds of seabuckthorn products made from the berries, oil, leaves, bark and the extracts from them have been developed. In Europe seabuckthorn juice, jellies, liquors, candy, vitamin C tablets and ice cream are readily available (Bernath and Foldesi, 1992; Wolf and Wegert, 1993). Seabuckthorn jams and jellies are produced on a small scale in Saskatchewan (PFRA, 1992). At present, the largest producers and consumers of seabuckthorn products are China, Russia, India, and Mongolia. They all have large scale processing facilities. Processed products include: oil, juice, alcoholic beverages, jam, biscuits, food colours, medicines, cosmetics and shampoos (Pan *et al.*, 1989; Huang *et al.*, 1991; Wu, 1991). Essential oil from seeds is the most valuable product and has medicinal value. High content of vitamin C in the fruits is another valuable natural product. More than ten different drugs have been developed from seabuckthorn in Asia and Europe and are available in different forms, such as liquids, powders, plasters, films, pastes, pills, liniments, suppositories and aerosols. Seabuckthorn oil from seeds is popular in cosmetic preparations, such as facial cream (Li and Wang, 1998).

In Europe and Asia, there are numerous products made from seabuckthorn, such as tea from leaves, beverages and jam from fruits, fermented products from pulp, and animal feeds from leaves, pulp, and seed residues. Various skin care, cosmetic and dietary products made of seabuckthorn are traditionally popular in different parts of the world. Besides seabuckthorn oil, extract obtained with olive oil, Floraleads GR produces seabuckthorn base anti-aging cream, and oil blend of seabuckthorn and Rose hip oils. Sustainol is a herbal dietary supplement combining nutritional benefits of seabuckthorn along with those of Bilberry and Rose hips. The following products are already available in China: i) Raw seabuckthorn such as unstrained juice, clear juice, concentrated juice, seabuckthorn fruit oil and seabuckthorn seed oil, seabuckthorn fruit residual oil, raw powder, seabuckthorn pigment etc; ii) Beverage with seabuckthorn such as soft drink, alcoholic drink (sweet wine, semi-fluid drink, wine beer), fruit juice, aerated fruit juice, powder, nutrient solution, jam; iii) Cosmetics with seabuckthorn such as hair shampoo, skin care cream, beauty cream, and body loation; iv) Medicines such as cough remedy, seabuckthorn flavanoid tablets and capsules. Healthcare medicines, including seabuckthorn oil, instant powder or granule preparation of seabuckthorn, and dried cream (Xu *et al.*, 1991). In addition, the UV-spectrum of the oils shows a moderate absorption in the UV-B range which makes seabuckthorn derived products attractive for sun care cosmetics (Quirin and Gerard, 1993). The body soap, a well-established cosmetic in Canada, is adding seabuckthorn oil to their sunscreen products as a sun-blocking and tan-enhancing agent. In Russia, seabuckthorn berries are often used in home made cosmetics. Recipes for moisturizing lotions, dandruff control and hair loss prevention are widely known and used in Russia (Pashina, 1993). The potential of seabuckthorn oils for dermatological applications such as facemasks, body lotions, sun lotions and shampoos is excellent.

Field Research Laboratory (DRDO), Leh, has successfully developed processing technology of seabuckthorn herbal beverage. It is rich in vitamins C, B, A, E, K, minerals, organic acids, antioxidants and antiageing properties, etc. The chemical composition of the herbal beverage is showing 25% Juice, 46% TSS, 200 mg vitamin C per 100 ml and 120 cal/100 ml energy values. The product, *Leh berry*, has made its impact at national and international level and has also been introduced in Indian Army, Railways and Airlines etc. It can be stored easily for a period of two years under Ladakh conditions and it will not freeze up to -20 ^{0}C (Dwivedi *et al.*, 2004). seabuckthorn provides a number of unique products that are widely popular among people, especially those that have proven health or nutritional benefit.

13.4 THERAPEUTIC USES OF OIL

Medicinal uses of seabuckthorn are well documented in Asia and Europe. Pulp oil can be used in a variety of cosmetic formulations. It is used for application of cell anti-ageing, enhancement of microcirculation, antioxidant protection, epidermal regeneration, anti-inflammation, natural UV-blocking and after sun cosmetics. The oil of the seabuckthorn effectively combats wrinkles, dryness and other symptoms of malnourished or prematurely aging skin and is utilized in anti-aging skin cream and lotions. Due to its overwhelming popularity in skin treatment, recent findings and further studies have substantiated its success rate when used internally. As an internal dietary supplement, it has shown to help promote healthy blood circulation, the rudimentary treatment of colitis, stomach ulcers, and as soothing agent for the gastro-intestinal tract. Seabuckthorn berry oil has UV-blocking activity as well as emollient properties and it is an aid in promoting regeneration of tissues (Lanev, 1995). Seabuckthorn seed oil actively reduces blood fat level on the one hand and nurture the blood vessels and improves the quality of the vessels. Linoleic acid regulates blood pressure and reduces serum cholesterol and prevents arrhythmia and linolenic acid reduces blood pressure, dissolves accumulated fat and promote metabolism. Sitosterol is one of the materials forming cell membrane and it stimulates metabolism. Sitosterol combines cholesterol with foods and thus reducing serum cholesterol and also effectively prevents arteriosclerosis. Seabuckthorn oils are known to regulate immune functions and antagonize the effects of immune suppressants (Yang, 2001). Seabuckthorn oils can be used as either food supplements or food ingredients. Seed and pulp oils in small bottles or as gelatin and vegetable capsules are among the best sellers of natural products in some European countries such as Finland and Sweden. Several encapsulated products containing mixtures of seabuckthorn oil and other vegetable oils and fish oils are also available in some European markets (Risto and Baoru, 2003). Since seabuckthorn is rich in omega-3-fatty acids, vitamin C, vitamin E, trace minerals, and other phytochemicals, it can certainly contribute nutritionally to our diets. The most important pharmacological functions attributed to seabuckthorn oil are: anti-inflammatory, antimicrobial, pain relief, and promoting regeneration of tissues. Seabuckthorn oil is also touted as a treatment for oral mucositis, vaginal mucositis, cervical erosion, radiation damage, burns, scalds, duodenal ulcers, gastric ulcers, skin ulcers caused by malnutrition and other skin damage (Li and Wang, 1998).

Seabuckthorn seed oil has tonic effects to cerebral cardiovascular systems. Total flavonoids increase blood flow of the coronary artery and nutrients in the blood supply to heart muscles, lower oxygen consumption, strengthen muscle contraction, thus improving the heart function and increase the anti-hypoxia under normal or below normal blood pressure (Xiao, 2003; Cheng, 2003). Seabuckthorn seed oil has anticarcinogenic factors. It blocks carcinogenic effects of NDMA, Hepatitis C virus, and aflatoxin, thus preventing growth of cancer cells. The content of phospholipids in the seabuckthorn ripe fruit (including lecithin, cephalin, phosphatidylinositol and phosphatidyl) is about 0.5%. These substances, as part of the membrane mitochondrion, participate in the electronic migration and the oxidative phosphorylation, can promote cellular metabolism, and have anti-fatty liver, anti-cirrhosis effect. There is 0.09% to 0.36% betaine in the ripe fruit, which is the methylating product of glycine and has anti-ulcer, and preventative and curative effects on arteriosclerosis. Also, there are the flavonoids, whose main components are the leucocyanidin, catechin, flavonol and trace flavanone (Chen, 1988). Seabuckthorn seed oil stimulates the immune functions and activities, and enhances phagocytosis of macrophage, thus strengthening the body resistance to cancerous mutations. Seabuckthorn seed oil increased microcirculation by increasing both the number of capillaries and their volume. Tricylglycerol is the major constituents of seabuckthorn seed oil. It can thus be used in a variety of cosmetic formulations. Seabuckthorn is traditionally used in the treatment of gastric ulcers and laboratory studies confirm the efficacy of the seed oil for this application (Zhou Yuanpeng, 1998; Xing, 2002). In China, a hemopoietic system after high dose chemotherapy in mice fed the seabuckthorn oil (Chen, 2003). The seed oil has been found to enhance non-specific immunity and to provide anti-tumor effects in preliminary laboratory studies (Yu, 1993; Zhong, 1989). Its function may be to normalize output of gastric acid and to reduce inflammation by controlling pro-inflammatory mediators. Seabuckthorn seed oil is already widely used alone or in various preparations topically applied for burns, scalds, ulcerations, and infections. Fiber in seabuckthorn helps to lower blood cholesterol, regulate blood sugar, protect against certain cancers and eliminate constipation. Components of the pulp or fruit likely contain some fiber. Seabuckthorn has potential as dietary supplement worldwide. In an era where nutrition and health are emphasized, seabuckthorn products could be promoted and consumed for its health advantages.

13.5 CONCLUSIONS

As regards to seabuckthorn very limited phytochemical work has been reported from India. The amount of experimental data evidencing important properties of many ingredients, and bioactive substances from *Hippophae* is vast and continues to increase rapidly. The Field Research Laboratory (DRDO) Leh has planned a programme to develop natural products from seabuckthorn with health promoting effects for the armed forces surviving in diverse climatic conditions. Since, seabuckthorn oil has been used since ancient times in common medicines for curing many diseases affecting humans and other animals, the commercialization of seabuckthorn oil based nutritious products would be a great achievement in alternative nutritional diet sources and health products.

References

AACC. 2000. Fat acidity-general method. In: 2000 *Approved Methods of the American Association of Cereal Chemists*, 10th edition, 1-2. St. Paul, MN: Ameican Association of Cereal Chemists.

Alam, Zeb. 2004. Chemical & nutritional constituents of seabuckthorn juice. *Pakistan Journal of Nutrition* 3(2), 99-106.

Alam, Zeb. 2004a. Important Therapeutic uses of Seabuckthorn (*Hippophae*): A Review. *Journal of Biological Sciences* 4(5), 687-693.

Bernath, J. & Foldesi, D. 1992. Seabuckthorn (*Hippophae rhamnoids* L.) a medicinal & food crop *J. Herbs, Spices & Medicinal Plants* 1(1/2), 27-35.

Beveridge, T., Li. T.S.C., Oomah.B.D. & Smith.A. 1999. Seabuckthorn products: manufacture and composition. *J. Agri. Food Chem.* 47, 3480-3488.

Beveridge,T.; Harison. J.E. & J.Drover. 2002. Processing effects on the Seabuckthorn juice from *H.rhamnoides* L. Cv. Indian Summer. *J. Agri. Food.Chem.* 50, 113-116

Beveridge, T. 2003. Chemical composition and some physical properties. In: Seabuckthorn: Production and Utilization, eds. Li T.S.C. and Beveridge T., 79-88, Ottawa, ON: NRC Research Press.

Cenkowski, S.; Yakimishen. R. Przybylski. R. & Muir. W.E. 2006. Quality of extracted Seabuckthorn seed and pulp oil, *Canadian Biosynthesis Engineering*, 48, 3.9-33.16.

Chen, T. 1988. Studies of the biochemical composition of *Hippophae* and its quality assessment in Gansu Province. *Hippophae* 1, 19-26.

Cheng, J. 2003. Inhibitory effects of total flavones of *Hippophae rhamnoids* on thrombosis in mouse femoral artery & *in vitro* platelet aggregation. *Life Sciences* 72(20), 2263-2271.

Chen, Y. 2003. Study on the effects of the oil from *Hippophae rhamnoides* in hematopoiesis. *Chinese Herbal Drugs* 26, 572-575.

Chen,Y,D.; Jiang, Z.R.; Quin, W.L.; Ni, M.N.; Li, X.L. & He, Y.R. 1990. Research on the chemical composition and characteristic of seabuckthorn berry & it oil *Chemistry and Industry of Forest Products* 10(3), 163-175.

Colquhoun, D.M.; Humphries, J.A.; Moores,D. & Somerset, S.M. 1996. Effects of a macadamia nut enriched diet on serum lipids & lipoprotein compared to a low fat diet *Food Aust.* 48, 216-222.

Derevich, Shindyapkin. 2004. Extraction of organic oil from seabuckthorn seeds with supercritical CO_2. *Theoretical Foundations of Chemical Engineering*, 38 (3), 274 - 283.

Dwivedi, S.K.; Attrey, D.P. and Chaurasia, O.P. 2004. Preparation herbal beverage from seabuckthorn *Indian Journal of Horticulture* 61(1), 78-80.

Eliseev, I.P. 1989. Evolutionary genetic aspects in assessment of achievements and perspectives of Seabuckthorn selection in the USSR. *Proc. Int. Symp. Seabuckthorn*, Xian, China, p. 184-193.

Franke, W. and Muller, H. 1983. Quantity and composition of fatty acid in the fat of fruit flesh & seed of seabuckthorn fruit. *Angew. Botanik* 57, 77-83.

Ge, Xiaoyan 1986. Preliminary research on the chemical composition of Seabuckthorn. *J. Chinese Herbs* 17, 42-44.

Gomez-Tubio, A.; Pita, M.L., Tavares, E.; Murillo, M.L.; Delgado, M.J. & Carreras, O. 1995. Change in the fatty acid profile of plasma & adipose tissue in rats after long term ethanol feeding *Alcohol. Clin. Exp. Res.* 19, 747-752.

Huang, Q.; Zhao, H. Wang. B. & Zong. D. 1991. The development and utilization of seabuckthorn in Mongolia. *Hippophae* 2, 43-46.

Jablczynska, R.; Krawczyk, U. & Minkowski, K. 1994. Fatty acid & tocopherol in some varieties of *Hippophae rhamnoids. Acta Poloniae Pharmaceutica – Drug Research* 51(3), 267-269.

Kallio, H.; Yang, B.R.; Tahvonen. R.; & Hakala.M. 1999. Composition of Seabuckthorn berries of various origins. *Pro. of Int. Symp. On Seabuckthorn* (IWS) Beijing, China.

Kallio, H.; Yang, B.R.; Peippo, P.; Tahvonen. R. & Pan. R. 2002. Triacylglycerols, glycerophospholipids, tocopherols, tocotrienols in berries and seeds of two species (ssp. Sinensis and mongolica) of Seabuckthorn. *Journal of Agri. & Food Chemistry* 50 (10), 3004-3009.

Lanev. 1995. The effect of an extract of Seabuckthorn on the healing of experimental skin wounds in rats. *Dermatology* 48(3), 30-33.

Li, T.S.C. and Wang. L.C.H. 1998. Physiological components and health effects of ginseng, Echinacea and Seabuckthorn. In: G.Mazza (ed.) Functional foods, biochemical & processing aspects. Technomic Publ.Co. Inc., Lancaster, PA.

Li, T.S.C. and Schroeder. W.R. 1999. A Grower's Guide to Seabuckthorn. Pacific Agri-Food Research Centre, Agriculture and Agri-Food Canada, Research Branch, Summerland, BC.

Liu, D.; Zhang, W.; Deng, X.J.; Li, C.Y.; Zhang, H.; Du, C.Y. & Li, Y.J. 1990. Study on dynamic changes in vitamins in Seabuckthorn berries, *Hippophae* 3, 32-35.

Lu, R. 1993. The chemical composition of *Hippophae* fruits in China. In: *Proceedings of International Symposium on Seabuckthorn* (*Hippophae rhamnoides* L.). Barnaul, Russia.

Ma, Z. & Cui.Y. 1989. Studies on the fruit character and biochemical composition of some forms within the Chinese Seabuckthorn in Shanxi, China. *Proc.Int. Symp. Seabuckthorn*, Xian, China. P. 106-112.

Mironov, V.A. 1989. Chemical composition of *Hippophae rhamnoides* of different populations of the USSR. In: *Proc. International Symposium on Seabuckthorn* (*H.rhamnoides* L.). Xian, China.

Moracova, J.; Filip, V.; Krlstkova, K.; Kubelka, V. & Jedinakova, V. 1995. The fatty acid composition in *Hippophae rhamnoids. Potrav. Vedy*. 13(4), 287-297.

Oomah, B.D. & Mazza. G. 1999. Health benefits of phytochemicals from selected Canadian crops. *Trends in Food Science and Technology* 10 (6-7), 193-198.

Pan, R.; Zhang, Z.; Ma, Y.; Sun Z. & Deng. B. (1989) The distribution characters of seabuckthorn and its research progress in China. *Proc. Int. Symp. Seabuckthorn* Xian, China. p. 1-16.

Parimelazhagan, T.; Chaurasia, O.P.; B. Ballabh & Ahmed, Z. 2004. Seabuckthorn: a source for highly versatile natural oil. In : the *First Indian Horticulture Congress*, The Horticultural Society of India, IARI, New Delhi. pp. 400-401.

Parimelazhagan, T., Chaurasia, O.P. & Ahmed, Z. 2005. Seabuckthorn: oil with promising medicinal value. *Current Science* 88(1), 8-9.

Parimelazhagan, T. & Zakwan Ahmed. 2005. Extraction and physico-chemical properties of Seabuckthorn oil from cold desert of Himalaya. In : the 2nd *International Seabuckthorn Association Conference*, Beijing, China.

Pashina Initials. 1993. Plants and cosmetics. Minsk, *Byelorussia*. p 130.

Perez-Jimenez, F.; Castro, P.; Lopez-Miranda, J.; Paz-Rojas, E.; Balnco, A.; Lopez-Segura, F.; Velasco, F.; Marin, C., Fuents, F.; & Ordovas, J. M. 1999. Circulating levels of endothelial function are modulated by dietry mono-unsaturated fats. *Artherosclerosis* 145(2), 351-353.

Pintea, A.; Marpeau, M.; Fave, C., & Socaciu., Gleizes. 2001. Polar lipids & fatty acid distribution in some carotenolipoprotein complex extracted from seabuckthorn fruits.*Phytochemical Analysis* 5, 293-298.

PFRA (Prairie Farm Rehabilitation Administration). 1992. Fruit bearing shrubs for mltiuse shelterbelts and orchards. *PFRA Shelterbelt centre Bulletin*, Indian Head, p. 15.

Quirin, K.; Gerard, W. & Sanddornlipide. D.; 1993. Interessante Wirkstoffe fur die Kosmetik, *Parfumerie kosmetik* 10, 618-625.

Quirin, K. & W. Gerard, D. 1993. Seabuckthorn pulp and kernel oils: valuable lipids for skin care. Unpublished report, flavex Natuuextrakte.

Risto Erkkola., and Baoru, Yang. 2003. Seabuckthorn oils: Towards healthy mucous membranes. *Agro Food Industry Hi-tech*. May-June. p. 53-57.

Rousi, A., & Aulin. H. 1977. Ascorbic acid content in relation to ripeness in fruit of six Hippophae rhamnoides clones from Pyharanta, SW Finland. *Ann. Agri. Fenn.*, 16, 80-81.

Schapiro, D.C. 1989. Bichemical studies on some hopeful forms and species of Seabuckthorn in USSR. *Proc. Int. Symp. Seabuckthorn* Xian, China, p.64-66.

Solonenko L.P., & E.E.Shishkina. 1989. Electrophorectic and amino acid. analysis of proteins in the Seabuckthorn fruit. *Proc.Int.Symp.Seabuckthorn*, Xian, China, p.91-95.

Solonenko, L.P.; Koshelev., Yu, A. & Ageeva, L.D. 2003. In: *Seabbuckthorn: A multipurpose wonder plant* (eds. Virender Singh), Vol. 1, pp. 504-507.

Tong, J.C. Zhang, Z. Zhao,Y. Yang, and K. Tian. 1989. The determination of the physical-chemical constants and sixteen mineral elements in raw Seabuckthorn juice. *Proc.Int.Symp. Seabuckthorn* Xian, China, p.132-137.

Ul'chenko, N.T.; Zhmyrko,T.G.; Glushenkova, A.I. & Murdakhaev, Yu, M. 1995. Lipids of *Hippophae rhamnoids* pericarp. *Chem.Nat.Comp.* 31(5), 565-567.

Wahlberg, K., Jeppsson. N. 1992. Development of cultivars and growing techniques for Seabuckthorn, black chokeberry and rowan. Sverigges Lantbruksuniversitedt Balsgard-Avdelningenfor Hortikulturell Vaxforadling Verksamhetsberattelse 190-1991, p.86-100.

Wolf, D., & Wegert. F. 1993. Experience gained in the cultivation, harvesting and utilization of Seabuckthorn. In: *Cultivation and utilization of Wild Fruit Crops*. Bernhard Thalacker Verlag Gmbh & Co., pp. 23-29.

Wu, F. 1991. A series of *Hippophae* drugs in the USSR. *Hippophae* 2, 38-41.

Xiao, Z. 2003. The inhibitory effect of total flavanoids of *Hippophae* on the activation of NF-kappa by stretching cultured cardiac myocytes, *Sichuan University Medical Journal* 34(2), 283-285.

Xing, J. 2002. Effects of Seabuckthorn seed and pulp oils on experimental models of gastric ulcer in rats. *Fitoterapia* 73(7-8), 644-650.

Xu Mingyu 1991. Present conditions and future research on the Seabuckthorn medicinal use. *J. Water and Soil Conservation of China* 5, 38.

Xurong, T. & Tigerstedt, P.M.A. 2001. Variation of vitamin C concentration and character correlation and between and within natural Seabuckthorn (*H. rhamnoids*. L) populations. *Acta Agriculture Scandinavica* 42, 12-17.

Yao, Y.; Tigerstedt., P.M.A. and Joy. P. 1992. Variation of vitamin C concentration and character correlation between and within natural Seabuckthorn populations. *Acta Agri. Scand.* 42, 12-17.

Yamori, Y.; Nara, Y.; Tsubouchi, T.; Sogawa, Y.; Ikeda, K. & Horie, R. 1986. Dietary prevention of stroke and its mechanism in stroke-prone spontaneously hypertensive rats, preventive effect of dietary fiber and palmitoleic acid. *J. Hypertens* 4(3), 449-452.

Yang, H.; S. Wang. & X.Su. 1998. A study on the dynamic variation of ascorbic acid in Hippophae. *Hippophae.* 4, 41-44.

Yang, B.R. 2001. Lipophilic components of Seabuckthorn seeds and beeries and physiological effect of Seabuckthorn oils. Ph.D. Thesis. University of Turku, Finland.

Yang, B.; Karlsson, R.M.; Oksman., P.H. & Kallio. H.P. 2001. Phytosterols in Seabuckthorn beeries: Identification and effects of different origins and harvesting times. *Journal of Agriultural and Food Chemistry* 49(11), 5620-5629.

Yang, B. & H. Kallio. 2002. Composition and physiological effects of Seabuckthorn lipids. *Trends in Food Science and Technology* 13(5), 160-167.

Yu Let. 1993. Effects of *Hippophae rhamnoides* juice on immunologic and antitumor functions. *Acta Nutrimenta Sinica* 15, 280-283.

Zadernowski, R.; Nowak-Polakowska, H.; Lossow, B. & Nesterowicz, J. 1997. Seabuckthorn lipids. *J. Food Lipids* 4, 165-172.

Zhang, W.; Zhang, Z.; Fan, J.; Yang, J.; Li, M.; Deng, Z.; Wang, G. & Zhang, F. 1988. The preliminary results of experimental obserbation and clinical application of acute radio-dermatitis treated with Seabuckthorn oil. *Hippophae* 1, 27-30.

Zhang, W.; Yan, J.; Duo, J.; Ren., B. & Guo. J. 1989. Preliminary study of biochemical constitutes of berry of Seabuckthorn growing in Shanxi province and their changing trend. *Proc. Int. Symp. Seabuckthorn*, Xian, China.

Zhang, P. 1989. The anti-cancer activities of *Hippophae* seed oil and its effect on the weight of the immunological organs. *Hippophae* 3, 31.

Zhong, Fei. 1989. Effects of the total flavonoid of *Hippophae rhamnoides* on nonspecific immunity in animals. *Shanxi Med.J.* 18, 9-10.

Zhou, Yuanpeng. 1998. Study on the effect of *Hippophae* seed oil against gastric ulcer. Institute of Medical Plants Resource Development, *The Chinese Academy of Medical Sciences*, Beijing, China.

□□□

14

Mechanical Harvesting of Fruits of Seabuckthorn

Gautam Biswas, Sanjai K. Dwivedi, Sanjeev Sagar and T. Stobdan

CONTENTS

14.1 INTRODUCTION

Seabuckthorn (*Hippophae rhamnoides* L.) is a unique and valuable hardy deciduous bushy shrub that grows well in the Himalayan hilly areas of Himachal Pradesh, Uttarakhand and Jammu & Kashmir. The plant consists of 5-7 branches. The height of the plant varies from 1-4 m. Its yellow-orange berries are potentially profitable. Juices and oil of this plant is a staple in many traditional medicines and cosmetics. The mass of the fruits varies from 0.20–0.35g and the diameter of the fruits ranges from 5–8 mm. Unfortunately, it is difficult to harvest the seabuckthorn fruits because the fruit is tightly clustered on thorn-covered branches. Due to the nature of the plant and lack of proper mechanism for harvesting the present harvesting method involves a range of problems, which includes (a) The berries are attached firmly to the branches and therefore it is not easy to remove it from the branches (b) The branches are densely covered with thorn making it difficult for manual operation (c) One more problem for the harvesting of the fruits is the lightweight of the fruits. The weight of the fruits also does not help in easy removal from the branches.

14.2 MANUAL HARVESTING METHOD

Manual harvesting for the seabuckthorn fruits as practiced in India is quite rough. In this method, the branches carrying the fruits are held with one hand and beaten with the help of a wooden stick. Due to impact of the wooden stick, the fruits get detached from the branches and fall on a piece of cloth lying under the plant and are collected. Another method of manual harvesting is to break off the branches from the mother plant and then remove the fruits. The capacity of harvesting by the former method is 2.5 kg/hr/man and by the latter method is 3.5-5 kg/hr//man (Dwivedi *et al.,* 2001). The major Problems in Manual Harvesting are (a) The branch of the plant is severely damaged due to the impact of the stick. Sometimes the branch gets totally damaged broken (b) Due to the thorny nature of the plant, it is tough to hold the branch by hand (c) The harvesting efficiency and capacity is quite low.

14.3 MECHANICAL HARVESTING METHODS: A REVIEW

Mechanical fruit harvesters can be classified as either direct harvesters or indirect harvesters. A direct harvester comes in direct contact of the fruits, while an indirect harvester causes the fruits to be removed without physically touching it. Indirect harvesting is usually accomplished by shaking a portion of the plant. Forces applied to the trunk, branch or foliage of the plant cause the fruit to be detached from the stem. Several indirect methods were applied for harvesting of seabuckthorn berries.

Stan *et al.* (1985) used a harvester to test seven cultivars of seabuckthorn. Only one cultivar could be successfully harvested with a vibration frequency of 18.5 Hz and an amplitude of 25 mm. Gaetke *et al.* (1991), tested a harvester at three different amplitudes, 13, 18 and 25 mm. At an amplitude of 25 mm, a frequency of 25 Hz was required for adequate removal. At smaller amplitudes, a frequency of 30 Hz was required. Quick freezing units were used by Wolf and Wegert (1993) to freeze the berries so that they could be easily removed by shaking. Best results were obtained when the whole branch was frozen at -36°C and the berries were removed by beating the branches. Whole branch harvesters were developed by Olander (1995) on the hypothesis that it is possible to expose every part of the branch to the same vibration if only a small portion of the branch is being shaken. But these kind of harvesters proved to be effective only in case of bushes that have one central trunk with small branches. Bushes with long branches are more difficult to harvest by shaking the trunk because much of the energy is lost before it reaches the berries. Manitoba Rural Adaptation Council, Canada, developed a hand-held and full-size harvester in the year 1999 which shakes or vibrates the branches of the plant for about a minute at 1 inch (25 mm) peak amplitude and at 25 hz. Mann *et al.* (2001) of Department of Biosystems Engineering, University of Manitoba, designed a branch shaker to harvest seabuckthorn of "Indian Summer" cultivar, consisted of a hydraulically powered slider-crank mechanism, with a rubber lined clamp attached to the slider. The slider functioned as a reciprocating body with a mass of 3.4 kg. The amplitude of motion could be varied from 6.4 to 44.5 mm by changing the eccentricity of the crank. The entire shaking apparatus was oriented in the vertical direction with the clamp inside an open bottomed hopper. The branches were shaken inside the hopper, and the fruits were collected in the hopper. They concluded that berries could be removed from the branches by shaking at a frequency of at least 20 Hz.

14.3.1 Conceptual Design of A Mechanical Harvestor

Field Research Laboratory, (FRL) Defence Research & Development Organisation, Leh, (India) contacted Agrosaw and discussed the issue of designing development of seabuckthorn harvester with Agro Oswa Industries Pvt. Ltd. Ambala and developed a concept on the issue. The concept was jointly examined by FRL and Osaw Agro about the mechanical harvesting of the seabuckthorn. It was decided to make a small machine comprising of:

a) An electrical motor,

b) A pipe assembly,

c) A hook assembly to hold the branch of the plant,

d) A cam assembly to shake the branch of the plant,

e) A generator to run the machine in the field and

f) A trolley for mounting the generator and the motor.

The drawing of the conceptual design is in Fig. 14.1.

14.3.2 Details of the Design

a. Electric Motor: 1 HP, single phase, 1440 RPM motor will be selected.

b. Generator: There is need for a generator, as the machine will be operated in the field where there will be no power supply. In the silent valley of the hilly area a silent generator will be needed, so the same is selected.

c. Cam assembly: The cam will be fabricated in mild steel. The amplitude of the cam will be variable. The exact diameter and amplitude of the cam will be decided at the time of actual fabrication.

d. Pipe assembly: It is planned to make the pipe assembly comprising of S.S pipes of different diameter so that the small diameter pipe can slide into the large diameter pipe and make the whole assembly folding. The operator can set the height of the pipe according to the height of the branch. The height of the pipe will vary from 2–4 m. There will be provision to increase the height of the pipe assembly by the operator standing on the ground by spring action. At the top of the pipe the hook assembly will be fitted.

e. Hook assembly: The hook is designed as two semi circular halves, one of them is fixed and the other one is movable in vertical direction. In Fig. 14.2, the lower half is fixed and the upper half is movable inside the hollow shaft of the lower half. The upper half will be normally in open position with spring action and can be tightened in the downward direction with the help of a string attached with the shaft. The concept of holding the branch of the plant with the hook assembly was to raise the hook at the level of the branch by increasing the length of the pipe, then place the hook around the branch and tight it by pulling the string.

f. Trolley: The trolley will be designed in such a way that it will accommodate the generator and the motor. The trolley will be portable.

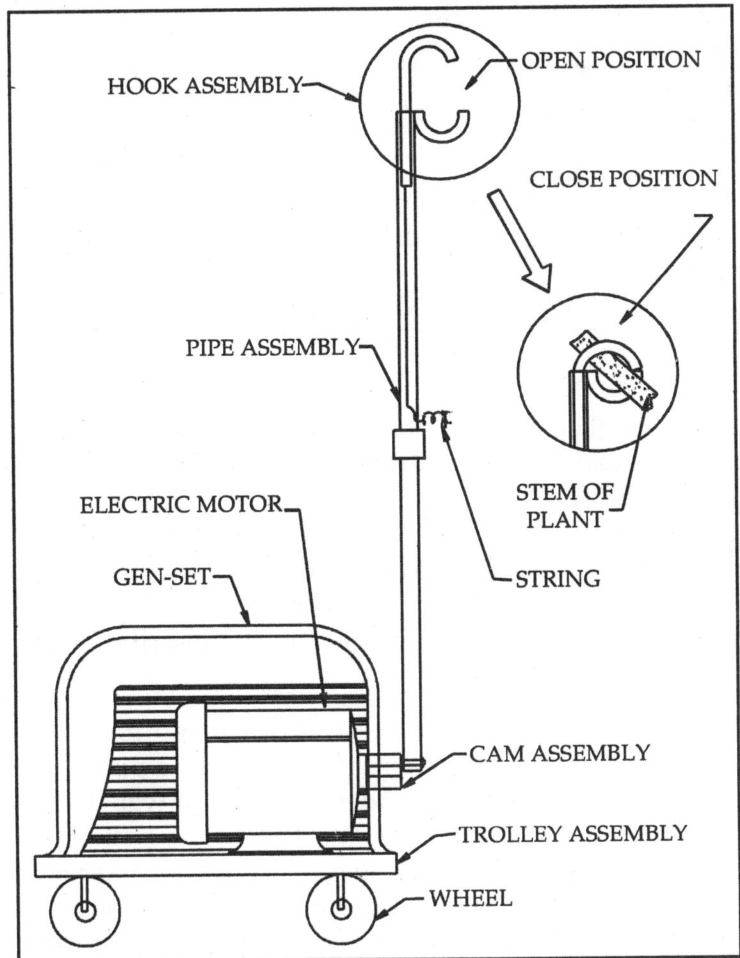

Fig. 14.1: Conceptual Drawing of Seabuckthorn Harvester

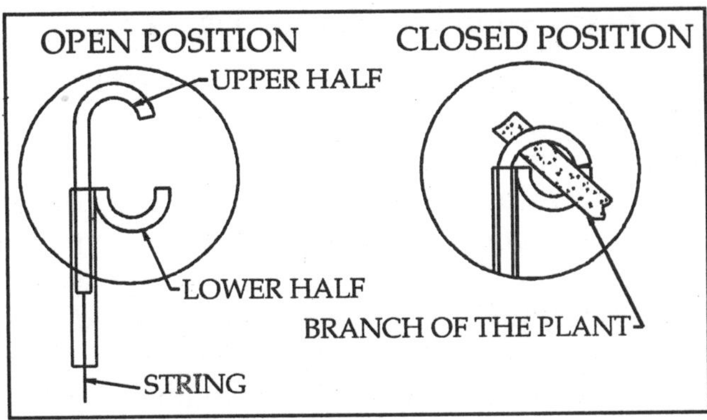

Fig. 14.2: Hook Assembly

14.3.3 Actual Fabrication Process

The specification of the machine during actual fabrication is mentioned below:

a. Electric motor: Power: 1 HP
Phase: Single
RPM: 1,440

b. Generator: Make: Honda
Model No: EXK-2,000 S
Type: Silent self start.
Fuel: Start by petrol, Run by kerosene.

c. Cam assembly: The diameter of the cam was 70 mm. The amplitude of the vibration can be varied from 0 to 70 mm. A M.S. axle was also fabricated to be fitted inside the slot cut on the body of the cam. This position of the axle inside the slot is responsible for the change in the amplitude of the motion. Universal joint bearing was attached on the end of the axle. On this bearing, the pipe assembly will be mounted. Different parts and assembly of the cam is described in Fig. 14.3 and Fig. 14.4.

d. Piston assembly: To convert the rotational motion of the cam into reciprocating motion, a piston was necessary. So, a piston was fabricated from cast iron along with a cylinder. The piston was designed to move inside the cylinder. The cam was attached to the bottom portion of the piston with the help of a connecting rod, the connecting rod was clamped with the piston by a pin. Thus the rotational motion of the cam was converted into reciprocating motion of the piston. On the upper portion of the piston one base plate was attached. On the two corners of the base plate two the pipe assemblies were fixed. The details of the piston assembly are described in Fig. 14.5 and Fig. 14.6.

e. Pipe assembly: The pipe assembly was fabricated from S.S. pipes. Pipes were selected having three different diameters: 12.7 mm, 15.87 mm and 19.05 mm and the thickness of all the pipes were 16 gauge (1.63 mm). The whole assembly was made foldable by inserting the smaller diameter pipe into the larger diameter pipe. The 15.87 mm pipe was inserted into the 19.05 mm pipe and the 12.7 mm diameter pipe was inserted into the 15.87 mm diameter pipe. To maintain the required height of the pipe locking arrangement was provided. Drilling was done at an interval of 150 mm in the 12.7 mm and 15.87 mm diameter pipes. Locking pins were fixed on 15.87 mm and 19.05 mm diameter pipes. The locking pins were spring operated having the same principle of the opening of umbrella. The locking pin inserts into the drill made on the surface of the pipe and lock the pipe at that certain position. When it is necessary to change the height of the pipes, pull out the pin and change the height of the pipe and after required height is reached, release the locking pin, it will lock the pipe at that position. The variable height of the pipe will be 2–4 m. A washer was fitted at the bottom of the pipe assembly. This washer was fitted with the vertical clamp fixed on the base plate with the help of a pin. Two similar pipe assemblies were fitted on two corners of the base plate. The pipe assemblies, fixed on the base plate can be inclined at any angle to make it easier to reach at every branch of the plant.

f. Hook assembly: Earlier it was decided to design the hook as two semi circular halves, one of them is fixed and the other one is movable in vertical direction. But the design was modified during actual fabrication process. Instead of one hook, two hooks were fixed on a frame and fixed on the top of the pipe. Metal string was fixed with the hooks. This string was guided by a pulley and reached at the ground level so that the worker standing on the ground can operate the closing and opening of the hooks. The inner surfaces of the hooks were coated with soft rubber so that it should not harm the branch of the plant.

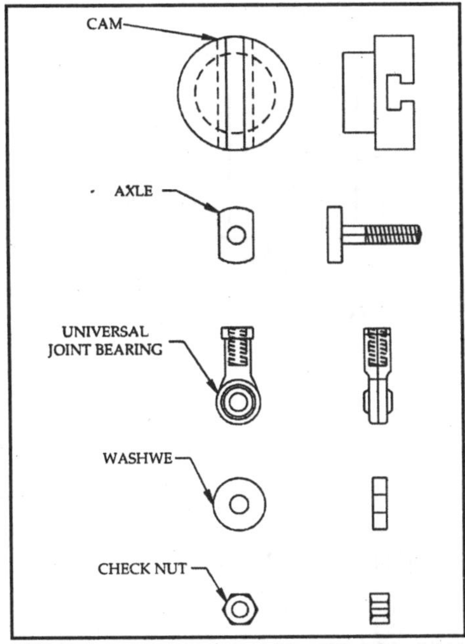

Fig. 14.3: Parts of Cam Assembly

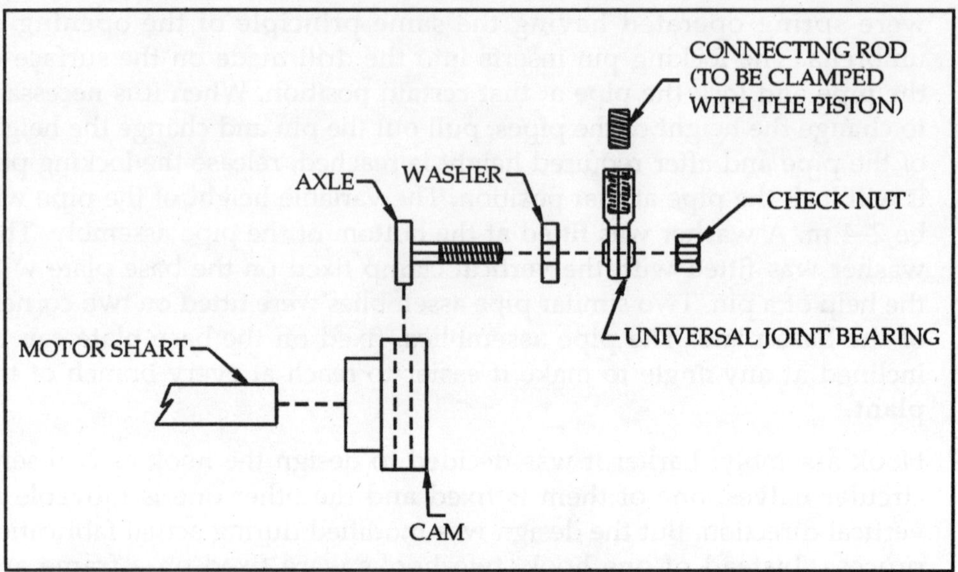

Fig. 14.4: Cam Assembly

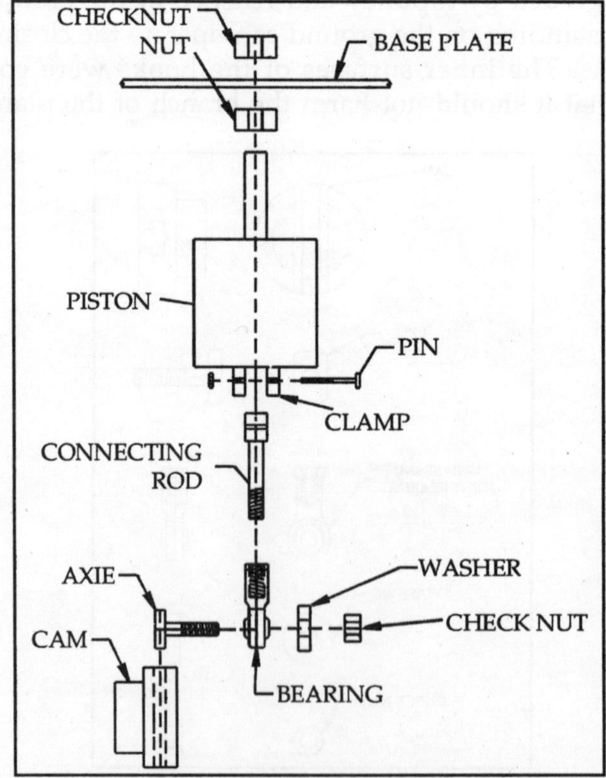

Fig. 14.5: Piston Assembly

g. *Trolley:* The trolley was made up of M.S. sheet of 3 mm thickness. The trolley was 1 m long and 75 m wide and it was strengthened by welding angle iron at its bottom. The trolley was fitted with castor wheels of 6" diameter and handle for pulling or pushing. The whole machine along with the generator and the motor was fitted on the trolley.

14.3.4 Trial of the Machine

The machine was tested at Leh (Ladakh) during fruiting season in the month of August. During the field trial, following points were observed:

A. The vibration created by the machine was sufficient to remove the fruits form the plant. But it was sensed that the vibration was not reaching up to the branch properly.

B. Due to more inclination of the pipe assembly, improper alignment caused overloading on the motor and the motor seemed to be stopped from time to time.

C. The high speed of the motor sometimes lead the removal of the pipe assembly from the base plate.

14.3.5 Modification of the First Prototype

The modifications of the first prototype includes:

A. The motor was replaced with 1 HP, 1 ph D.C motor with variable speed controller to maintain the speed of the motor according to the situation.

B. The diameter of the cam was increased to 75 mm in order to increase the amplitude of the vibration.

C. The shaft, joining the cam with the motor, was supported by two pillow block bearings to overcome the overloading on the motor.

D. The pipe assembly was made in single piece and the diameter of the pipe was also increased to make it robust so that it could deliver full vibrations to the branch of the plant.

E. The clipping arrangement was also modified. This time the hook was made in the shape of human fingers so that it could hold the branch more firmly.

14.3.6 Trial of the Second Prototype

The modified machine was again tested in Ladakh in the month of September. But its performance was almost similar to the first prototype. The result was not satisfactory.

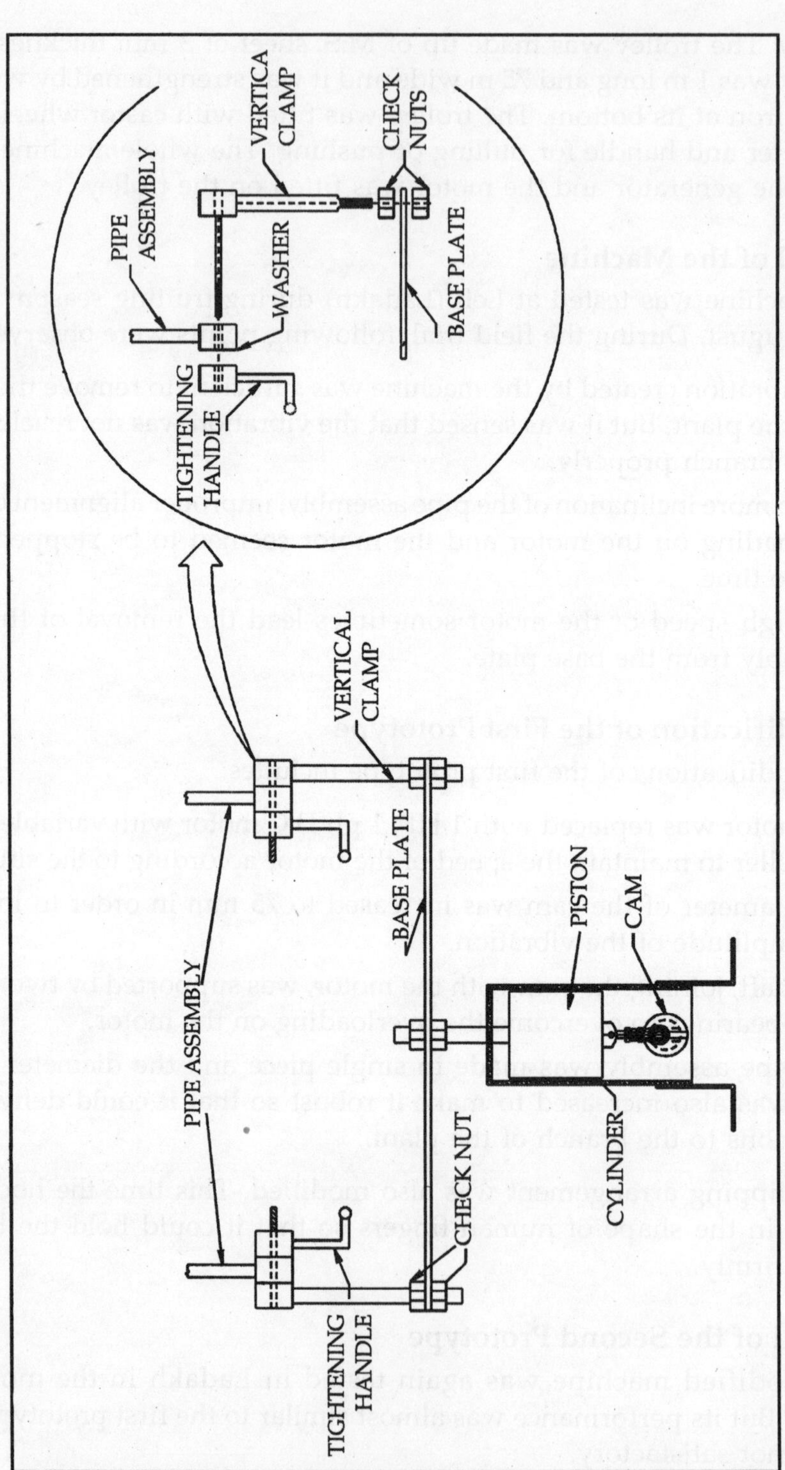

Fig. 14.6: Piston assembly with Base Plate

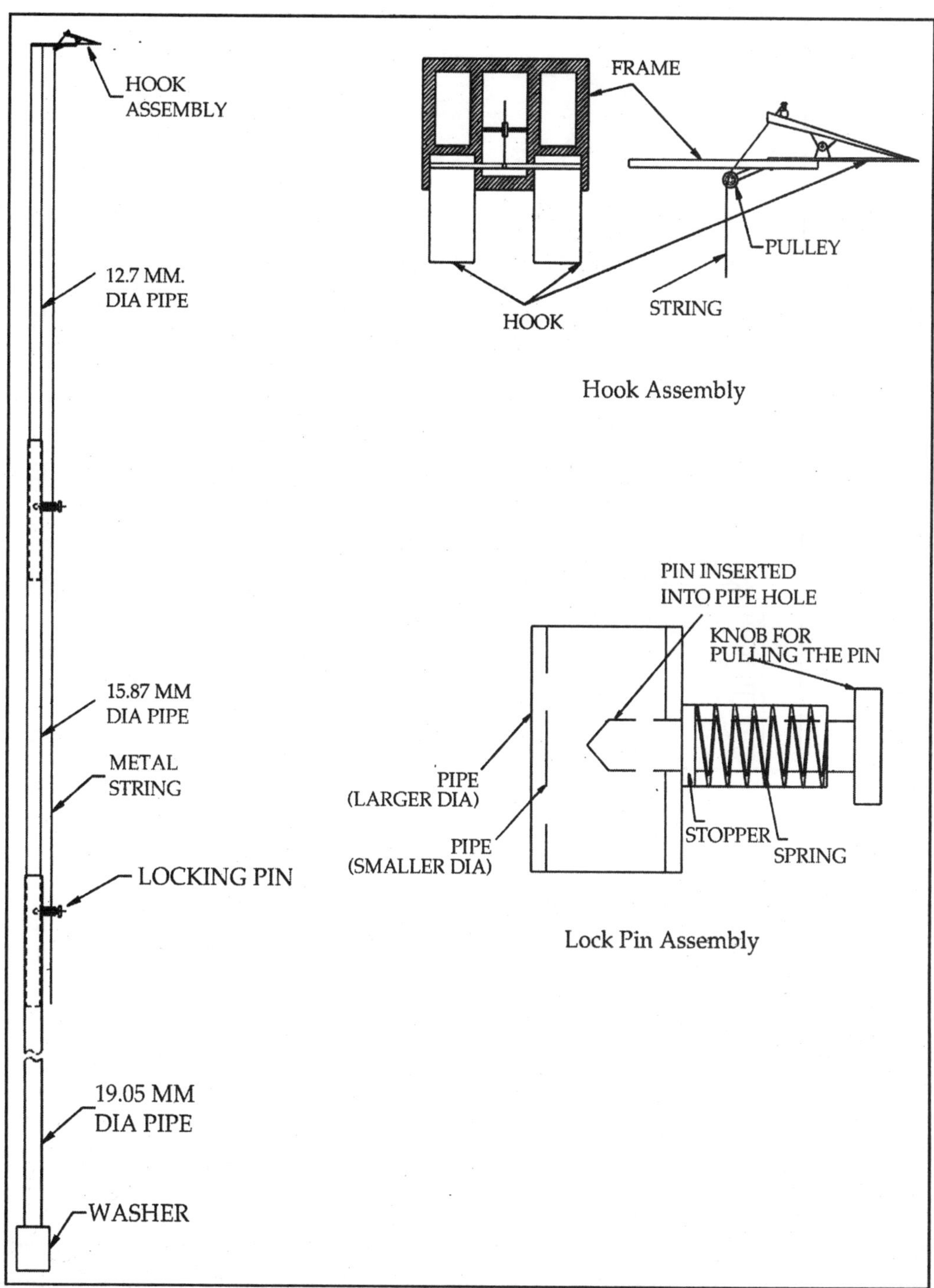

Fig. 14.7: Pipe and Hook Assembly

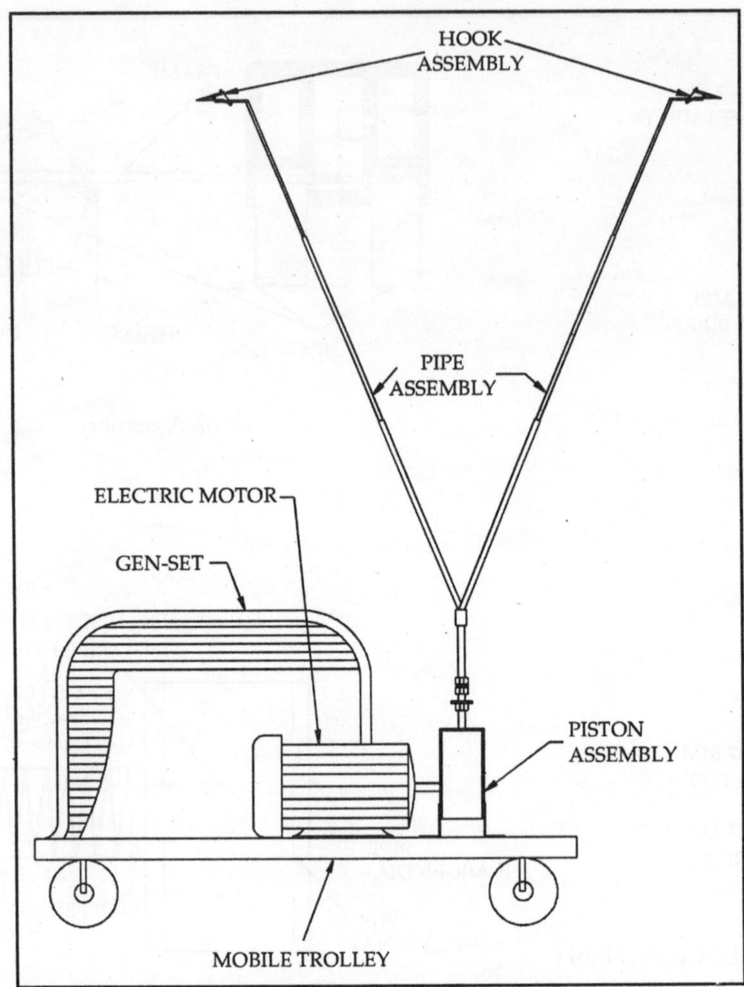

Fig. 14.8: Complete Assembly

14.4 DESIGN OF HAND HELD HARVESTER

After the failure of the two prototypes, the problems were re-analysed. Several other methods for harvesting of the fruits like using a rotating brush or using a vacuum cleaner were also tested. The physical appearance as well as the structure of the plant was also studied deeply. Depending on the overall scenario, the need for a hand-held harvester was strongly recommended by FRL by using the same vibration principle. The Agrosaw team developed a harvesting assembly with hook which was fitted on a simple cutting tool wherein to and fro motion was available. Even the frequency of the operation was as per the requirement and interestingly the power requirement was very much less. This machine was compact in design, robust and user friendly. There were no bulky motors, piston and cylinder, long pipes, trolley etc. The inside surface of the hook was coated with rubber to save the branch from getting damaged. The method of operation of this machine was to clip it to the branch of the plant and to start the machine. After harvesting of the branch, remove the clip from that branch, fix it to another branch and continue the operation.

14.4.1 Modification of Hand Held Harvester

The hook was fabricated in two parts: two semi circular halves among which one was fixed and the other was movable. The hook was designed in such a way that it would hold the branch firmly and complete vibration without damaging the branch.

14.5 CONCLUSIONS

The hand held harvester was the most successful machine developed for mechanical harvesting of seabuckthorn during the course of this project. This machine has several new features like, the machine is small, handy as well as robust has less power consumptive, machine is truly portable, Two types of same machines can be run at the same time through the generator and the machine is light in weight user friendly. The joint effort of Agrosaw team and FRL (DRDO) will lead to open up new horizons in the field of seabuckthorn fruit harvesting.

References

Dwivedi, S.K. and Attrey, DP and Chaurasia, OP 2001. Seabuckthron – A hope f or greening of Ladakh. Field Research Laboratory (DRDO) – Leh.

Gaetke, R.,M Schmidt, E. Triquart and F. Wegert. 1991. Ernteverfahern *sanddorn. Erwerbsobstbau 2:49-51.*

Olander, S. 1995. Mechanical harvesting of seabuckthorn. In: *proceedings of the International Sea Buckthorn Workshop*. Beijing, China. December.

Stan, G., M. Botez and S. Stan. 1985. Possibilities of using vibration to harvest sea buckthorn. *Togungsbericht, Akademie der Landwirtschaftwissenschaften der Deutschen Demokranschen Republik 232:217-223.*

Wolf, D. and F. Wegert, 1993. Experience gained in the harvesting and utilization of sea buckthorn. *In Cultivation and Utilization of Wild Fruit Crops, 23-29.* Braunschweig, Germany: ernhard Thalacker Verlag Gmbh & Co. (in German)

www. mrac.ca/index.cfm/fuseaction

❑❑❑

15

Frankia, Actinomycete Symbiont of Seabuckthorn

T. Stobdan, Sanjai K. Dwivedi and Zakwan Ahmed

CONTENTS

15.1 INTRODUCTION

Actinomycetes belonging to the genus *Frankia* comprise heterotrophic Gram-positive bacteria. *Frankia* is characterized by their ability to induce nitrogen-fixing root nodules on more than 200 species of non-leguminous woody plants in 25 genera of angiosperms (Schwintzer and Tjepkema, 1990). The first strain of *Frankia*, strain CpI1 (also known as HFPCpI1), was reported from *Comptonia peregrina* nodules in 1978 (Callaham *et al.*, 1978). The report demonstrated that *Frankia* strains constituted a previously undescribed group of soil actinomycetes that are facultatively symbiotic with higher plants. The strain strongly resembled the previously isolated strain from *Alnus glutinonosa* nodules by E.H. Pommer in 1959 (Pommer, 1959). Pommer's strain was lost, but, based on his morphological description; his was probably the first true strain. The plant nodulated by *Frankia* strains are known as actinorhizal plants. Unlike the rhizobium-legume symbiosis, in which the host plants, with a few exceptions, belong to a single large family, actinorhizal plants are distributed among eight families and 25 genera (Benson and Silvester, 1993). These plants have a common characteristic to grow in marginally fertile soils and often serve as pioneer plant species in successional plant community development. Because these plants can often thrive on marginal soil in a variety of ecosystem, it has potential application in reclaiming soil and reforestation. Seabuckthorn is one such actinorhizal plant belonging to the family Elaeagnaceae. It is native to central Asia, distributed to the North Sea to the Black Sea and to the Himalayas.

15.2 TAXONOMY AND PHYLOGENY OF *FRANKIA*

Actinomycete taxonomy has traditionally been based on morphology, chemotaxonomy and more recently on phylogenetic analysis. *Frankia* strains are readily recognized by their appearance in liquid culture. They form extensive hyphae and bear sporangiospores in multilocular sporangia located either terminally or in an intercalary position on the hyphae. They produce three cell types: vegetative hyphae, vesicles and spores.

The G+C content of *Frankia* strains is around 66-75 mol% which is typically of the filamentous Actinobacteria. *Frankia* strains have a Type III cell wall containing mesodiaminopimelic acid, alanine, glutamic acid, muramic acid and glucosamine (Lechevalier and Lechevalier, 1990) commonly found amongst aerobic actinomycetes. Whole cell sugars include fucose, ribose, xylose, madurose, mannose, galactose, glucose, rhamnose, and 2-O-methyl-D-mannose. The presences of 2-O-methyl-D-mannose and hopanoid lipids have been proposed as diagnostic phenotypic features for the genus *Frankia*.

At present, the genus *Frankia* is classified as follows in Bergy's Manual:

Phylum BXIV	Actinobacteria
Class I	
Subclass V	Actinobacteridae
Suborder	Frankineae
Family	Frankiaceae
Genera	*Frankia*

The phylogeny of the genus *Frankia* has been deduced by comparative sequence analysis of the 16S rRNA gene, the genes for nitrogen fixation (*nif*) and by other genes (Benson and Clawson, 2000). Strains generally fall into three major clusters, referred to as cluster 1, 2 and 3, each having different and sometimes overlapping plant specificity, physiological properties and symbiotic interactions. Within each group are definable subgroups. In general cluster 1 *Frankia* strains form nodules on member of the higher Hamamelidae, in the order Fagales, including the Betulaceae, Myricaceae and Casuarinaceae. Cluster 2 strains infect only members of Coriariaceae, Datiscaceae, Rosaceae and Rhamnaceae. These strains have not been isolated in pure culture despite many attempts to do so and may therefore be obligate symbionts (Benson *et al.*, 2004). Cluster 3 strains form effective nodule on member of Myricaceae, Rhamnaceae, Elaeagnaceae and Casuarinaceae. On occasion, they have been isolated from nodules of the Betulaceae and Rosaceae. Thus, cluster 3 strains can be considered to be cosmopolitan. Of the three major clusters, most ecological information, including distribution and diversity measurements, is available for cluster 1 strains that are commonly known as alder and casuarinas strains. Less is known about member of cluster 3 and even less about member of cluster 2.

15.3 ISOLATION AND GROWTH CHARACTERISTICS

Frankia is difficult to isolate directly from soil, so most strains originate from root nodules. Major factors limiting success in isolating *Frankia* are the slow growth of the strains, the lack of a clearly defined growth media, and the presence of fast-growing contaminants. Successful isolation protocol should begin by using greenhouse or fresh field collected nodules in the spring or early summer when *Frankia* is actively growing (Carú, 1993). Nodules must be cleaned with tap water to eliminate most of the microorganisms present on the nodule surface. Sterilization treatments with hypochlorite with or without detergents, $HgCl_2$, have been employed on most isolation procedures. The surface-sterilized nodule lobes are fragmented and the dilutions of the resulting vesicle cluster suspension are plated on solid media or dilute to extinction in liquid media. Vesicle clusters consist of a hyphae mass plus vesicles and plant cell cytoplasmic debris, and sometimes *Frankia* sporangia. Outgrowth of hyphae from vesicle cluster is monitored microscopically and thereby improves the chance of obtaining a pure culture.

The medium used for isolating *Frankia* strains is important but till date universal or selective media have not been reported (Benson and Silvester, 1993). Media used range from simple tap water, agar to the complex Q Mod medium. Strains need only a minimal medium for growth. They use mainly short fatty acids (acetic, pyruvic, succinic, propionic) as a carbon source or some sugars, such as glucose. They can use a variety of nitrogen sources, such as amino acids, urea or nitrate; ammonia salts are generally employed. All *Frankia* strains are able to use atmospheric N_2 as nitrogen source. Richer medium tend to favour the growth of contaminants. Antifungal agents are often incorporated to minimize fungal contaminants. *Frankia* strains are slow growing heterotrophic aerobes with doubling time of 15 h or more. Some strains prefer lower O_2 levels than others. The growth kinetics of *Frankia* strains are complex primarily because of their filamentous growth habit and abundant sporangia found in liquid culture that impedes the homogeneous and constant access to nutrients, including O_2, to all parts of mycelia. Growth kinetics consists of a stationary phase followed by a short exponential phase and then by a slower increase in biomass over time (Benson and Silvester, 1993). Most strains are grown and maintained in liquid culture and generally grow slowly on solid media, which are used mainly for purifying cultures and for some specific requirements.

15.4 INFECTION OF HOST

Frankia infects host plants by root hair infection (RHI) or by direct intercellular penetration (IP) of root epidermis cells and cortex depending upon the host plants. A single strain of *Frankia* can use different infection pathways when nodulating different host species. RHI is preceded by root hair branching and curling induced by an unknown *Frankia* signal. *Frankia* penetrates the curled root hair and infection proceeds intercellularly in the root cortex. At the same time cell division is stimulated in the hypodermis and cortex leading to the formation of a small external protuberance called the prenodule. Nodule lobe primordia arises in the pericycle, endodermis, or cortex during prenodule development. *Frankia* enters the developing nodule lobe primordial cells to form the infected nodule. During IP infection root hair are either not present or are not infected. *Frankia* becomes encapsulated in a pectin capsule extending through the intercellular spaces. As *Frankia* hyphae enter the root hair they become encapsulated with plant cell wall-like material and remain outside the host plasmalamma. Encapsulated surrounds hyphae during invasion and vesicles in the mature nodule (Lalonde *et al.*, 1977). The invasive phase of nodule development is characterized by hyphae which pass through both the middle lamella of cell wall and the host cytoplasm. The hyphae differentiated into colonizing hyphae, which penetrate cell walls, and proliferating hyphae, which grow and branch in the center of host cell.

Studies on molecular response to *Frankia* infection have shown that a set of genes is activated in developing nodules (called the actinorhizal nodulin genes) (Mullin and Dobritsa, 1996). By homology to legume nodulins, two major types of actinorhizal nodulin genes have been defined by their pattern of expression and function. Early nodulin genes are expressed before the beginning of nitrogen fixation; they are thought to be involved in plant infection or in nodule organogenesis. The late nodulin genes comprise sequences involved in different metabolic activities necessary for the functioning of the nodule (Obertello *et al.*, 2003). The nodulation genes are expressed in particular cell types. The earliest symbiotic genes such as *cg*12 were found to be expressed in young infected cells of prenodules and in the nodule infection zone just before onset of nitrogen fixation (Svistoonoff *et al.*, 2003). The role of Cg12 is suggested to be in processing of extracellular or the processing of cell wall proteins during early stages of *Frankia* infection (Svistoonoff *et al.*, 2003). Other early actinorhizal genes such as *ag*164 and *agNt*84 have been observed to express in young infected cells. Ag164 and agNt84 may be involved in transport/storage of metal during the infection process (Pawlowski *et al.*, 1997). Plant hemoglobin Hb is the moist studied late nodulins and have been found in actinorhizal nodules. The role of Hb in symbiosis is to supply oxygen to the bacterial respiration chain while preserving the oxygen intolerant nitrogenase enzyme complex. The largest amount of *hb* mRNA was found to present in cells completely filled by *Frankia* (Gherbi *et al.*, 1997). Although our understanding of molecular mechanisms of actinorhizal symbiosis has increased considerably in the last few years, more work is needed to understand the regulation of actonorhizal nodule development.

15.5 NITROGEN FIXATION

Frankia strains have the ability to fix nitrogen in non-leguminous woody plant species including seabuckthorn. Root nodule of Seabuckthorn have been found to fix 180 kg of nitrogen per ha per year, which generally improves soil fertility (Jike and Xiaoming, 1992). In Egypt, a nitrogen-fixing potential of

288 kg of nitrogen per ha per year has been reported for *Casuarina* (Diem and Dommergues, 1990). *Frankia* strains make vesicles in nitrogen deficit culture, and often in symbiosis. The vesicle interior provides the enzyme with an anaerobic activity. Evidence from structural and physiological studies indicates that a variety of mechanisms exists for limiting O_2 diffusion to nitrogenase in symbiosis. The process of nitrogen fixation requires lots of energy in the form of ATP. The enzymatic reaction of biological nitrogen fixation can be represented by the following equation:

$$N_2 + 8e^- + 8H^+ + ATP \rightarrow 2\,NH_4^+ + H_2 + ADP + Pi$$

The reaction is carried by nitrogenase enzyme of *Frankia*. Hydrogen is evolved as a byproduct. At least one H_2 is produced per N_2 reduced to ammonia. A number of physical and chemical parameters influence the nitrogen fixation capacity of *Frankia*. The effect of temperature on the nitrogen fixing capabilities in the excised root nodule of non-legume woody trees and shrubs including seabuckthorn showed maximum nitrogenase activity at 20°C and no significant change was observed up to 30°C. At 40°C there was high temperature induced irreversible damage of nitrogenase. At temperature near 0°C, the nitrogenase activity was found to be very low (McNeil and Carpenter, 1980). The effect of nitrate-nitrogen on the nodule symbiosis of nodulated seabuckthorn using labeled N^{15} shows that applied nitrogen promotes plant growth in seabuckthorn while nodule growth and rate of nitrogen fixation per plant is depressed (Bound and Mackintosh, 1975). The root nodular nitrogen fixation capacity in seabuckthorn fluctuated with stage of plant growth, plant age, plant variety, environmental factors, and seasonal variation as well as between morning and evening. Plant sex did not influence the nitrogenase enzyme activity (Wang *et al.*, 1999). Nitrogen fixing capacity by *Frankia* is also influenced by infected root nodular ultrastructure. A very elegant relationship was established between nitrogen fixing capacity and nodular ultrastructure in seabuckthorn during early to multilobbed root nodule nitrogenase development (Andreeva *et al.*, 1980). The nitrogen fixing activity is proportional to the number of endophytic vesicles per unit cell area. Deficiencies of molybdenum severely affect the formation of vesicles which reduces the nitrogen fixing activity.

15.6 DISTRIBUTION AND DIVERSITY OF *FRANKIA* STRAINS

Studies on the presence of *Frankia* in soil are important for understanding its ecological and population dynamics. Also, in view of the optimization of the symbiotic prior to increase plant production, it may be crucial to identify the existing *Frankia* population before the introduction of specific, selected strains for nodulating actinorhizal plants (Schwencke and Carú, 2001). Many studies have addressed the diversity and distribution of *Frankia* strains in nature and some have dealt with the biogeographic distribution of strains and the host plants. The existing pattern of *Frankia* strain distribution can be viewed as resulting from adaptation by both host plants and *Frankia* strains within a geographic mosaic of environment development over millions of years (Benson *et al.*, 2004). Several issues need to be considered while discussing *Frankia* strain diversity in nature. Conceptual difficulties arise when studying the biogeography of actinorhizal symbiosis. The major problem, which is common to all bacteriological studies, is of defining a *Frankia* strain. A variety of markers have been used to study ecological diversity of *Frankia* strains but each has different levels of resolution. These markers range from simple phynotypic traits to protein pattern and isozyme analysis to PCR-RFLP and DNA sequencing (Benson *et al.*, 2004). Differences in resolution have resulted in a varied distribution and diversity of *Frankia* strains. Earlier studies have focused on nodulation-dependent approach, which has low resolution while in recent years molecular based approaches detecting minor differences in DNA sequences are used.

15.6.1 Molecular Approach for Assessing *Frankia* Diversity and Populations

Frankia population inhabits two distinct ecological niches, the root nodules and the soil. Studies on *Frankia* population in both niches, however, have been impeded by problems encountered in the isolation and identification. Limited studies of *Frankia* population in soil have until recently been based solely on plant bioassays. Bioassays are performed by diluting soil samples, inoculating a test plant, and then calculating nodulation units based on the number of nodules formed on plants per gram or cm^3 of soil. The major limitation of this approach is that it underestimates the number of *Frankia* present since only those strains capable of infecting the test plant and that forms a root nodule are counted. In recent years, molecular approaches have increasingly supplemented nodulation-dependent detection method for studying *Frankia* diversity and population in nature. The new methods are revealing much about the genetic diversity and distribution of *Frankia*, as well as refining and expanding knowledge about endophyte-host specificities. The molecular approaches open door to more sophisticated studies of environmental influences on the dynamic of indigenous or introduced *Frankia* populations in plants and soil. One disadvantage of this method is the amplification of dead cells or free-DNA which results in an overestimation of the microbial populations in environment. Molecular approaches target sequences of DNA or ribosomal RNA (rRNA) and thus unaffected by the limitations of culturability. The most widely used molecular techniques in accessing *Frankia* population is the polymerase chain reaction (PCR), which enabled researchers to amplify minute quantities of target sequences of DNA and to rapidly retrieve sequence information from uncultured organisms in the environment. The potential of this technique was initially exploited for the phylogenetic analysis of isolates and uncultured endophytes in nodules but has also been used for the detection and analysis of specific *Frankia* population in nodules and soil as specific sequence information becomes available (Hahn *et al.*, 1999).

The assignment of isolates to the genus *Frankia* is usually based in a PCR-assisted sequence retrieval from isolated DNA, subsequent cloning and sequencing of the amplicons and finally comparative analysis of sequence derived from pure cultures or uncultured populations with those of existing reference collection. Generally, genes with a high phylogenetic significance such as the gene encoding rRNAs have been targeted for such analysis. Other PCR based approach includes comparison of the size of the intergenic spaces (IGS) between the nitrogenase *nif*H and *nif*D (*nif*H-D) genes or the *nif*D and *nif*K (*nif*D-K) genes and the 16S-23S rRNA operon. *Frankia* have also been differentiated from other actinomycetes by using the glutamine synthetase II (*gln* II) gene as the target for PCR. In recent years, the quantification of *Frankia* populations usually employed PCR-MNP using nested or booster PCR (Myrold and Huss-Danell, 1994). Comparison of this method of quantification with that of plant bioassays revealed that the fraction of *Frankia* cabable of nodulation, which ranged between 0.2 and 2940 nodulation unit g^{-1} soil was just a small portion of the total population of *Frankia* as measured by genomic units, which ranged from 200 to 92000 genomic units g^{-1} soil (Myrold and Huss-Danell, 1994). The discrimination of *Frankia* at a lower taxonomic level has often been based on restriction fragment length polymorphism (RFLP) analysis. RFLP analysis is done in digested DNA extracted from *Frankia* strains and subsequent hybridization with DNA probes. Today, RFLP analysis are usually performed on PCR-amplified nucleic acids such as 16S rDNA, the *gln* II, the IGS of the 16S-23S rRNA operon, the IGS of *nif*H-D or the IGS of *nif*D-K. At the sub-species level, arbitrary primers, as well as intervening sequence of *Frankia* genomic DNA amplified by PCR with oligonucleotide primers targeting consensus motif of repetitive elements common to prokaryotic genomes such as repetitive extragenic palindromic (REP) or enterobacteria repetitive intergenic consensus (ERIC) elements (rep-PCR), provides and effective means to distinguish closely related strains (Hahn *et al.*, 1999). Such markers are more variable than 16S rDNA gene sequence and have resolution at strains level.

15.6.2 *Frankia* in Soil

Studies on the occurrence of *Frankia* in soil are limited and have mainly focused on the infecting *Frankia* population using a combination of plant bioassay and subsequent analysis of *Frankia* in nodule formation. The nodulation assays showed that *Frankia* are able to survive in soil, even in an arid environment. The cellular mechanisms that permit *Frankia* to avoid or to tolerate desiccation in soil are not well understood. As sporulation is a general response of bacteria to environmental stress, it is assumed that *Frankia* spores allow for dispersal through time and space. Studies have shown that *Frankia* hyphae may also become drought tolerant using cellular trehalose as an osmotic protectant. Presence of *Frankia* population infective on *Elaeagnus angustofolia* L. was found in different depths of same soil collected at a site devoid of Elaeagnaceae. A decrease in density of *Frankia* was correlated with a decrease of the soil organic matter content with the soil depth (Nalin *et al.*, 1997). Bartzli *et al.* (2004) studied abundance and distribution of *Frankia* in sand dune ecosystem differing in successional stage, vegetative cover, edaphic characteristics and topography. Infective *Frankia* were found to present in soils of young dunes prior to the establishment of any actinorhizal host. Submerged soils had no infectious capacity, whereas soils with greater in situ moisture content and soils subject to intermittent saturation tends to have lower infective capacity overall. The results suggested that soilborne, infective *Frankia* genotypes are not only host specific but are also associated with spatially and chronologically distinct sets of ecological conditions. The first attempt to quantify *Frankia* population directly from soil dealt with rRNA as the target molecule. In RNA extracts from 1 g of soil, genus-specific hybridization with oligonucleotide probes resulted in the detection of *Frankia* with an estimated detection limit of 10^4 cells (Hahn *et al.*, 1990).

15.6.3 *Frankia* in Nodules

Diversity of *Frankia* in nodules has most often focused on isolates and uncultured endophytes in nodules. Molecular approaches using rRNA as the target for hybridization, nodules of about 1 mg provided sufficient target molecules for a reliable detection of *Frankia* strains (Hahn *et al.*, 1990). *In situ* hybridization for analysis of *Frankia* population in nodules of various *Alnus* spp. on the subgroup level allowed detection limit of one cell. The analysis revealed the presence of only one *Frankia* population in every nodule homogenate (Zepp *et al.*, 1997). Low genetic diversity of cluster 1 *Frankia* spp. strains was observed in root nodules of actinorhizal species of Rosaceae, Ceanothus (Rhamnaceae) and *Datisca glomerata* (Datiscaceae) in California by using three different DNA regions as markers (Hauvel *et al.*, 2004). Maggia *et al.* (1992) found close relationship between 60 *Frankia* isolates obtained from root nodules of *Casuarina equistofolia* and the reference strain CeD using a combination of RFLP analysis of 16S-23S rDNA IGS amplicons with hybridization to oligonucleotide probe of PCR-amplified *nif*H-D. In contrast to this study which suggests a small genetic diversity among *Frankia* strains that infect members of the family Casuarinacaeae, five distinct genetic groups were recognized in *Casuarina* by Rouvier *et al.* (1996).

RFLP analysis of PCR-amplified *nif*D-K IGS region with the endonuclease *Hae*III has resulted in the generation of seven different profile of uncultured *Frankia* population in nodule of *Elaeagnus angustifolia* L. (Nalin *et al.*, 1997). Similarly, RFLP analysis of PCR-amplified *nif*D-K IGS region resulted in clustering of 22 *Frankia* strains of the *Elaeagnus* host infective group into seven genomic groups (Lumini *et al.*, 1996). *Frankia* microsymbionts in root nodules from *Colletia hystrix* showed high genetic diversity. Fifteen hapltypes were recognized on the basis of PCR-RFLP (Cháver and Caàu, 2006). With the advancement of molecular detection analysis strains can be differentiated at sub-species level. Genomic fingerprinting on DNA of root nodules of several *Caenothus* spp. using rep-PCR differentiated twelve distinct groups, indicating a considerable genetic diversity of *Frankia* in the sampled nodules (Murry *et al.*, 1997). A similar high degree of diversity was obtained from isolates from nodules of *A. incana* which could not be differentiated by a comparative sequence analysis of 23S rRNA insertion but which all exhibited different rep-PCR patterns (Maunuksela *et al.*, 1999).

15.7 *FRANKIA* GENOME MAPPING

Beginning in 2003, three strains of *Frankia* - CcI3, Ean1pec and ACN14a- were put into the genome sequencing pipeline. Two of them, CcI3 and Ean1pec, were sequenced at the Joint Genome Institute (JGI) in collaboration between D. Benson at the University of Connecticut and L. Tisa at the University of New Hampshire together with Dr M. Pilar Francino at JGI. The third was initiated independently by P. Normand at the University Lyon in collaboration with scientists from Genoscope in France and several other groups. Details of the genome sequencing can be accessed at the Integrated Microbial Genome site: http://img.jgi.doe.gov/cgi-bin/pub/main.cgi.

15.8 CONCLUSIONS

The studies on interaction of host plants and *Frankia* has provided a great insight into the mode of action, the development of root nodules, the functional relationship of nodule ultra structure and nitrogen-fixation, effect of various physical, chemical and biological factors on biological nitrogen-fixation in woody plant spp. In the past decade, considerable advances have been made in the identification and characterization of plant genes involved in the development and functioning of actinorhizal nodules. Studies on natural populations and ecology of *Frankia* by using molecular techniques without prior isolation have allowed and will provide significant advances in obtaining reliable information about the diversity of the symbionts, including the factors that affect their survival.

Despite the advances made in understanding the nitrogen fixation in non-leguminous plants by *Frankia* and its diversity in nature, relatively little is known about the mechanisms involved in its symbiotic interaction with actinorhizal species. Many aspects of the molecular biology of *Frankia* have been hampered by the lack of methods for the genetic manipulation by the actinomycetes. Many questions about the genes and mechanisms involved in nodulation and host specificities most likely will be answered after gene knockout and genetic transformation systems become established. There is a need to search and domesticate efficient *Frankia* strains with higher nodular capacity, broad host range, and easy adaptability to tolerate extremes of physical, chemical and biological factors. Actinorhizal tree plantation appears to be good candidate for establishment of large plantations aimed not only at the production of wood, but also at improving soil fertility. Such efforts will help in improving overall biomass and yield of plant and improving soil fertility especially in cold arid region that is characterized by nutritionally poor soil and harsh environmental conditions. Further, it will reduce our dependence on the costly synthetic fertilizers and other toxic agrochemical.

References

Andreeva, I.N., Tibilov, A.A., Il'-Yosava, V.B. & Zhiznevskaya, G.Y. 1980. Ultrastructure of nitrogen-fixing nodules in seedlings of Seabuckthorn (*H. rhamnoides* L.). *Soviet Plant Physiol* 29:598-605. Translated from *Fiziologica Rastenii* 27, 791-799.

Bartzli, J.M, Zimpfer, J.F ; Huguet, V ; Smyth, C.A., Fernandez, M. & Dawson, J.O. 2004. Distribution and abundance of infective, soilborne *Frankia* and host symbionts *Shepherdia, Alnus,* and *Myrica* in a sand dune ecosystem. *Can. J. Bot.* 82, 700-709.

Benson, D.R, & Clawson, M.L. 2000. Evolution of the actinorhizal plant symbioses, p. 207-224. *In* Triplett E.W. (ed.), Prokaryotic nitrogen fixation: A model system for analysis of biological process. Horizon Scientific Press, Wymondham, UK.

Benson, D.R. & Silvester, W.B. 1993. Biology of *Frankia* strains, actinomycete symbionts of actinorhizal plants. *Microbiol. Rev.* 57, 293-319.

Benson, D.R; Heuvel, B.D.V, & Potter, D. 2004. Actinorhizal symbiosis: diversity and biogeography, p 99-129. *In* Gillings M and Holmes, A (ed.) Plant Microbiol. Garland Science/BIOS Scientific Publishers, Oxford.

Bound, G. & Mackintosh, A.H. 1975. Effect of nitrate-nitrogen on the nodule symbiosis of *Coriaria* and *Hippophae*. *Proc. Royal Soc. London* B 190, 1099, 199-209.

Callaham, D.; DelTredici, P. & Torrey, J.G. 1978. Isolation and cultivation in vitro of the actinomycete causing root nodulation in *Comptonia*. *Science*, 199, 899-902.

Carú, M. 1993. Characterization of native *Frankia* strains isolated from Chilean shrubs (Rhamnaceae). *Plant and Soil*, 157, 137-145.

Chávez, M. & Carú, M. 2006. Genetic diversity of *Frankia* microsymbionts in root nodules from *Colletia hystrix* (Clos.) plants by sampling at a snall-scale. *World J. Microbiol. Biotech.* 22, 813-820.

Diem, H.G. & Dommergues, Y.R. 1990. Current and potential uses and management of Casuarinaceae in the tropics and subtropics, p. 317-342. *In* : Schwintzer C. R. and Tjepkema J. D. (ed.), The Biology of *Frankia* and Actinorhizal Plants. Academic Press, San Diego.

Gherbi, H; Duhoux, E; Franche, C; Pawlowski, K; Nassar, A; Berry, A; & Bogusz, D. 1997. Cloning of a full-length symbiotic hemoglobin cDNA and *in situ* localization of the corresponding mRNA in *Casuarina glauca* root nodule. *Physiol. Plant*. 99, 608-616.

Hahn, D.; Kester, R.; Starrenburg, M.J.C. & Akkermans, A.D.L. 1990. Extraction of ribosomal RNA from soil for detection of *Frankia* with oligonucleotide probes. *Arch. Microbiol*, 154, 329-335.

Hahn, D.; Nickel, A. & Dawson, J. 1999. Assessing *Frankia* populations in plants and soil using molecular methods. *FEMS Microbiol. Ecol*, 29, 215-227.

Hahn, D.; Starrenburg, M.J.C. & Akkermans, A.D.L. 1990. Oligonucleotide probes that hybridize with rRNA as a tool to study *Frankia* strains in root nodules. *Appl. Environ. Microbiol*, 56, 1342-1346.

Heuvel, B.D.V.; Benson, D.R.; Bortiri, E. & Potter, D. 2004. Low genetic diversity among *Frankia* spp. strains nodulating sympatric populations of actinorhizal species of Rosaceae, *Ceanothus* (Rhamnaceae) and *Datisca glomerata* (Datiscaceae) west of the Sierra Nevada (California). *Can. J. Microbiol*, 50, 989-1000.

Jike, Z. and Xiaoming, Z. 1992. Progress of study on Frankia in nodules of Seabuckthorn . *Hippophae* 2, 4-10.

Lalonde, M. & Quispel, A. 1977. Ultrastructural and immunological demonstration of the nodulation of the European *Alnus glutinosa* (L.) Gaertn. host plant by the North-American *Alnus crispa* var. *mollis* Fern. root nodule endophyte. *Can. J. Microbiol*, 23, 1529-1547

Lechevalier, M. P. & Lechevalier, H.A. 1990. Systematics, isolation, and culture of *Frankia*, p. 35-60. *In* Schwintzer C. R. and Tjepkema J. D. (ed.), The biology of *Frankia* and Actinorhizal Plants. Academic Press, New York.

Lumini, E.; Bosco, M. & Fernandez, M.P. 1996. PCR-RFLP and total DNA homology revealed three related genomic species among broad-host-range *Frankia* strains. *FEMS Microbiol. Ecol*, 21, 303-311.

Maggia, L.; Nazaret, S. & Simonet, P. 1992. Molecular characterization of *Frankia* isolates from *Casuarina equisetifolia* root nodules harvested in West Africa (Senegal and Gambia). *Acta Oecol*, 13, 453-461.

Maunuksela, L.; Zepp, K.; Koivula, T.; Zeyer, J.; Haahtela, K. & Hahn, D. 1999. Analysis of *Frankia* populations in three soils devoid of actinorhizal plants. *FEMS Microbiol. Ecol.* 28, 11-21.

McNeil, R.E. & Carpenter, P.L. 1980. Effect of temperature of acetylene-reducing activity associated with rot nodules on woody trees and shrubs. *J. American Soc. Hort. Sci*, 105, 543-546.

Mullin, B.C. & Dobritsa, S.V. 1996. Molecular analysis of actinorhizal symbiotic system: Progress to date. *Plant Soil*, 186, 9-20.

Murry, M.A.; Konopka, A.S.; Pratt, S.D. & Vandergon, T.L. 1997. The use of PCR-based typing methods to assess the diversity of *Frankia* nodule endophytes of the actinorhizal shrub *Ceanothus*. *Physiol. Plant* 99, 714-721.

Myrold, D. D. & Huss-Danell, K. 1994. Population dynamics of *Alnus*-infective *Frankia* in a forest soil with and without host trees. *Soil Biol. Biochem*, 26, 533-540.

Nalin, R.; Normand, P. & Domenach, A.M. 1997. Distribution and N2-fixing activity of *Frankia* strains in relation with soil depth. *Physiol. Plant.* 99, 732-738.

Obertello, M.; Sy, M.O.; Laplaze, L.; Santi, C.; Svistoonoff, S.; Auguy, F.; Bogusz, D. & Franche, C. 2003. Actinorhizal nitrogen fixing nodules: infection process, molecular biology and genomics. *African J. Biotech.*, 2, 528-538.

Pawlowski, K.; Twigg, P.; Dobritsa, S.; Guan, C. & Mullin, B.C. 1997. A nodule-specific gene family from *Alnus glutinosa* encodes glycine- and histidine-rich proteins expressed in the early stages of actinorhizal nodule development. *Mol. Plant Microbe Interact.*, 10, 656-664.

Pommer, E.H. 1959. Uber die Isolierung des Endophyten aus den Wurzelknollchen *Alnus glutinosa* Gaertn. und uber erfolgreiche re-infektions versuche. *Ber. Dtsch. Bot. Ges*, 72, 138-150.

Rouvier, C.; Prin, Y.; Reddell, P.; Normand, P. & Simonet, P. 1996. Genetic diversity among *Frankia* strains nodulating members of the family Casuarinaceae in Australia revealed by PCR and restriction fragment length polymorphism analysis with crushed root nodules. *Appl. Environ. Microbiol*, 62, 979-985.

Schwencke, J. & Carú, M. 2001. Advances in actinorhizal symbiosis: Host plant-*Frankia* interactions, biology, and application in arid land reclamation. A review. *Arid Land Res. Manag.*, 15, 285-327.

Schwintzer, C.R. & Tjepkema, J.D. 1990. The biology of *Frankia* and actinorhizal plants. Academic Press, Inc., New York.

Svistoonoff, S.; Laplaze, L.; Auguy, F.; Runions, J.; Duponnois, R.; Haseloff, J.; Franche, C. and Bogusz, D. 2003. cg12 expression is specifically linked to infection of root hairs and cortical cells during *Casuarina glauca* and *Allocasuarina verticillata* actinorhizal nodule development. *Mol. Plant. Microbe. Interact.*, 16, 600-607.

Wang, Z.; Guo, B. & Yan, L. 1999. Study of Seabuckthorn root nodule nitrogen fixation. In: *Proceedings of International Workshop on Seabuckthorn*, China.

Zepp, K.; Hahn, D. & Zeyer, J. 1997. Evaluation of a 23S rRNA insertion as target for the analysis of uncultured Frankia populations in root nodules of alders by whole cell hybridization. *Syst. Appl. Microbiol*, 20, 124-132.

□□□

16

Seabuckthorn Community Structure, *Frankia* Diversity and Nitrogen Fixation in the Zones of North Sikkim

Bipin Kumar Sharma, Hridip Kumar Sharma, A.K. Shukla and S.C. Tiwari

CONTENTS

16.1 INTRODUCTION

The study of microbes and their ecology under natural habitats has gained tremendous importance in recent years due to their intricate interrelationship with the environment. One primary area in the last three decades that has contributed to a considerable constructive understanding on the nature of the microbial world is the plant-microbe interaction. It is only through our careful observations that today we know how microbes fix atmospheric nitrogen and subsequently enhance soil fertility. Incidentally, the identification of nitrogen fixing bacteria in root nodules of both leguminous and non-leguminous plants mesmerized our senses to wonder how the delicate balance between nature and the environment is regulated. Indeed, we cannot stop but keep on appreciating the tremendous array of microbial diversity as more and more gets discovered every year and with every discovery a new hope of unraveling the mystifying world of microbes is re-established.

16.1.1 Soil and Rhizosphere Microbes

Soil is a dynamic habitat for an enormous variety of life-forms. It not only provides mechanical support to plants but also support sustenance to habitats colonized by a staggering variety of microorganisms. All these forms of life interact with one another and with the soil to create continually changing biotic conditions. Microbial communities constitute an essential component of biological characteristics in soil ecosystems. The important functional attributes of soil microorganisms include population number, biomass, respiratory efficiency and diversity patterns. The distribution of microbes throughout the soil is not even. Microorganisms tend to cluster around the roots of higher plants. This phenomenon is referred to as the rhizospheric effect. The majority of microorganisms found in the rhizosphere are bacteria, micro-fungi and actinomycetes that congregate together and constitute a complex ecosystem. Microorganisms are thought to gain nutrients from soil, plants and decomposed substances. However, auxotrophic mutants requiring various amino acids have also been isolated from the rhizosphere. Plants may also derive benefit from such arrangement. Certain bacteria and actinomycete fix nitrogen in an inorganic form that can be taken up and used by plants. Under certain circumstances, the association between microorganisms and higher plants can become very intimate giving rise to a symbiotic relationship for mutual benefit termed 'symbiosis'.

Microorganisms are responsible for breakdown of organic matter and nutrient transformations (Tiwari *et al.*, 1988), influencing plant growth and are considered crucial to the functioning of soil ecosystems. Apart from their indispensable role in biogeochemical cycling of nutrients they are reported to be useful indicators of fertility and quality of soil since they are sensitive to changes occurring in the soil environment. Hence they also act as biosensors constantly monitoring the environment and indicating shifts between degradation and restoration. In fact, soil microorganisms play very important role in soil fertility not only because of their ability to carry out biochemical transformations but also due to their importance as a source and sink for mineral nutrients (Jenkinson and Ladd, 1981). The populations of soil microbes are especially dense in the rhizospheric region, where root exudates, lysates, and mucigels provide microbes with vital nutrients for growth, in the form of sugars, amino acids, organic acids and fatty acids, (Brady, 1995). The rhizosphere forms a heterogeneous ecosystem, where a large variety of microorganisms exist together. It has been shown that many of these microorganisms interact with *Frankia* and/or its host. Some soil organisms, like *Pseudomonas cepacia* (Knowlton *et al.*, 1980, Knowlton and Dawson, 1982) facilitate infection of the host root system by *Frankia*, probably through induction of root hair deformation. Rojas *et al.* (1992) depicted as to how several soil actinomycete isolates had a positive effect on *Frankia acetylene* reduction rates with *Alnus rubra*. Probanza *et al.* (1996) found positive effects of some *Bacillus* spp. on the growth of alder, whereas some *Pseudomonas* spp. negatively influenced alder development.

16.1.2 Soil Physicochemical Parameters Determine Microbial Activity

Soil physicochemical characteristics influence the level of biomass and the activity of microorganisms. Seasonal changes in soil moisture, soil temperature and C input from crop roots, rhizosphere products (i.e., root exudates, mucilage, sloughed cells, etc.), and plant residues can have a large effect on soil microbial biomass and its activity (Bornemen and Triplett, 1997), which in turn, affect the ability of soil to supply nutrients to plants through soil organic matter turnover (Coyne, 1999). Microbial biomass has been reported to vary seasonally in European soils Singh *et al.* (1989) have also reported a seasonal variation in the microbial C, N and P in forests and Savanna. Generally, a negative relationship between soil moisture content and microbial biomass is observed (Ross, 1987). Short-term fluctuations of moisture and temperature conditions have been shown to influence the amount of microbial biomass carbon (Bossio and Scow, 1995). Soil organic carbon levels, too, has been reported to be governed by climatic conditions. Limited data, however, exists on the magnitude of seasonal changes in soil microbial biomass and enzyme activity in different ecosystems.

Explanations for variations in N_2 transfers within soils lie in the litter decomposition process itself. It is known that the release of mineral N_2 from decaying litter results from complex interactions between microbial populations and chemical compositions of leaves. Hence, the substrate quality, as defined by its C/N and its lignin and phenolic contents are critical factors determining the rate of litter decay (Swift and Posner, 1972; Jenkinson, 1965). Sparling *et al.* (1981) found that rates of liter decomposition were negatively correlated with the ratio of lignin concentration to initial N_2 concentration in hardwood leaves. The effect of actinorhizal plants on associated non-fixing species was found to depend on the considered species (Dawson, 1986).

16.1.3 Actinorhizal Symbiosis and N_2 Fixation in Soils

Two groups of bacteria participate in N_2 fixing symbioses with plants. One group consists of diverse proteobacteria possessing the *rhizobia* nodulation genes (called the *nod* genes) that confer the ability to nodulate legumes of the family Fabaceae and the single genus *Parasponia* of the family Ulmaceae (Clawson *et al.*, 2004) while the other group consists of actinomycetes high molecular G+C content that nodulate plants from 8 families (Benson and Silvester, 1993). While it is well established that the symbioses of *rhizobia* (i.e. *Rhizobium, Bradyrhizobium, Sinorhizobium*) are confined only to the family Leguminosae and to the far-related genus *Parasponia* (Akkermans and Van Dijk, 1981), interestingly the actinomycete *Frankia* has been found to be associated with a large number of higher plant genera. Till date, more than 200 species representing 25 genera of actinorhizal plants have been identified (Benson and Silvester, 1993). Unlike the *Rhizobium*-legume symbiosis, in which the host plants with a few exceptions belong to a single large family, actinorhizal plants are distributed among 8 families and 7 orders of angiosperms. These plants have in common a predilection to grow in marginally fertile soils and they often serve as pioneer species early in successional plant community development. Representatives can be found in most climatic zones, and they inhabit a variety of ecosystems including the arctic tundra (*Dryas species*), coastal dunes (*Casuarina, Hippophae, Myrica* and *Elaeagnus* species) riparian areas (*Alnus* and *Myrica* species), forest areas (*Alnus, Casuarina, Coriaria* and *Shepherdia* species) dry chapparal and xeric regions (*Casuarina, Purshia, Ceanothus, Cercocarpus, Comptonia,* and *Cowania* species) and high altitude alpine zones (*Alnus, Hippophae, Myrica* species). The input of fixed nitrogen by these plants is considerable, especially in colder temperate areas where indigenous legumes are absent or rare.

16.1.4 The Actinorhizal Plant Seabuckthorn (*H. salicifolia* D. Don)

Seabuckthorn (*Hippöphae* spp.) is a group of deciduous, spinescent, non-leguminous, angiospermic, actinorhizal woody shrubs distributed throughout the temperate zones of Asia and Europe, belongs to the family Elaeagnaceae and bears the microsymbiont *Frankia* in their root nodules in a symbiotic association that fixes huge quantities of atmospheric nitrogen (Ambaye, 1971). The genus is represented by 4 species and a number of sub-species among which the most important species are *H. rhamnoides* and *Hippöphae salicifolia* D. Don. *H. rhamnoides* is widely distributed throughout Europe and Asia but not profoundly available along the eastern Himalayas of India. In comparison, the species i.e. *H. salicifolia* D. Don is profoundly available and distributed along high altitude fragile mountain slopes of the eastern Himalayas (Sikkim and Arunachal Pradesh) of the Indian sub-continent better known as the Hindu-Kush Himalayan mountain ranges (Singh, 1998). These mountain ranges along with the Qinghai mountains of China, the Tibetan Plateau and the Burmese mountain ranges constitute a large contiguous region in Asia, where resource base has started diminishing at an alarming rate showing distinct signs of unsustainability, decreasing soil fertility and a high degree of instability. The rapid population growth coupled with urbanization along these fragile high altitude areas has immensely contributed to the degradation of limited productive land. Because of the profound availability, superior growth characteristics, wood quality with high calorific value and presence of high nitrogen content in leaves and branches, *H. salicifolia* D. Don. is widely utilized by the people dwelling in such mountain areas. *H. salicifolia* D. Don. thus has become a major source for meeting the increasing demand for fuelwood and fodder requirements which has resulted into rapid natural resource extraction of the plant both from forests and farmlands resulting into degradation of its natural ecosystems.

H. salicifolia D. Don., a hard, woody actinorhizal shrub is abundantly observed to grow along high altitude slopes of the Eastern Himalayas of Sikkim, North-Eastern India. The distribution of the plant in Sikkim ranges between 2,100 m asl to 3,500 m asl under harsh and stress natural conditions. The availability of the plant along the fragile mountain slopes of North Sikkim is thought to have marked influence on soil enrichment. *H.* is well adapted under stressed conditions, thereby contributing remarkably to recuperation and soil improvement in and along fragile landscapes (Rousi, 1965; Yao and Zhu 1985; Salo, 1989; Yao and Tigerstedt, 1995). In recent years, a lot of research on actinorhizal plants contributing to high levels of nitrogen assimilation in soils has attracted scientists and ecologists towards understanding the ecological dynamics of such plants under variable microclimatic conditions and their subsequent influence on soil characteristics.

16.1.5 Definition of the Problem

Biological nitrogen fixation, important on a global basis is essential to the development of degraded ecosystems. Over the past several decades, a lot of research groups have focused stress on understanding symbiotic nitrogen fixation in leguminous plants with good success. However, nitrogen fixation in non-leguminous plants has been a subject of inquisitiveness for researchers in the field of symbiotic nitrogen fixation in recent years. A characteristic feature of seabuckthorn is its ability to fix atmospheric nitrogen in soil. It has an outstanding ability to fix enormous amount of nitrogen (180 kg of nitrogen/ hectare/annum) thereby increasing soil fertility. Researchers around the globe have been trying to address the possibilities of enhanced nitrogen fixation with seabuckthorn (*H. salicifolia*). The plant *H. salicifolia* bears nodules in its root system and these nodules harbour *Frankia*, a gram-positive actinomycete responsible for symbiotic nitrogen fixation. Uninterrupted anthropological interventions in the name of sustainable agricultural practices along high altitude areas of these regions has not only led to devastation of natural *Hippophae* forest lands but also has contributed to decreased soil fertility resulting into erosion and landslides whose consequences will be devastating with time. Under such circumstances, sensible exploitation of the actinorhizal plant *H. salicifolia* D. Don and the study of their symbioses with the soil microsymbiont *Frankia* becomes critical and important for a number of reasons among which the most important are reforestation and reclamation efforts of depauperate, nitrogen limiting soils along high altitude zones of the Indian Eastern Himalayas. The majority of the actinorhizal plant species identified so far are usually recognized as pioneer species during early ecological successions (Bond, 1983). It is also established that these plants are particularly distributed along the temperate zones of the world, most of which are either highlands or mountainous terrains. Climatic conditions in these areas are therefore harsh and life inhospitable. Most of the temperate zones are snow covered throughout the year and may be referred to as cold deserts characterized by extremely poor and infertile soils. It is known that soils from a natural forest ecosystem harbour a great deal of microbial population and their activity. Being one of the most important components of soil environment, microorganisms exert considerable influence on soil fertility and plant growth

(Tiwari *et al.*, 1991). These microbial communities are inherently stable but at the same time are also dynamic structures (Campbell, 1983). In a natural seabuckthorn forest root nodules die and decomposition takes place and the debris from leaves and branches on the soil surface enters into the disturbed soil serving as food for microbial growth. Also there are very few reports on the effects of soil microbes on nitrogen fixation along temperate zones of the Indian subcontinent. In the last four decades a lot of research has been focused on the assessment of nitrogen fixing ability of actinorhizal plants and in determining the genetic diversity of *Frankia* strains associated with them. *Frankia* is a filamentous, gram-positive, nitrogen-fixing endosymbiont that belongs to the order *Actinomycetales,* family *Frankiaceae* and genus *Frankia* (Becking, 1970) together with the related genus *Geodermatophilus* (Hahn *et al.,* 1989). However, very few studies on other microbial communities associated with rhizospheric soils of actinorhizal plants and their probable interaction with *Frankia* has been accomplished. The present study aims at enumerating and assessing the tree and plant diversity associated with growth of *H. salicifolia* natural strands of North Sikkim. In addition *Frankia* strains associated with *H. salicifolia* were isolated and studied in terms of their N_2 fixing ability in an attempt to understand their probable mechanistic role in enhancing soil fertility in poor fragile soils of North Sikkim.

16.2 MATERIALS AND METHODS

16.2.1 Study Site

The study sites for the present investigations are situated in the Eastern Himalayas along the temperate belt of mount Kangchendzonga near Lachen (Zema) areas in North Sikkim, North-Eastern India. The study site lie a part of lesser Himalayas, the youngest, tallest and the most fragile mountain system of the world, covering an area of approximately 5,91,000 km^2 and extending for about 2,500 km from the east to the west. The study sites are situated between longitudes 27° 75' N and 27° 81' N and latitudes 88° 54' E and 88° 56' E, respectively. The study area ranges in elevation from 2,127 to 3,150 m asl, a difference of 1023 meters. The rugged configurations of monolithic cliffs, high slopes, cold dry climate and extreme temperatures along these higher Himalayan regions allow a unique array of biological diversity to survive. The study sites were divided into three different elevation zones based on availability of the plant species *H.salicifolia* at different altitudinal variations which are termed as Lower Elevation Zone (LEZ) at an altitude ranging from 2,127 m asl to 2,279 m asl, Middle Elevation Zone (MEZ) at an altitude from 2,660 m asl to 2,810 m asl and Upper elevation zone (UEZ) at an altitude from 2,998 m asl to 3,150 m asl where gregarious growth of the species was observed. All the three elevation zones are characterized by the occurrence of mixed vegetation along with natural stands of *H. salicifolia*. Each elevation zone was further divided into two cover types termed as control plots or CTLwithout *H. salicifolia* D. Don growth and sample plots or SBT supporting growth of *H. salicifolia*.

16.2.2 Climate

Data on weather conditions for the present study were measured at Lachen (2,735m asl). Relative humidity was measured suing a hygrometer (Barigo, Germany), atmospheric temperature was measured using a air-thermometer (Hicks, Japan), and altitude was measured using an altimeter (Barigo, Germany), respectively. Data for mean average rainfall and photosynthetically active radiation was collected from Gangtok weather station's Metrological and Agricultural Department.

16.2.3 Ambient Temperature

Mean monthly air temperature was recorded to be highest during the month of June-July (Min.-19.04°C, Max. 27.81°C to 30.79°C) and lowest in January (Min.-2.37°C, Max. 8.94°C). Air temperature increases progressively from January till June after which there is gradual reduction from July to October. However, during the months of November-January there a sudden decline in air temperatures was observed.

16.2.4 Relative Humidity

The relative humidity varied form 65% to 98% through out year with lowest in the month of December (65.44%) and comparatively higher in the month of July (92.62%). Summer/Rainy season have higher relative humidity followed by autumn, winter and spring. Slight diurnal variation in relative humidity was also observed which might be attributed to the obstruction of the horizontal flow of air by mountains.

16.2.5 Rainfall

The average annual rainfall was recorded between 3,100mm to 4,100mm, temporal variation was recorded in monthly rainfall, low rainfall was received between November to February while fairly good rainfall was received every month from March to October and highest was recorded in the month of June (820 mm). Most of Sikkim does not experience high intensity winds. However, at many hilltops and passes, high speeds winds and sometimes during winters, snow blizzards do occur.

16.2.6 Sunshine (Photosynthetically Active Radiation)

Photosynthetically active radiation (PAR) was recorded at Gangtok weather station (1,750 m asl). Highest (638.32 μ mol s^{-1} m^{2-1}) mean PAR was recorded in the month of April while lowest (365.97μ mol s^{-1} m^{2-1}) was recorded in February, 2003-2004.

16.2.7 Tree Diversity and Associated Vegetation

The phytosociology of the tree vegetation higher then 1.5 m was studied by quadrate method (n=10, 25 m x 50m) for trees (Magurran, 2004; Kent and Coker, 1992; Mishra, 1968). Ten quadrats of 25 m^2 x 50 m^2 for trees were laid randomly in each site. All the individuals of tree species within the quadrate were measured were listed and their CBH (circumference at breast height, 1.3 m above ground level) were measured & grouped as trees (~ 30 cm girth at breast height of 1.3 m), for each tree species density (trees ha^{-1}) and basal area values were calculated. Quantitative analysis of vegetation for frequency, density and dominance was calculated following Mishra (1968). The associated vegetation under seabuckthorn shrubs are listed below (Table 16.1).

Table 16.1: Associated Vegetation in LEZ, MEZ and UEZ

S. No.	Botanical Name	Family
1	*Anaphalis cortorta*	Asteraceae
2	*Artemesia* spp.	Asteraceae
3	*Clinopodium umbrosum*	Chenopodiaceae
4	*Deschampsia* spp.	Poaceae
5	*Fagaria nubicola*	Roseaceae
6	*Galium asperifolium* Wallich	Rubiaceae
7	*Impatiens urticifolia*	Chenopodiaceae
8	*Meconopsis horridula*	Papaveraceae
9	*Pieris formosa*	Ericaceae
10	*Polygonum* spp.	Polyconaceae
11	*Primula dendriculata*	Premulaceae
12	*Primula sikkimensis*	Premulaceae
13	*Prunella vulgaris*	Asteraceae
14	*Ranunculus* spp.	Ranunculaceae
15	*Rhododendron arboreum.1*Ericaceae	
16	*Rhododendron barbatum*	Ericaceae
17	*Rhododendron beesian*	Ericaceae
18	*Rhododendron calophytum*	Ericaceae
19	*Rhododendron wightii*	Ericaceae
20	*Sempervivum montanum*	Asteraceae
21	*Senecio* spp.	Asteraceae

16.2.8 Nodule and Stem Collection, Surface Sterilization and Nursery Propagation of *H. salicifolia* Plants

Fresh root nodules, fruits and stem cuttings of *H. salicifolia* were collected from the above mentioned elevation zones of Lachen (North Sikkim) and were transported to the laboratory in an icebox containing solid ice. Also soil from the study sites were carried for nursery propagation of *H.* plants. The root nodules were kept in a deep freezer (-20ºC) for 24 h. and subjected to acetylene reduction activity (ARA) for estimation of nitrogenase activity. The stem cuttings collected were surface sterilized with 5% sodium hypochlorite solution, dipped into solution containing different concentrations of Indole Butyric Acid (IBA), gibberellic acid (GA_3) and planted into pots containing soil from study sites for root initiation. After almost 65 days root development was distinct with emergence of fresh root nodules. These fresh young nodules were collected from the nursery-grown plants and were separated into individual lobes with the help of a disinfected scalpel. A total of 27 young, fresh individual lobes were separated from single nursery-grown *H. salicifolia* plant and treated for homogenate preparation.

16.2.9 Nodule Disinfection and Preparation of Nodule Homogenate for Inoculation

Individual lobes were washed in tap water 3 times to remove extraneous debris, treated in 1 ml of 1:1 ratio of sodium hypochlorite (5.25%) solution and 0.1% Tween 80 for 20 minutes in 1.5 ml eppendorf tubes by centrifuging at 2000 rpm in a microcentrifuge (Model-Spinwin Tarsons, India). The lobes were then thoroughly washed with double distilled water in the eppendorf tubes using a 10-100 μl micropipette (Labsystems Finnpipette, Finland) to ensure minimum contamination. Individual lobes were then soaked into 1.0 ml of 5% sodium bicarbonate solution for 10 minutes and thawed several times in eppendorf tubes. The lobes were then placed in sterilized petridishes in wetted filter papers (Whatman no. 1) and the outer epidermis peeled off in a dissecting microscope olympus (The Netherlands) in order to remove root phenolics and contaminants that might be available in the epidermal layers. After eventual peeling, the individual lobes were re-sterilized again in 1.0 ml of 30% hydrogen peroxide (v/v H_2O_2) by centrifuging the effendorf tubes at 2000 rpm for 10 minutes. After treating the lobes with H_2O_2 these were further washed twice with double distilled water and the individual lobes were then crushed in a 2 ml tissue homogenizer (HiMedia) for isolation of clonal population of *Frankia* cultures *in vitro* following the methodologies of Lalonde *et al.* (1981), Benson (1982), respectively. The nodule homogenates obtained were then mixed with 1 ml of phosphate buffered saline (137 m M NaCl,

2.7 m M Na_2HPO_4, 1.8 m M KH_2PO_4; pH 7.4) supplemented with 1% cetyl trimethylammonium bromide (CTAB) and 0.5% polyvinyl pyrrolidone (PVP) in eppendorf tubes (Sarma *et al.*, 1998). The tubes were then centrifuged at 2000 rpm for 5 minutes and in microcentrifuge and the supernatant were aspirated out with the help of sterilized Pasteur pipettes. The pellets of the homogenates then recentrifuged at 10,000 rpm for 5 minutes in 1.5 ml eppendorf tubes with autoclaved distilled water. This suspension was used for inoculation.

16.2.10 Nodule Homogenate Inoculation and Isolation of Clonal Population of *Frankia*

10 µl of homogenates obtained from each of the 27 disinfected lobes were inoculated into petriplates containing solid Qmod growth media for growing *Frankia* strains following the procedures of Lalonde and Calvert (1979). The inoculated plates were then incubated in a BOD incubator at 29 ±0.5 C for 14 weeks to obtain *Frankia* growth. Once growth of *Frankia* colonies was obtained, these were streaked individually into Petri plates containing solid Qmod growth media in order to separate the individual colony types from a consortium. After successful sub-culturing, the strains were transferred to 500 ml Erlenmeyer flasks containing 200 ml liquid BAP medium of Murry *et al.* (1984) containing 10 mM sodium pyruvate as sole carbon source and ammonium chloride (NH_4Cl) as sole nitrogen source to obtain pure growth of the isolates. The flasks were incubated in a BOD incubator at 29 ±0.5 C for 8 weeks in order to obtain *Frankia* mass culture.

16.2.11 Identification of *Frankia* Strains from H. *salicifolia*

After successful isolation of *Frankia* axenic cultures, the isolates were grown through entrapment of *Frankia* spores in Ca^{+2} alginate beads following the procedure of Sarma *et al.* (1998) to obtain genetically pure cultures of *Frankia*. The single spore cultures obtained were subjected to phase contrast micrography using a LEICA DMRX Image processing system. For accomplishing the same, 5 ml of the single spore isolates grown in BAP and DPM medium were carried in to Regional Sophisticated Instrumentation Centre (RSIC, Shillong, India) for preliminary identification of the strains and subsequent microphotography.

16.2.12 Morphological Studies and Photomicrography of *Frankia* strains

Microscopic examination of single spore cultures obtained was prepared from 7 days, 14 days, 21 days and 30 days old cultures. 5ml of the single spore cultures of all the strains were transferred to 1 ml of 3% glutaraldehyde solution and fixed for 3 hours to which 50% glycerol was added following the methodology of Mansour *et al.* (1990). The sample was shaken gently in a vortex mixer (REMI, India) for 10 min. in order to ensure a homogenous solution and to dismantle the filamentous network of *Frankia*. However, precaution was taken so that filaments do not break and are intact. Single drop of the prepared media solution were taken in slides with the help of sterilized Pasteur pipettes, smeared and observed directly in a phase contrast LEICA DMRX image processing system for microscopical observation of *Frankia* and resolution of morphological characters like colony colour, colony size, hyphal diameter, spores, vesicles and sporangia forms as was described by Racette *et al.* (1989). Observations were made in triplicate sample preparations and recorded at 400X, 800X and 1000X magnification respectively. Photographs at different resolutions were taken to obtain clarity. Measurements for colony size and diameter of hyphae, spores, vesicles and sporangia forms were done in "Leica Qwin" image analyzing software supplied with the LEICA DMRX System.

16.2.13 Nitrogenase Activity of *Frankia in vitro* (Acetylene Reduction Assay)

Nitrogenase activity of the isolated *Frankia* strains and the reference strains was estimated by the reduction of acetylene to ethylene following the methodology of Lopez *et al.* (1986) Replicate 2 ml samples of BAP grown *Frankia* cells were incubated at 29 C ± 0.5 C for 14 days (2 weeks), aseptically removed from culture medium by inserting a sterilized syringe of 16 mm gauge and introduced into sterilized 6 ml serum stoppered vials. Determination of conversion of acetylene to ethylene during incubation of the samples was analysed in a gas chromatograph (Chemito, model-8610) fitted with a flame ionization detector (FID) and Porapak N column (stainless steel, 180 cm long). The column, detector and injector temperatures were maintained at 80 °C, 130 °C and 155 °C, respectively. Nitrogen was used as the carrier gas, hydrogen as fuel gas and zero air as supporting gas with flow rates of 20 ml min^{-1}, 30 ml min^{-1} and 250 ml min^{-1} respectively. Proper conditioning of the columns was done periodically after analysing each gas sample injected by heating the columns to 250 °C. The columns for the purpose were calibrated and standardized using gas mixtures of acetylene and ethylene concentrations from 2-100 µl l^{-1} and 2 ml standard mixture of, 8 ppm C_2H_2 and 5 ppm C_2H_4 , respectively. Using a gas tight disposable syringe (Dispovan), a 1 ml gas sample was withdrawn from each of the incubated stoppered tubes (in triplicates) and was injected into injecting ports of the gas chromatograph for analysis of C_2H_2 reduced and C_2H_4 produced. Gases injected were analysed for a time interval of 10-15 minutes. Chromatograms obtained were analysed by comparing their retention time with authentic compounds obtained from Chemtron Inc., India. Nitrogenase activity of *Frankia* cultures and the quantity of acetylene reduced/ethylene produced was measured in nmol C_2H_4 formed per mg protein per hour was calculated using the formula:

$$\text{N.A.} = \frac{\text{nmole } C_2H_4\text{/ unit area of the standard X area of } C_2H_4 \text{ in the sample X Volume of the vial}}{\text{mg protein/ ml X incubation time}}$$

where, N.A. is nmole of ethylene produced/per mg protein/hour.

16.2.14 Nitrogenase Activity of *Frankia in vivo* inside Nodules (Acetylene Reduction Assay)

The nitrogenase activity of nodules collected from the study sites, which were stored at –20 ºC, were also measured within 5 days of collection following the methodology of Lopez *et al.* (1986). The stored root nodules were washed several times with deionized water to remove all extraneous debris and then subjected to treatment with 5% sodium hypochlorite solution for 10 minutes followed by 30% hydrogen peroxide (H_2O_2) wash for 5 minutes. This was followed by repeated washing with double distilled sterile water in sterilized petridishes inside a laminar flow hood. The washed nodules were dried on sterilized tissue paper. 10 nodule lobes were separated from each individual nodule collected from the three elevation zones (viz. LEZ, MEZ and UEZ), inserted into 10 ml sterilized serum stoppered vials and properly sealed with parafilm to ensure that the vials are airtight. One vial contained one nodule lobe each and a total of 150 nodule lobes collected from 30 trees were sampled for performing acetylene reduction assay. For performing acetylene reduction assay, air in the vials was removed by the derived gas phase, which contained 10% (v/v) acetylene. In order to check the purity of gas, a control containing only acetylene without any nodules was also taken for the assay. The stoppered vials containing nodule lobes were incubated for 3 hours at 28 ± 1 ºC. To perform the assay, 1 ml of incubated gas samples from the vials were injected directly into the injection port of gas chromatograph (Chemito-8610). The injector, oven and detector temperatures were maintained at 120, 190 and 175 ºC respectively. The gases used and their flow rates were similar to ARA performed on the *Frankia* pure cultures mentioned above. In order to confirm the retention time 1 ml of standard ethylene was injected into the gas chromatograph and the ethylene peaks of the standard were compared with sample peaks. Under the mentioned conditions, the retention time for ethylene was approximately 1.53 minutes and that of acetylene was 2.44 minutes, respectively. After recording the peak area of retention for ethylene gas of all the samples, the nodules from each vial were weighed using a micro Sartorius balance and their respective fresh weights recorded. During the experiments the conditions for the performance activity of all the samples were kept constant by maintaining the incubation time, thus making comparative inference valid.

The nitrogenase activity and the quantity of acetylene reduced/ethylene produced per gram nodule fresh weight per hour was estimated in terms of nmoles of C_2H_4 produced per gram fresh weight per hour using the formula:

$$N.A. = \frac{\text{nmole } C_2H_4/\text{unit area of the standard X area of } C_2H_4 \text{ in the sample X volume of the vial}}{\text{nodule fresh weight X incubation time}}$$

where, N.A. is nmole of ethylene produced/gram fresh weight of nodules/hour.

16.3 RESULTS

16.3.1 Plant Distribution

H. salicifolia is a deciduous, usually spinescent, shrub. Usually the plant grows to a height of 1-5 metres. The plant enjoys a widespreased but dispersed distribution along the mountain slopes of North Sikkim. We could observe dense distribution along with *rhododendron* spp. along steep mountain slopes of Lachen and Lachung. Interestingly, the persistence of *H. rhamnoides* could not be observed during our frequent field visits. The presence of thick vegetation as well as *H. salicifolia* forests could be observed along the slopes of hill streams in all the study sites.

16.3.2 Tree Structure, Composition and Diversity in *H. salicifolia* Natural Forest

The natural forest stand and the study sites (i.e. LEZ, MEZ and UEZ) have showed uneven distribution of different tree stands (above 1.5 m). Forests located at Zema area of Lachen were surveyed to understand forest structure and species diversity. A total of 17 identified tree species were found in tree stratum of LEZ (Table 16.2a) having DBH in Between 10 to 50 cm whereas a total of only 7 and 5 identified species were found in MEZ and UEZ (Table 16.2 b, c) forest sites having DBH in between (10 to 40 cm). *H. salicifolia* accounted alone for 30.88%, 64.71% and 46.45%, respectively for LEZ, MEZ and UEZ. In case of UEZ a lesser number of tree species diversity was recorded then MEZ and UEZ. Abundance, Basal area m^2/ha and Individual Value Index (IVI) of all three stands are calculated and listed in Table 16.2 (a,b & c).

The index of dominance was recorded highest for UEZ (0.183) (Table 16.1) depicting that the stand is shared by a few species only. Species diversity and Index of dominance are presented in Table 3 (a, b, and c). Species diversity was estimated as 2.824, 1.998 and 1.812 for LEZ, MEZ and UEZ,

Table 16.2 (a) : Tree species density, basal area, abundance and IVI in UEZ

S. No.	Species	Local name	Family	Frequency (%)	Density (no./ha)	Abundance	Basal area (m^2/ha)	IVI
1	*Hippöphae salicifolia* D. Don.	Tarubo/ chuma /Durchuk	Eleagnus	100.000	318.400	3184.000	0.008	87.366
2	*Oak* spp.	Dewadar	Fagaceae	100.000	40.800	408.000	0.100	60.019
3	*Pinus* spp.	Cheer	Pinaceae	100.000	66.400	664.000	0.037	43.916
4	*Criptomeria* spp.	Dhupi	Taxodiaceae	80.000	19.200	240.000	0.013	22.504
5	*Rhododendron* spp.	Gurnas	Ericaceae	80.000	13.600	170.000	0.017	22.635
6	*Ostodes paniculata*		Euphorbiaceae	70.000	12.000	171.429	0.086	43.788
7	Un-identified-I			40.000	4.000	100.000	0.035	19.769

Table 16.2 (b) : Tree species density, basal area, abundance and IVI in MEZ

S. No	Species	Local name	Family	Frequency (%)	Density (no./ha)	Abundance	Basal area (m^2/ha)	IVI
1	*Hippöphae salicifolia* D Don	Tarubo/ chuma /Durchuk	Eleagn aceae	100	568.80	5688.00	0.008	99.700
2	*Quercus* spp.	Arkholo	Fagaceae	40	7.20	180.00	0.100	27.910
3	*Pinus* spp.	Cheer	Pinaceae	50	8.00	160.00	0.036	16.902
4	*Criptomeria* spp.	Dhupi	Taxodiaceae	70	12.80	182.86	0.012	16.160
5	*Rhododendron* spp.	Gurnas	Ericaceae	90	64.80	720.00	0.015	27.559
6	*Oak* spp.	Dewadar	Fagaceae	50	8.00	160.00	0.098	29.414
7	*Castanopsis* spp.	Katus	Fagaceae	70	12.00	171.43	0.098	33.375
8	Un-identified-I			80	13.60	170.00	0.085	32.646
9	Un-identified-II			40	5.60	140.00	0.044	16.327

Table 16.2 (c): Tree species density, basal area, abundance and IVI in LEZ

S. No.	Species	Local name	Family	Frequency (%)	Density (no./ha)	Abundance	Basal area (m^2/ha)	IVI
1	*Hippöphae salicifolia* D. Don.	Tarubo/ chuma /Durchuk	Eleagn aceae	100	145.60	1456.00	0.008	43.784
2	*Pinus* spp.	Cheer	Pinaceae	70	12.00	17.14	0.037	11.346
3	*Criptomeria* spp.	Dhupi	Taxodiaceae	50	7.20	144.00	0.012	6.577
4	*Rhododendron* spp.	Gurnas	Ericaceae	90	29.27	325.20	0.022	15.793
5	*Symplocos* spp.	Kholmein	Symplocaceae	80	21.60	270.00	0.028	13.682
6	*Quercus* spp.	Arkholo	Fagaceae	90	27.20	302.22	0.204	30.969
7	*Acer* spp.	Kapashe	Acerraceae	70	10.40	148.57	0.025	9.941
8	*Alianthus* spp.		Meliaceae	60	8.80	146.67	0.037	9.815
9	*Castanopsis* spp.	Katus	Fagaceae	90	21.60	240.00	0.100	20.685
10	*Macaranga* spp.	Malata		70	20.00	285.71	0.023	12.139
11	*Alnus nepalensis*	Utrish	Betulaceae	90	16.80	186.67	0.095	19.038
12	*Artocarpus* spp.	Badhar	Moraceae	50	11.20	224.00	0.155	19.858
13	*Schima wallichii*	Chilaune	Theaceae	50	4.80	96.00	0.058	9.930
14	*Cinnamon* spp.	Sissi	Lauraceae	80	25.60	320.00	0.039	15.658
15	*Michilus* spp.	Punsi	Lauraceae	50	9.60	192.00	0.027	8.471
16	*Sapium* spp.		Euphorbiaceae	60	14 .40	240.00	0.058	12.998
17	*Shorea robusta*	Sakhuwa	Dipterocarpaceae	70	10.40	148.57	0.100	16.458
18	Un-identified-I			60	7.20	120.00	0.071	12.423
19	Un-identified-II			50	5.60	112.00	0.062	10.436

respectively. Where as *H. salicifolia* diversity alone in the sites were 0.281, 0.366 and 0.359 out of the total diversity of 2.824, 1.998 and 1.812 for LEZ, MEZ and UEZ conforming that the spp. is the most diverse and dominating spp. among others. (Table 16.3 a, b and c).

The presence of *H. salicifolia* in the study sites was predominant with an increase in altitude. The soil pH of the study sites varied from 5.46 ± 0.04 in the lower elevation zone, 5.27 ± 0.04 in the middle elevation zone and 6.19 ± 0.07 in the upper elevation zone while the average temperatures of the study sites recorded during the present investigations varied between 18 °C in the upper elevation zone, 21 °C in the middle elevation zone and 22 °C in the lower elevation zone during summer and –5 °C in the upper elevation zone

Table 16.3 (a) : Tree species diversity and dominance in UEZ

S. No.	Species	Local name	Family	distribution pattern	Sp. diversity	Sp. dominance
1	*Hippöphae salicifolia* D. Don.	Tarubo/ chuma / Durchuk	Eleagn aceae	31.840	0.359	0.085
2	*Oak* spp.	Dewdar	Fagaceae	4.080	0.322	0.040
3	*Pinus* spp.	Cheer	Pinaceae	6.640	0.281	0.021
4	*Criptomeria* spp.	Dhupi	Taxodiaceae	3.000	0.194	0.006
5	*Rhododendron* spp.	Gurnas	Ericaceae	2.125	0.195	0.006
6	*Ostodes paniculata*		Euphorbiaceae	2.449	0.281	0.021
7	Un-identified-I			2.500	0.179	0.004

Table 16.3 (b) : Tree species diversity and dominance in MEZ

S. No.	Species	Local name	Family	Distribution pattern	Sp. diversity	Sp. dominance
1	*Hippöphae salicifolia* D. Don.	Tarubo/ chuma / Durchuk	Eleagn aceae	56.880	0.366	0.110
2	*Quercus* spp.	Arkholo	Fagaceae	4.500	0.221	0.009
3	*Pinus* spp.	Cheer	Pinaceae	3.200	0.162	0.003
4	*Criptomeria* spp.	Dhupi	Taxodiaceae	2.612	0.157	0.003
5	*Rhododendron* spp.	Gurnas	Ericaceae	8.000	0.219	0.008
6	*Oak* spp.	Dewdar	Fagaceae	3.200	0.228	0.010
7	*Castanopsis* spp.	Katus	Fagaceae	2.449	0.244	0.012
8	Un-identified-I			2.125	0.241	0.012
9	Un-identified-II			3.500	0.158	0.003

Table 16.3 (c) : Tree species diversity and dominance in LEZ

S. No.	Species	Local name	Family	Distribution pattern	Sp. diversity	Sp. dominance
1	*Hippöphae salicifolia* D. Don.	Tarubo/ chuma / Durchuk	Eleagn aceae	0.281	0.021	14.560
2	*Pinus* spp.	Cheer	Pinaceae	0.124	0.001	0.245
3	*Criptomeria* spp.	Dhupi	Taxodiaceae	0.084	0.000	2.880
4	*Rhododendron* spp.	Gurnas	Ericaceae	0.155	0.003	3.613
5	*Symplocos* spp.	Kholmein	Symplocaceae	0.141	0.002	3.375
6	*Quercus* spp.	Arkholo	Fagaceae	0.234	0.011	3.358
7	*Acer* spp.	Kapashe	Acerraceae	0.113	0.001	2.122
8	*Alianthus* spp.		Meliaceae	0.112	0.001	2.444
9	*Castanopsis* spp.	Katus	Fagaceae	0.184	0.005	2.667
10	*Macaranga* spp.	Malata		0.130	0.002	4.082
11	*Alnus nepalensis*	Utrish	Betulaceae	0.175	0.004	2.074
12	*Artocarpus* spp.	Badhar	Moraceae	0.180	0.004	4.480
13	*Schima wallichii*	Chilaune	Theaceae	0.113	0.001	1.920
14	*Cinnamon* spp.	Sissi	Lauraceae	0.154	0.003	4.000
15	*Michilus* spp.	Punsi	Lauraceae	0.101	0.001	3.840
16	*Sapium* spp.		Euphorbiaceae	0.136	0.002	4.000
17	*Shorea robusta*	Sakhuwa	Dipterocarpaceae	0.159	0.003	2.122
18	Un-identified-I			0.132	0.002	2.000
19	Un-identified-II			0.117	0.001	2.240

and -2 °C in the middle as well as in the lower elevation zones, respectively. For isolating *Frankia* strains and conducting laboratory experiments, matured root nodules, ripened fruits and stem cuttings from the actinorhizal plant *H. salicifolia* D. Don and fresh soils from the rhizosphere of the plants were collected randomly from a number of selected quadrates in the lower, middle and the upper elevation zones during four seasonal visits to the study sites situated at Zema, North Sikkim in the years 2002-2003. The nodules, fruits and the leaves were stored in an ice box supplied with dry ice and brought back to the laboratory. The nodules were immediately stored in at -20 °C in a deep freezer for performing acetylene reductase activity. The seeds collected were allowed to dry in sunlight for a period of 2 weeks and stored at -20 °C, which were later used for germinating plantlets for performing infectivity and effectivity tests *in-vitro* under greenhouse conditions. The stem cuttings were disinfected with $HgCl_2$ (5% v/v) and were planted into earthen pots to allow root germination. The stem cuttings were treated with different doses of plant hormones to observe the effects of growth on germinating roots and shoots (Table 16.4).

Table 16.4: Effect of plant hormones on root and shoot initiation of stem cuttings of *Hippöphae salicifolia* D. Don

Treatments	Initial initiation (days)	Complete initiation (days)
IBA (1 ppm)	11	13
IBA (150-200 ppm)	08	11
IBA (300-600 ppm)	–	–
GA3 (1 ppm)	18	22
GA3 (150-200 ppm)	12	15
GA3 (300-600 ppm)	09	13
GA3 (700 ppm)	–	–
NAA (5 ppm)	09	11
NAA (10 ppm)	08	11
NAA (100 ppm)	–	–
Control	17	25

Based on the results obtained, the stem cuttings of *H. salicifolia* were treated with a dosing regime of 150 ppm IBA (Indole Butyric Acid) and 300 ppm GA_3 (Gibberlic acid) and were allowed to grow in soil collected from study sites, under greenhouse conditions and at an optimum temperature of 22 °C. Successful growth of the stem cuttings were obtained in 8 weeks which grew into 6-7 inches tall plantlets and distinct young root nodules could be observed which were collected with sterilized scalpel and treated for isolation of *Frankia* strains. A selective incubation procedure with much poorly developed nutrient Qmod media was used for isolation of *Frankia* strains. Also antibiotics like Cyclohexemide and Nalidixic acid were added to the autoclaved media to inhibit growth of gram-negative bacteria and fungus that contaminated the Petri dishes during subsequent incubation. A total of 70 solid Qmod petridishes inoculated with 27 nodule homogenates were incubated inside a BOD incubator at 29 °C. All the 70 petridishes did not show development of *Frankia* growth. Observable colonies could be obtained in petriplates which were overlaid with a layer of Qmod agar. The other group of petriplates did show some contamination and hence were discarded. Out of 70 incubated petriplates, 14 plates showed emergence of small colonies, perfectly oval in shape and similar in morphology but different in colouration ranging from pink and red to opaque white. Sub-culturing in solid media was done four times in order to achieve pure forms of the colonies. The strains were then transferred to liquid BAP medium with different formulations of carbon sources and an additional nitrogen source common to all. It was observed that almost all the strains responded positively to simple sugars and growth was convincing although the propensity of growth to all the sugars was not the same (Table 16.5).

16.3.3 Isolation of Single Spore *Frankia* Isolates

The most presumable fact regarding isolation of *Frankia* is extreme slow growth of the genus (generation time 2-5 days, Diem and Dommergues and sometimes in some strains it is 15 days (Benson and Silvester, 1993) which often antecedes growth overtaken by faster growing contaminating microorganisms, which thereby causes the greatest problem in obtaining pure cultures. Furthermore, isolating *Frankia* strains by crushing nodule lobes in suitable culture media as inoculants does not ensure the isolation of genetically pure cultures because a single nodule lobe may harbor consortium of several *Frankia* strains. The use of calcium alginate beads using immobilization of *Frankia* strains in agar plates ensures faster growth of the organism as well as reduction of contaminants (Sarma *et al.*, 1998). All the isolates from *H. salicifolia* D. Don were subjected to single spore isolation by the plating method.

Table. 16.5: Utilization of different carbon sources by all 31 probable Frankia isolates obtained during sub-culturing

		Simple and complex sugars										Organic acids				
S. No.	Strain	Glucose*	Fructose*	Maltose*	Xylose	Madurose	Trehalose*	Mannitol*	Mannose*	Lactose*	Sucrose*	pyruvate'	propionate'	succinate'	acetate'	Tween 80'
1	Hsli1	1	1	1	0	0	1	1	1	1	1	1	0	0	0	0
2	Hsli2	1	1	1	0	0	1	1	1	1	1	1	1	1	0	0
3	Hsli3	1	1	1	0	0	1	1	1	1	1	1	1	0	0	0
4	Hsli4	1	1	1	0	0	1	1	1	1	1	1	1	1	0	0
5	Hsli5	1	1	1	0	0	1	1	1	1	1	1	0	0	0	0
6	Hsli6	1	1	1	0	0	1	1	1	1	1	1	1	0	0	0
7	Hsli7	1	1	1	0	0	1	1	1	1	1	1	1	1	0	0
8	Hsli8	1	1	1	0	0	1	1	1	1	1	1	1	0	0	0
9	Hsli9	1	1	1	0	0	1	1	1	1	1	1	1	1	0	0
10	Hsli10	1	1	1	0	0	1	1	1	1	1	1	0	0	0	0
11	Hsli11	1	1	1	0	0	1	1	1	1	1	1	0	1	0	0
12	Hsli12	1	1	1	0	0	1	1	1	1	1	1	0	0	0	0
13	Hsli13	1	1	1	0	0	1	1	1	1	1	1	1	0	0	0
14	Hsli14	1	1	1	0	0	1	1	1	1	1	1	1	0	0	0
15	Hsli15	1	1	1	0	0	1	1	1	1	1	1	0	0	0	0
16	Hsli16	1	1	1	0	0	1	1	1	1	1	1	0	0	0	0
17	Hsli17	1	1	1	0	0	1	1	1	1	1	1	1	0	0	0
18	Hsli18	1	1	1	0	0	1	1	1	1	1	1	0	0	0	0
19	Hsli19	1	1	1	0	0	1	1	1	1	1	1	0	0	0	0
20	Hsli20	1	1	1	0	0	1	1	1	1	1	1	0	0	0	0
21	Hsli21	1	1	1	0	0	1	1	1	1	1	1	0	0	0	0
22	Hsli22	1	1	1	0	0	1	1	1	1	1	1	1	0	0	0
23	Hsli23	1	1	1	0	0	1	1	1	1	1	1	1	0	0	0
24	Hsli24	1	1	1	0	0	1	1	1	1	1	1	0	0	0	0
25	Hsli25	1	1	1	0	0	1	1	1	1	1	1	0	0	0	0
26	Hsli26	1	1	1	0	0	1	1	1	1	1	1	1	0	0	0
27	Hsli27	1	1	1	0	0	1	1	1	1	1	1	1	0	0	0
28	Hsli28	1	1	1	0	0	1	1	1	1	1	1	0	0	0	0
29	Hsli29	1	1	1	0	0	1	1	1	1	1	1	0	0	0	0
30	Hsli30	1	1	1	0	0	1	1	1	1	1	1	1	0	0	0
31	Hsli31	1	1	1	0	0	1	1	1	1	1	1	0	0	0	0

1= positive growth observed, 0= no growth observed, *denotes simple sugars, 'denotes organic acids.
Growth was measured by estimating soluble protein in culture observed at 660 nm in a Spectrophotometer (SPECTRONICS, India)

Altogether, 27 different single spore isolates were identified with the help of an inverted microscope; a few of which were photographed and these 27 single spore isolates were subjected to growth in liquid BAP medium during repeated sub-culturing. Thus, genetically pure forms of single spore isolates were obtained which were further subjected to morphological identification with the help of LEICA DMRX Phase contrast light photomicrography.

16.3.4 Morphological Identification and Photomicrography of *Frankia* Strains

One of the preliminary methods to delineate *Frankia* strains based on morphological differences was simple microscopic observation. Many successful attempts have revealed the structures to be filamentous with extensive branching, presence or absence of sporangia and spores as well as presence of vesicular structures which are sites of the enzyme nitrogenase. The *Frankia* isolates from *H. salicifolia* were all filamentous in appearance with vesicular diameter varying between 0.4-0.6 mm. However, diameter of vesicles in all the isolates could not be measured due to limitations in resolution of DMRX Phase contrast imaging. The hyphal diameter and diameter of sporangia forms in all the isolates were measured in µm per m^2 (µm m^{-2}) at three different magnifications. The observed characteristics for hyphae, filaments and sporangia forms are provided below (Table 16.6). The morphology of the filamentous forms was observed to be of two types. Type I were characterized by very sinuate hyphae with extensive branching while Type II strains were characterized by smooth hyphae that were rarely branched and were slightly sinuate. Based on microscopical observations all the 27 single spore isolates were identified as 11 distinct strains that differed in morphological characteristics. The morphological features observed for the isolates support previous findings (Weber *et al.*, 1998; Benson, 1982; Racette and Torrey, 1989; Weber *et al.*, 1998).

16.3.5 Nitrogenase Activity of *Frankia* Isolates and Root Nodules of *H. Salicifolia*

The most important aspect of *Frankia* research is the ability of the actinomycete to fix huge amounts of atmospheric nitrogen. All pure cultures of *Frankia* tested till date differentiates specialized structures called vesicles under conditions of nitrogen limitation that induce nitrogenase activity. It has been well established that *Frankia* vesicles protect the enzyme nitrogenase from oxidation by molecular oxygen by virtue of highly specialized laminated

Table 16.6: Frankia isolates from *Hippophae salicifolia.*

Diameter µm/m2	HsIi2	HsIi4	HsIi5	HsIi7	HsIi8	HsIi9	HsIi10	HsIi11	HsIi12	HsIi13	HsIi14
Filament-1	7.24 ± 0.21	13.38 ± 0.26	9.98 ± 0.46	9.42 ± 0.66	9.99 ± 0.38	6.90 ± 0.30	16.94 ± 0.89	5.02 ± 0.22	10.12 ± 0.42	11.82 ± 0.59	5.02 ± 0.22
Filament-2	8.73 ± 0.39	18.40 ± 0.42	12.24 ± 0.80	8.39 ± 0.64	9.81 ± 0.32	6.60 ± 0.34	19.00 ± 1.03	6.46 ± 0.20	9.88 ± 0.33	12.46 ± 0.56	6.46 ± 0.20
Filament-3	10.46 ± 0.52	10.22 ± 0.51	11.62 ± 0.57	8.10 ± 0.78	9.58 ± 0.28	7.35 ± 0.49	20.00 ± 0.76	6.01 ± 0.30	9.54 ± 0.32	12.76 ± 0.49	6.01 ± 0.30
Sub-hyphae-1	4.91 ± 0.18	8.61 ± 0.24	8.63 ± 0.19	6.64 ± 0.88	5.36 ± 0.21	2.98 ± 0.17	10.05 ± 0.48	3.08 ± 0.11	7.40 ± 0.32	4.14 ± 0.31	3.13 ± 0.12
Sub-hyphae-2	3.45 ± 0.14	10.74 ± 0.56	8.28 ± 0.51	6.55 ± 0.49	7.09 ± 0.18	3.28 ± 0.23	10.16 ± 0.66	3.05 ± 0.17	6.54 ± 0.38	4.57 ± 0.18	2.37 ± 0.20
Sub-hyphae-3	3.86 ± 0.10	9.22 ± 0.35	9.28 ± 0.29	6.49 ± 0.55	7.24 ± 0.26	3.40 ± 0.23	6.30 ± 0.31	2.53 ± 0.19	7.24 ± 0.26	4.55 ± 0.19	2.19 ± 0.15
Sub-hyphae-4	3.95 ± 0.11	9.77 ± 0.47	8.77 ± 0.25	6.83 ± 0.47	7.40 ± 0.32	3.90 ± 0.12	9.77 ± 0.46	2.98 ± 0.12	7.09 ± 0.18	4.27 ± 0.18	2.52 ± 0.20
Sub-hyphae-5	3.21 ± 0.13	8.71 ± 0.58	9.30 ± 0.28	7.49 ± 0.78	6.54 ± 0.38	3.24 ± 0.19	8.62 ± 0.49	2.42 ± 0.21	5.36 ± 0.21	5.25 ± 0.24	2.14 ± 0.18
Sporangia-1	21.43 ± 1.01	38.60 ± 1.26	34.66 ± 1.79	27.60 ± 1.12	23.25 ± 1.47	13.27 ± 0.44	55.34 ± 1.22	10.77 ± 0.33	17.20 ± 0.92	32.51 ± 1.80	11.53 ± 0.38
Sporangia-2	26.88 ± 1.38	46.99 ± 2.46	32.99 ± 1.70	27.72 ± 1.31	24.84 ± 1.73	12.72 ± 0.31	42.44 ± 1.03	10.69 ± 0.37	12.70 ± 0.86	33.48 ± 2.68	10.91 ± 0.42
Sporangia-3	26.41 ± 1.36	36.89 ± 1.87	32.91 ± 1.78	28.87 ± 1.31	21.60 ± 1.00	12.76 ± 0.48	56.31 ± 1.38	10.50 ± 0.37	17.32 ± 1.29	30.22 ± 1.98	11.16 ± 0.43

n=7; NA = ± SE

envelope that present a diffusion barrier to gases. The oxygen sensitivity of nitrogen fixation in *Frankia* may also be influenced by the carbon substrate being utilized for the growth of the isolates. The objectives of the present study were to observe the nitrogenase activity of *Frankia* isolates from *H. salicifolia* in BAP medium supplemented with sodium pyruvate and in both presence and absence of nitrogenous source (10 mM NH_4Cl) and their comparison to the effect of oxygen tension on nitrogenase activity in B medium without any carbon and nitrogenous sources, respectively (Table 7). The nitrogenase activity of the *Frankia* strains *in vivo* inside root nodules of *H. salicifolia* were also evaluated by conducting acetylene reduction assay. The observations thus obtained are summarized in a tabular format in Table 7. Results show a subsequent increase in nitrogenase activity of nodules from plants with an increase in altitude i.e. nodules collected from plants of the Upper Elevation Zone (UEZ) of the study sites showed the highest nitrogenase activity compared to nodules collected from plants from Lower Elevation Zone (LEZ) and the Middle Elevation Zone (MEZ), respectively. The variability of nitrogenase induction in root nodules and nitrogen fixation is diverse, which of course may be attributed to the type of plants from where nodules were collected. The UEZ of the study sites harbour larger plants of *H. salicifolia* compared to plants of the MEZ and the LEZ, which harbour comparatively smaller shrubs of the plant.

16.4 DISCUSSIONS

The high altitude areas of Zema in North Sikkim exemplified a tremendous diversity of gymnosperms as well as angiosperms. The regions show typical temperate forest cover types with fair distribution of actinorhizal species like *Alnus* and *Hippophae*. The lower elevation zone harbour the maximum number of plant species with a concurrent decrease in plant communities with an increase in altitude. In addition, Oaks, laurels, pines, cedars, chestnuts, and *Rhododendrons* exist together but the frequency of *Rhododendrons* and *Catanopsis* spp. was found to be higher compared to other plant spp. All these observations suggest an intricate relationship between co-existence of stress tolerant plant species and their probable role in symbiotic N_2 fixation. The Shannon diversity index for all the plant species was observed to be highest in LEZ and lowest in UEZ (Table 2) which definitely explain the role of microclimatic parameters in the overall distribution of plant species which seems to be restricted to LEZ. *H. salicifolia* (0.021 m^2/ha) and *Quercus* spp. (0.011m^2/ha) are the dominant and sub-dominating species in tree stratum for LEZ. In MEZ *H. salicifolia* (0.110m^2/ha) and *Castanopsis* spp. (0.012m^2/ha) showed dominance and sub-dominance over other available species while in the UEZ *H. salicifolia* (0.085m^2/ha) and Oak spp. (0.040m^2/ha) were

Table 16.7: Nitrogenase activity in root nodules of *H. salicifolia* based on Acetylene Reduction Assay (ARA)

S No.	Root nodule sample	Av. no. of nodules per root system	Nitrogenase activity in nmole C_2H_4 reduced gram fresh $wt^{-1} h^{-1}$
1	LEZ-A1	3.30 ± 0.23	21.59 ± 1.11
2	LEZ-A2	2.76 ± 0.16	20.16 ± 2.56
3	LEZ-A3	3:20 ± 0.47	21.50 ± 1.24
4	LEZ-A4	2.71 ± 0.30	18.33 ± 1.52
5	LEZ-A5	2.69 ± 0.25	27.86 ± 1.91
6	LEZ-A6	3.23 ± 0.13	33.37 ± 2.44
7	LEZ-A7	2.95 ± 0.18	26.77 ± 1.87
8	LEZ-A8	3.23 ± 0.13	26.71 ± 0.72
9	LEZ-A9	3.31 ± 0.12	40.38 ± 1.10
10	LEZ-A10	2.90 ± 0.09	36.57 ± 1.27
11	MEZ-A1	2.26 ± 0.04	10.25 ± 1.93
12	MEZ-A2	1.83 ± 0.08	10.44 ± 1.08
13	MEZ-A3	2.66 ± 0.13	15.69 ± 1.03
14	MEZ-A4	1.99 ± 0.13	22.76 ± 0.90
15	MEZ-A5	3.13 ± 0.14	18.33 ± 2.17
16	MEZ-A6	3.90 ± 0.13	24.96 ± 1.43
17	MEZ-A7	3.53 ± 0.13	33.78 ± 2.93
18	MEZ-A8	2.34 ± 0.06	32.72 ± 2.38
19	MEZ-A9	1.20 ± 0.06	27.70 ± 2.27
20	MEZ-A10	1.94 ± 0.08	32.20 ± 2.81
21	UEZ-A1	3.39 ± 0.14	53.39 ± 3.13
22	UEZ-A2	3.17 ± 0.11	42.21 ± 2.88
23	UEZ-A3	3.06 ± 0.07	29.95 ± 2.31
24	UEZ-A4	2.63 ± 0.17	40.03 ± 2.62
25	UEZ-A5	3.81 ± 0.20	55.02 ± 2.34
26	UEZ-A6	2.94 ± 0.13	23.57 ± 1.82
27	UEZ-A7	2.62 ± 0.26	32.28 ± 3.51
28	UEZ-A8	3.13 ± 0.12	49.16 ± 2.15
29	UEZ-A9	3.89 ± 0.14	29.44 ± 1.76
30	UEZ-A10	3.50 ± 0.19	23.57 ± 1.82

predominant. Although the possible link between co-existence of certain species with *H. salicifolia* is difficult to interpret, it can however be assumed that these plants evolved almost parallely and hence their existence under harsh cold climatic conditions is virtually dependent on one another.

Identification and characterization are the two most important aspects of any bacterial classification. The Bergeys Manual of Systematic Bacteriology, which taxonomically defines different groups of bacteria, is primarily based on morphological and biochemical characteristics of the microbial world. In view of the increasing awareness of the importance of biological nitrogen fixation in non-legumes, the study of *Frankia* associated with high altitude actinorhizal plant of the Eastern Himalayas i.e. *H. salicifolia* is of great importance to ecology, biology and agriculture which has not been reported till date. Initial growth medium used for isolation of *Frankia* strains from *H. salicifolia* root nodules often resulted into contamination (Schwintzer and Tjepkema, 1990). The method of selective incubation of poor media was thought to be necessary to obtain *Frankia* growth (Burggraaf *et al.*, 1981). Taking into consideration of the fact that initial failures to isolate *Frankia* strains from *H. salicifolia* was quite justified until the media requirements for proper growth could be optimized, certain changes in formulating media requirements became inevitable. Initial isolation was tried in solid Qmod media and apreciable growth of micro-colonies that could be obtained. Qmod media for *Frankia* isolation was supplemented with á-Lecithin as lipid supplement in addition to antibiotics like Cyclohexemide and Nalidixic acid. Schwencke (1991) described the growth behaviour of three different types of *Frankia* isolates for the actinorhizal plant *Casuarina* to grow under defined conditions in liquid BAP medium showing a rapid exponential growth without any sporangia formations and no lysed hyphae. It was also reported that the lipid supplements like lecithin and phosphatidyl cholines result into delayed appearance of sporangia during *Frankia* growth (Schwencke, 1991), a contradictory remark to what was reported by Quispel *et al.* (1983) who documented that diglycerides and phospholipids from *Alnus* root extracts stimulate growth of some *Frankia* strains *in vitro* when grown in Qmod media of Lalonde and Calvert (1979). Phosphatidylcholines may improve the growth characteristics of *Frankia* in culture and as a consequence, it may result into delayed sporangia appearance, provided that cholines are utilized for membrane biosynthesis (Schwencke, 1991). The *Frankia* strains isolated from *H. salicifolia* during the present invstigations also showed a marked reduction in sporangia formation which supplements the observations of Schwencke (1991). Earlier a lot of workers had reported the culture conditions for rapid growth of *Frankia* strains and vesicle formation in liquid BAP and DPM medium (Zhongze *et al.*, 1986). *Frankia* cultures grown under nitrogen-limiting conditions in BAP supplements also showed concomitant nitrogenase activity *in vitro* (Tjepkema *et al.*, 1981). Weber *et al.* (1988) had earlier reported the growth of *Frankia* strains from *Alnus glutinosa* and *A. incana* root nodules in Finland in Tween, Propionate, Casamino acid (TPC) medium containing basal solution with $CaCl_2$ but incidentally the same carbon sources die not result into growth of the *Frankia* strains in the present studies. Weber *et al.* (1988) had also reported that *Frankia* isolates originating from the same nodule were generally identical but they did not confirm their phylogenetic identity. Lumini and Bosco (1996) had reported that individual nodule lobes may host more than one *Frankia* strain and further generalized the essentiality of culture purity and identity as essential prerequisites to be ensured while characterizing *Frankia* strains *in vitro*. Benson and Silvester (1993) had suggested that once a *Frankia* strain has been isolated, it should be purified and identified to ensure that a co-culture of different strains is not obtained. On the contrary, the number of isolates (31 based on morphological characteristics in culture) obtained in the present studies was quite few in terms of enormous diversity of the genus reported by a number of workers (Wall, 2000). However, to ensure their purity status from a consortium, generation of single-spore isolates became necessary and which was achieved with success. Sarma *et al.* (1998), Prin *et al.* (1991) and Lumini and Bosco (1996) have earlier reported methods for generation of single-spore cultures of *Frankia*. Based on these facts the 31 isolates obtained from *H. salicifolia* was subjected to single-spore isolation following the methodology of Sarma *et al.* (1998) which resulted in the identity of actually 27 different *Frankia* strains rather than the original 31 types obtained, a few of which were mere duplications during identification. Generation of pure cultures also ensured the genetic purity of the isolates obtained (Dobritsa and Stupar, 1989; Prin *et al.*, 1991). Incidentally, the *Frankia* isolates from *H. salicifolia* did not grow under nitrogen limiting conditions as was earlier reported by Lumini and Bosco (1996) who sub-cultured wild type *Frankia* strain UFI 132715 (E15) in BAP medium without the addition of exogenous nitrogen source. This suggests that nitrogen is a prime requirement for *in vitro* growth of *Frankia* strains isolated from *H. salicifolia*. In another attempt to establish the identity of the 27 single spore isolates under the present investigation, the isolates were subjected to light photomicrography using a DMRX Phase Contrast Imaging. On comparing the hyphal diameter, shape and diameter of sporangiophores, it was concluded that all the 27 single-spore isolates actually represented 11 distinct strains. Observations based on similar morphological types provided conclusive evidences that isolates HsIi1, HsIi2, HsIi3, HsIi18, HsIi19 and HsIi25 were similar strains or rather same strains because they also depicted similar hyphal and sporangia diameter, all showed pinkish colouration in BAP medium, did not secrete any pigments in culture and showed oval sporangia in culture when observed through photomicrography. Therefore, only one of the above representatives was chosen for further studies. Similarly, the strains HsIi6, HsIi7 and HsIi22 were similar strains with similar hyphal morphology, dermatophilus type sporangia, dirty white colonies in appearance and secrete dirty-opaque pigments in culture. The representative chosen from these three strains for further studies was HsIi7. Strains HsIi9, HsIi20 and HsIi27 were similar in appearance during microscopic observations with stalked sporangia, pinkish-white in appearance and secrete dirty-opaque pigments in culture. The representative chosen for further studies was HsIi9. Strains HsIi10 and HsIi15 showed pinkish colouration in media with dirty-opaque pigment secretion and stalked sporangia. The representative chose from this group was HsIi10. Isolates HsIi12, HsIi16, HsIi23 and HsIi24 showed pinkish colouration in media, round sporangia and did not secrete any pigments in culture. The representative chosen from this group included isolate HsIi12. Isolates HsIi14, HsIi21 and HsIi26 showed yellowish colony formation in culture, dermatophilus sporangia and yellowish pigment secretion in culture. The representative chosen from this group was HsIi14. The two other isolates HsIi13 and HsIi17 showed reddish colouration in media, oval sporangia and did not secrete any pigments in culture. The representative chosen for further studies was isolate HsIi13. The other *Frankia* strains were different in some form or the other, either colony morphology, colouration, secretion of pigments and even hyphal diameter were all chosen for the studies. Finally a total of 11 strains were identified as distinct pure culture strains based on photomicrographic observations.

The nitrogenase activity of the isolates was determined by estimating the reduction of acetylene to ethylene. It was evident from the results that the isolates HsIi7, HsIi8, HsIi11 and HsIi14 showed remarkable nitrogenous activity compared to other isolates. Among the isolates HsIi2, HsIi12 and HsIi13 were observed to show the least nitrogenase activity in presence of carbon sources. Not much significant differences for nitrogenase activities of all the isolates in BAP and B medium supplemented with carbon sources could be observed, although the *Frankia* strains grown in B medium showed lesser nitrogenase activities compared to strains grown in BAP medium. The most significant differences could be observed for oxygen tension (pO_2) of the strains grown in B medium without any exogenous carbon and nitrogenous sources, which depicted a marked decrease (approximately five fold decrease) in nitrogenase activities in all the strains tested. This reflects that the metabolic activity for enzyme nitrogenase resides in the vesicles and not the filaments or hyphae (Lopez *et al.*, 1986) of the *Frankia* strains in the present investigations.

16.5 CONCLUSIONS

The studies performed provided substancial evidences regarding the existence and sustenance of N_2 fixing bacteria in the plant *H. salicifolia*, their tremendous phenotypic diversity and their role in enhacing soil fertility. The important concern is the role of these microbes in the growth and sustenance of other plants co-existent with *H. salicifolia* is still a mystery. While no studies on our part have been conducted on the isolation and identification of *Frankia* strains from other actinorhizal spp. like *Alnus* in the region, it is therefore felt that actinorhiza infecting *Alnus* root nodules might not belong to same phylogenetic clades as that in *H. salicifolia*. Nonetheless, it is evisent from the studies that there exists a diversified distribution of *Frankia* spp. in root nodules of *H. salicifolia* although their probable existence in soil or their possible role in sustenance of other plant species is still unknown and needs further exploration.

References

Akkermans, A. D. L., & Van Dijk C. 1981. Non-leguminous root-nodule symbioses with actinomycetes and *Rhizobium*, 57-103. In: W. J. Broughton (ed.) *Nitrogen Fixation* Vol. 1: Ecology. Clarendon Press, Oxford.

Ambay L. 1971. The Genus *Hippophae* L : "A taxonomic study'. In *Ann Bot. Fennici.*

Becking J. H. 1970. *Frankiaceae* fam. nov. (*Actinomycetales*) with one new combination and six new species of the genus *Frankia* Brunchorst 1886, 174. *International Journal of Systematic Bacteriology* 20, 201-220.

Benson D. R. 1982. Isolation of *Frankia* strains from alder actinorhizal root nodules. *Applied and Environmental Microbiology* 44, 461-465.

Benson D. R., & Silvester W. B. 1993. Biology of *Frankia* strains, actinomycete symbionts of actinorhizal plants. *Microbiological Reviews* 57, 293-319.

Bond G. 1983. Taxonomy and distribution of non-legume nitrogen-fixing systems, 55-87. In: J. C. Gordon and C. T. Wheeler (ed.) *Biological nitrogen fixations in forest ecosystems: foundations and applications*. Martinus Nijhoff / Dr W. Junk Publishers, The Hague.

Borneman, J., & Triplett, E.W. 1997 Molecular microbial diversity in soils from Eastern Amazonia: evidence for unusual microorganisms and microbial population shifs associated with deforestation. *Applied and Environmental Microbiology* 63, 2647-2653.

Brady, N.C. 1995, *The Nature and Properties of Soils.* Tenth Edition, Prentice-Hall of India Pvt. Ltd., New Delhi, pp-621.

Burggraaf A. J. P., Quispel, A., & Tak T., Valstar J. 1981. Methods of isolation and cultivation of *Frankia* species from actinorhizas. *Plant and Soil* 61, 157-168.

Campbell, R. 1983. Microbial Ecology. In: Wilkinson, J.F. (ed.). *Basic Microboiology*. Vol.5, Second Edition, Blackwell Scientific Publications, Oxford.

Clawson, M.L.; Caru, M. and Benson, D. R., 1998. Diversity of *Frankia* strains in root nodules of plants from the families *Elaeagnaceae* and *Rhamnaceae*. *Appl. Environ. Microbiol* 64, 3539-3543.

Coyne, M. 1999. *Soil Microbiology: An explanatory approach*. Delmer Publicsher, New York, pp-462.

Dawson J. O. 1986. Actinorhizal plants: their use in forestry and agriculture. *Outlook on Agriculture* 15, 202-208.

Dobritsa S. V., & Stupar O. S. 1989. Genetic heterogeneity among *Frankia* isolates from root nodules of individual actinorhizal plants. *FEMS Microbiologial Letters* 58, 287-292.

Hahn D.; Dorsch M.; Stackebrandt E., & Akkermans A.D.L. 1989a. Synthetic oligonucleotide probes for identification of *Frankia* strains. *Plant and Soil* 118, 211-219.

Jenkinson D. S. 1965. Studies on the decomposition of plant material in soils. *Journal of Soil Science* 16, 104-115.

Jenkinson, D.S., & Ladd, J.N. 1981. Microbial biomass in soil: measurement and turnover. *In:* Paul, E.A and.Ladd, J.N. (eds), *Soil Biochemistry Vol.* 5, Dekker, New York, 415-471.

Kent, M., & Coker, P. 1992. Vegetation description and analysis. International Book distributors, Dehra Dun, India, 363pp.

Knowlton S., & Dawson J. O. 1982. Effects of *Pseudomonas cepacia* and cultural factors on the nodulation of *Alnus rubra* roots by *Frankia*. *Canadian Journal of Botany* 61, 2877-2882.

Knowlton S.; Berry A., & Torrey J. G. 1980. Evidence that associated soil bacteria may influence root hair infection of actinorhizal plants by *Frankia*. *Canadian Journal of Microbiology* 26, 971-977.

Lalonde M., & Calvert H. E. 1979. Production of *Frankia* hyphae and spores as an infective inoculant for *Alnus* species, 95-110. In: J. C. Gordon, C. T.Wheeler and D. A. Perry (ed.). *Symbiotic nitrogen fixation in the management of temperate forests*. Forest Research Laboratory, Oregon State University, Corvallis.

Lalonde, M.; Calvert, H. E., & Pine, S. 1981. Isolation and use of *Frankia* strains in actinorhizae formation. In A. H. Gibson and W. E. Newton (eds) Symbiotic nitrogen fixation in the management of temperate forests. Forest Research Laboratory. Oregon State University, Corvallis.

Lopez, M. F.; Murry, M. A., & Torrey, J. G. 1986. Effect of oxygen on substrate utilization for nitrogen fixation and growth in *Frankia* sp. *Arch. Microbiol.* Vol 145: 209-214.

Lumini, E., and Bosco, M. 1996. PCR-Restriction Fragment Length Polymorphism identification and host range of single-spore isolates of the flexible *Frankia* strain UFI 132715. *Applied and Environmental Microbiology* 62 (8), 3026-3029.

Magurran, A.E. 2004. Measuring biological diversity. Blackwell publishing, UK.

Mansour, S.R.; Dewedar, A., & Torrey, J. G. 1990. Isolation, culture and behaviour of *Frankia* strain *HFPCgI4* from root nodules of *Casuarina glauca*. *Bot. Gaz.* 151(4), 490-496.

Mishra, R. S. 1968. Ecology Work Book. Oxford and IBB publishing, Co. Pvt. Ltd. New Delhi, pp.

Murry M. A.; Fontaine M. S., & Tjepkema J. D. 1984. Oxygen protection of nitrogenase in *Frankia* sp. HFPArI3. *Archives of Microbiology* 139, 162-166.

Prin, Y.; Maggia, L.; Picard, B.; Diem, H. G., & Goullet, P. 1991. Electrophoretic comparison of enzymes from 22 single-spore cultures obtained from *Frankia* strain ORS140102. *FEMS Microbiol. Lett* 77, 223-228.

Probanza, A.; Lucas J.A.; Acero, N., & Gutierrez Mañero F.J. 1996. The influence of native rhizobacteria on european alder (*Alnus glutinosa* (L.) Gaertn.) growth. *Plant and Soil* 182, 59-66.

Quispel, A. 1954a. Symbiotic nitrogen fixation in non-leguminous plants. I. Preliminary experiments on the root nodule symbiosis of *Alnus glutinosa* (L.) Gaertn. *Acta Botanica Neerlandica* 3, 495-511.

Racette, S., & Torrey, J. G. 1989. The isolation, culture and infectivity of a *Frankia* strain from *Gymnostoma papuanum* (Casuarinaceae). *Plant and Soil* 118, 165-170.

Rojas, N. S.; Perry, D.A.; Li, C.Y., & Friedman, J. 1992. Influence of actinomycetes on *Frankia* infection, nitrogenase activity and seedling growth of red alder. *Soil Biology and Biochemistry* 23, 1043-1049.

Rousi, A. 1965. Observation on the cytology and variation of European and Asiatic populations of *Hippophae rhamnoides*. *Ann. Bot. Fenn.* 2, 1-18.

Salo, K. 1989. The site requirements of seabuckthorn (*Hippophaë rhamnoides* L.) on some islands in the Gulf of Bothnia, Finland. Luonnon Tutkija 93 (2), 40-45.

Sarma, G.; Sen, A.; Verghese, R., & Misra, A. K. 1998. A novel technique for isolation and generation of single-spore cultures. *Can. J. Microbiol.* 44, 490-492.

Schwencke, J. 1991. Rapid, exponential growth and increased biomass yield of some *Frankia* strains in bufferd and stirred mineral medium (BAP) with phosphatidylcholine. *Plant Soil* 137, 37-41.

Schwintzer, C.R., & Tjepkema, J.D. 1990. The Biology of *Frankia* and Actinorhizal plants (eds). Academic Press Inc.Harcourt Brace Jovanovich Publishers, San Diego, California 92101. USA.

Singh, V. 1998. Geographical adaptation of seabuckthorn (*Hippophae L.*) in dry temperate Himalayas. In: *Proc. Int. Symp. On Seabuckthorn*, Ulan-Ude, Russia.

Sparling, G.P.; Ord, B.G., & Vaughan, D. 1981. Changes in microbial biomass and activity in soil amended with phinolic acid. *Soil Biology and Biochemistry* 13, 455-460.

Swift R.S., & Posner A.M. 1972. Nitrogen Phosphorous and sulpher content of humic acid fracsoined with respect to molecular weight. *Journal of Soil Sciences* 23, 50-57.

Tiwari, S.C.; Tiwari, B.K., & Mishra, R.R. 1991. Seasonal variation in microfungal population in pineapple (*Ananas comosus* L.) orchard soils. *Acta Botanica Indica* 19, 55-61.

Tiwari, S.C.; Tiwari, B.K., & Mishra, R.R. 1988b. Phosphatase activity in Pineapple (*Annas comosus* L.) orchard soils. *Journal of Soil Biology and Ecology* 8, 83-89.

Tjepkema, J. D.; Ormerod, W., & Torrey, J.G. 1981. Factors affecting vesicle formation and acetylene reduction (nitrogenase activity) in *Frankia* sp. CpI1. *Canadian Journal of Microbiology* 27, 815-823.

Wall, L.G. 2000. The Actinorhizal Symbiosis. *Journal of Plant Nutrition and Soil Science* 19, 167-182.

Weber, A.; Smolander A.; Nurmiaho-Lassila, E.L., & Sundman V. 1988. Isolation and characterization of *Frankia* strains from *Alnus incana* and *Alnus glutinosa* in Finland. *Symbiosis* 6, 97-116.

Yao, Y.C. Zhu. 1985. Investigation on natural resources of *Hippöphae rhamnoides* L. in West-Shanxi Plateau. Shansxi, *Science and Technology of Forestry (China)*, 4, 105-119.

Yao, Y., & Tigerstedt, P. M. A. 1995. Geographical variation of growth rhythm, height, and hardiness, and their relations in *Hippöphae rhamnoides*. *Journal of the American Society for Horticultural Science* 120(4), 691-698.

Zhonge, Z.; Murry, M. A., & Torrey, J. G., 1986. Culture conditions influencing growth and nitrogen fixation in *Frankia* sp. *HFPCcI3* isolated from *Casuarina*. *Plant Soil* 91, 3-15.

□□□

17.1

Neuro-chemical and Neuro-psychophysiological Modulation by "Seabuckthorn" with Special Reference to Senile Dementia of Alzheimer Type and Parkinson's Disease

Aruna Agrawal, G. Victor Rajamanickam and G.P. Dubey

CONTENTS

17.1.1 INTRODUCTION

Recently global attention has been directed towards the utilization of eco-friendly, bio-friendly plant based medical products to replace the synthetic chemicals in the prevention and management of various mental and physical disorders. Indian subcontinents rely on traditional medicine mainly based on plants. Traditional medicine could not get global acceptance due to lack of scientific evidence. Considering the facts, W.H.O. has provided guide lines for the validation of safety and efficacy profile of traditional form of practice.

The incidence of neuro-degenerative disorders is increasing throughout the world due to increase in life expectancy. Neuro-degeneration is an out come of aging process. Several environmental and genetic factors play a role in the early neuro-degeneration. The currently available synthetic drugs are inadequate, costly and have high risk profile. In view of the above, attempt has been made to evaluate the therapeutic potentials of seabuckthorn (*Hippophae rhamnoides*) in the prevention and management of Senile Dementia of Alzheimer Type (SDAT) and Parkinson disease. The test drug containing the hydroalcoholic extract of fruits and leaves of seabuckthorn was studied in animal models of Alzheimer's disease. It has been observed that seabuckthorn acts on cholinergic, nor-adrenergic and serotonergic receptors and thus, prevents the loss of cholinergic neurons following Colchicine induced neuro-toxicity. Further, a long-term study of this drug in human model demonstrated the beneficial role of this drug among cases suffering from (SDAT). In one of the series of investigations, the organic extract of seabuckthorn showed prevention of loss of dopaminergic and nor-adrenergic neurons following MPTP neurotoxin in experimental animals. In preliminary clinical trial, a significant clinical improvement like reduction in tremors and associated neuro-psychiatric problems were observed on treating with seabuckthorn based medicine. It is concluded that a long term application of plant based drug seabuckthorn may serve as useful remedial measure in the prevention of early brain aging, as well as prevention and management of neuro-degenerative disorders and associated problems.

Age associated neurological impairment can take a variety of forms including memory loss and senile dementia. Dementia is a general term for disease involving nerve cell deterioration. It is defined as a loss of complex behaviour including language memory, visual and spatial abilities. Dementia, the most serious form of age associated mental impairment, is often a slow, gradual process that may take months or even years to reach noticeable level. Symptoms vary according to the areas of the brain is affected (Cummings *et al.*, 1992). Dementia can be considered as a global impairment of intelligence, memory and personality. It can occur at any age but appear more frequent with advancing age, with a prevalence of 5-10% to those over 65 in age and 20% to those over 80s in age (Saunder, 1993). It is seen more frequent among women, due to their increased longevity. Dementia is an acquired, progressive impairment of overall mental function affecting memory, cognition and personality without any clouding of consciousness. Impairment of personality is the most distressing factor to those who cover them because the patient becomes a different person. Emotional liability is often seen (Chatterjee, 1992). Neuro-psychological evaluation and cognitive testing remain as the most effective differential diagnostic methods in discriminating pathophysiological dementia from age-related cognitive decline (La Rue, 1992; Lezak, 1995). The dementia is described as involving multiple progressive cognitive deficits among older persons in the absence of disturbance in the consciousness, presence of psychoactive substance or any other medical, neurological or psychiatric conditions (Rediess and Caine 1996; Smith *et al.*, 1996). In dementia, the rate of processing the information become slower. The patterns of thinking become regressive and rigid. Impairment of memory is well noticed and often wrongly considered to be the only symptom of a dementia. The short-term memory is affected first with preservation of long term memory some time. Poor concentration due to other problems such as depression or anxiety can give a false impression of severity of memory loss (Ferris and Kluger 1996; Kaszniak and Christenson, 1994; Larrabee, 1985; Ballard *et al.*, 1996). The principal cause of dementia is the degenerative condition of the brain known as Alzheimer's disease (Evans *et al.*, 1989). Formerly, when it affected people under 65 years, it was called as pre-senile dementia whereas to those over 65 as senile dementia. The second most common cause of vascular disease of the brain which goes under the term multi-infarct dementia, often referred to in the past as arteriosclerotic dementia. In population studies, higher dietary intake of fat and calories was associated with an increased risk for Alzheimer's disease, whereas high intake of fish oil rich in Omega-3 fatty acids, such as DHA was associated with a decreased risk (Grant 1997; Smith *et al.*, 1997; Kalmijn *et al.*, 1997; Miyanaga *et al.*, 1995). The main cause of Alzheimer's disorder is still not known, although the problem appears to involve abnormal breakdown of acetylcholine. Some studies suggest that it may be related to an accumulation of aluminium in the brain (Priest, 1993). People with higher levels of non-occupational activities, such as playing a musical instrument, gardening, physical exercise or playing board games were less likely to develop Alzheimer's disease later in life (Friedland, 2000).

Alzheimer's disease is the major cause of dementia in later life. Alzheimer's disease is characterized by severe, progressive problems in moving newly learned information into long-term memory (Gao *et al.*, 1998; Buschke *et al.*,

1999). Alzheimer's include a pattern of forgetfulness, short attention span, difficulty in performing routine tasks, language problems, disorientation, poor judgement problems with thinking, misplacing things, depression, irritability, paranoia, hostility and lack of initiative. Alzheimer develops progressive loss of memory and gradually loses the ability to function, and to take care of themselves (Albert and Drachman, 2000; Hebert *et al.*, 1995). Early identification of Alzheimer's disease provides an opportunity to enhance quality of life and long term care. Medical intervention, such as a trial of donepezil hydrochloride (Aricept), sulpiride and thioridazine or other memory-enhancing medications can be given when results are likely to be optimal (Copper *et al.*, 1991). Phosphatidylserine, which is related to lecithin, is a naturally occurring compound present in the brain. Studies showed 100 mg of phosphatidylserine taken three times per day has been shown to improve mental function, in the people with Alzheimer's disease, (Crook *et al.*, 1992; Engel *et al.*, 1992). In another double-blind trial, 200 mg of phosphatidylserine, taken twice daily has produced short term improvements in mental function among the patients with Alzheimer's disease (Heiss et al 1994). Research shows that higher blood levels of vitamin E correlate with better brain functioning in middle aged and older adults (Schmidt *et al.*, 1998). The possible protective effect of antioxidants may be explained by the observation that oxidative damage appears to play a role in the development of dementia (Lethem and Orrell 1997).

Keeping the facts in view, it was decided to validate the beneficial role of seabuckthorn, a high altitude growing plant which has shown a multi-dimensional utility particularly in the prevention and management of neuro-psychophysiological deterioration. In the present series the hydroalcoholic extract of fruits and leaves have been taken and are evaluated on WHO guidelines as regards its safety and efficacy profile is concern. An ethical approval has been obtained before commencing the studies both experimental and clinical. For clinical study, a prior informed consent of the subjects has been taken.

17.1.2 PRE-CLINICAL STUDY (EXPERIMENTAL MODEL)

Before going for clinical trails, the safety and efficacy profile of the drug seabuckthorn was evaluated. The LD50 and effective human dose was determined.

Acute and Sub-Acute Toxicity

The acute toxicity was studied in albino rats. The average weight of rats varied from 150-200 gm. The animals were introduced the drug in increasing manner of the doses upto 1000 mg/day. From the table, it is evident that there is no lethality even in highest dose of 1000 mg/day.

Acute Toxicity of Seabuckthorn

Part used (fruits): Experimental animal Charles Foster strain rats

Dose level mg/100 gm	Sample size	Mortality observed	
		After 24 hours	After 72 hours
100	10	0	0
250	10	0	0
500	10	0	0
750	10	0	0
1000	10	0	0

No lethality is found at the highest does of 1000 mg / 100 gm body weight.

Neurological Observations

Factors	Normal control	After 24 hours	After 72 hours
Flexing reflex	+	+	+
Grasping reflex	+	+	+
Corneal reflex	+	+	+
Auditory startle	+	+	+
Head shaking	+	+	+
Body equilibrium	+	+	+

Effect of Hydro-Alcoholic Extract of Seabuckthorn on Histamine and Histaminase Level to Determine Allergic Reaction of the Drug

Parameters	Normal control	After 24 hours	After 72 hours
Histamine (µg/ml)	0.196 ±0.025	0.187 ±0.029	0.193 ±0.027
Histaminase (µg/ml)	32.92 ±3.72	34.75 ±4.36	33.88 ±3.98

It is evident from the present experimental study that seabuckthorn did not demonstrate any lethality even after oral administration of 1000 mg/gm of the body weight. Similarly, the drug did not exert any allergic reaction following oral administration of seabuckthorn. The histamine level did not alter even after ten times higher effective doses.

Chronic Toxicity of Seabuckthorn

Chronic toxicity studies were carried out among Albino rabbits of Holland strain upto 12 months continuously and the effect of drug was assessed on various organ systems of the body. The organic (hydro-alcoholic) extract of seabuckthorn was 100 mg/ day.

Neurological Observations

Dose (100 mg / day): Animal model- Holland strain Rabbits

Factor	Normal control	After 3 months	After 6 months	After 12 months
Flexing reflex	+	+	+	+
Grasping reflex	+	+	+	+
Corneal reflex	+	+	+	+
Auditory startle	+	+	+	+
Head shaking	+	+	+	+
Body equilibrium	+	+	+	+

No neurological abnormality could be observed even after one year of oral administration of seabuckthorn in higher doses (100 mg/kg/day by oral route).

Effect of Drug on Hemopoetic Function

Factor	Normal control	After 3 months	After 6 months	After 12 months
Total WBC (Thousand / cu mm)	6870.55 ±1010.85	6685.64 ±1120.75	5975.88 ±985.72	6587.95 ±1127.55
Polymorph (%)	62.86 ±5.23	61.42 ±4.33	62.36 ±5.28	63.25 ±4.12
Lymphocytes (%)	28.75 ±4.85	27.26 ±3.28	30.52 ±2.86	28.77 ±3.25
Eosinophils (%)	5.19 ±1.25	5.26 ±1.35	4.46 ±1.26	5.28 ±1.32
R.B.C. (millions/cu mm)	2.36 ±0.44	2.43 ±0.38	2.53 ±0.40	2.51 ±0.42
Platelet count (lacs/cu mm)	1.38 ±0.25	1.49 ±0.28	1.42 ±0.26	1.46 ±0.23

No adverse effect on Hemopoetic system could be seen.

Chronic Toxicity Study of Seabuckthorn on Liver Function

Factor	Normal control	After 3 months	After 6 months	After 12 months
Alkaline phosphatase (kings Au/100 ml)	6.28 ±0.79	5.02 ±0.75	5.44 ±0.72	5.89 ±0.76
G.G.T.P. (I.U.)	21.32 ±4.45	19.75 ±5.56	21.42 ±3.78	19.69 ±4.20
S.G.O.T. (I.U.)	19.52 ±5.02	20.86 ±5.85	24.80 ±7.62	21.77 ±6.32
S.G.P.T. (I.U.)	24.80 ±3.05	18.52 ±5.85	17.85 ±4.55	20.92 ±4.82
Serum billirubin (mg %)	0.42 ±0.04	0.46 ±0.07	0.44 ±0.06	0.36 ±0.05
Total protein (gm %)	5.41 ±1.21	5.69 ±1.02	4.72 ±2.48	5.67 ±1.32

No adverse effect on liver function tests could be noticed.

Chronic Toxicity Study of Seabuckthorn on Renal Function

Factor	Normal control	After 3 months	After 6 months	After 12 months
Blood urea (mg%)	19.36 ±1.99	18.98 ±3.15	21.68 ±2.42	19.25 ±2.25
Serum creatinin (mg %)	1.64 ±0.22	1.56 ±0.24	1.34 ±0.25	0.98 ±0.23
Serum sodium (MEQ / L)	138.55 ±6.52	140.20 ±5.95	136.87 ±8.77	144.60 ±7.85
Serum potassium (MEQ / L)	4.20 ±0.17	4.45 ±0.18	4.60 ±0.35	4.55 ±0.55
Creatinin clearance (ml/min.)	52.85 ±4.77	-	-	54.75 ±6.85
Plasma volume (CC)	46.36 ±7.85	-	-	49.28 ±6.50

No toxic effect of the drug could be noticed on kidney function among the experimental rabbits.

Efficacy Profile: Study has been divided into two groups.

1. Experimental Study
2. Clinical Study

17.1.3 EXPERIMENTAL STUDY

Materials and Method

10 Albino rats of Charles Foster strain with an average body weight of 218.95±20.92 gm were taken for the experiment. The age of the rats varied from 6-8 weeks. The rats were housed in polypropylene cages at room temperature. All animals were kept in one room and no other species being housed with them. The room was well ventilated with fresh air and normal temperature. The animals had free access to palliated feed of standard composition. Aqua Guard water was provided ad-libitum. Animals were marked on their back with 0.5% picric acid. The animals were distributed randomly into different groups as per protocol.

Doses

Name of the drug	Prescribed effective dose for human consumption	Experimental dose/kg/day average Indian weight 50 kg
Seabuckthorn (Part used fruits leaves)	Hydro-alcoholic extracts of the drug 900 mg/day	18 mg/kg

Following tests were performed to assess the beneficial role of test drug on Alzheimer's disease.

Experimental Design

The study was divided into four groups :

Group-I: This group served as normal control. In this group the artificial cerebrospinal fluid (ACSF) in the dose of 5 ml to all the 10 rats were given continuously for 14 days.

Group-II: 10 rats of this group were treated with neurotoxin colchicine in the dose of 15 μgm through intra-cerebroventricular (ICV) administration, following the method described by Emerich and Walsh (1990).

Group-III: The test drug prepared out of extract of the plant Seabuckthorn (fruits and leaves) 8 mg/kg was administered for a period of 14 days.

Group–IV: This group of rats were treated with colchicine along with test drug (seabuckthorn 18 mg/kg).

Biochemical Analysis

Dopamine estimation was done by method developed by Chang (1964). Acetylcholine, nor-adrenaline and 5-HT estimations were done by fluorometric method of Speeg (1974). ChAT activity was determined by the radiometric method of Haba *et al.*, (1988). Active avoidance test was performed by method developed by Sen and Bhattacharya (1991). At the end of 14 days of treatment the animals were sacrificed and the following enzymatic, biochemical and anti-oxidant profiles were studied. The anti-oxidant profiles to the brain, heart, liver and kidney were also determined.

Brain Dissection

Rats were decapitated by cervical dislocation after anesthetization with other. Brain of each animal was quickly removed and placed on a glass plate over ice and dissected as described by Glowinski and Iverson (1966). The regions were subjected to biochemical and histopathological studies.

Soft Tissues Dissection

Liver, lung, kidney and heart were dissected for biochemical and histopathological studies.

Table 17.1.1: Changes in acetylcholine concentration with administration of seabuckthorn- an experimental study

Treated groups (mg/kg po)	Acetylcholine concentration (n mol/gm.)			
	Frontal cortex		Hippocampus	
	Day 7	Day 14	Day 7	Day 14
Vehicle (ACSF) * (N=10)	22.84 ±3.45	26.92 ±5.06	30.44 ±4.98	28.94 ±3.75
Treated with Colchicine * * (N=10)	11.84 ±2.11	6.92 ±1.82	15.87 ±2.65	12.99 ±2.87
Treated with seabuckthorn * * * (N=10)	24.94 ±3.45	26.22 ±4.04	29.84 ±3.11	31.75 ±4.14
Treated with Colchicine + seabuckthorn * * * *(N=10)	17.87 ±2.85	16.92 ±3.01	23.97 ±3.75	22.20 ±2.97

Comparison:

* vs **	$p<0.001$	$p<0.001$	$p<0.001$	$p<0.001$
** vs ***	$p<0.001$	$p<0.001$	$p<0.001$	$p<0.001$
*** vs ****	$p<0.001$	$p<0.01$	$p<0.01$	$p<0.01$

Table 17.1.2: Effect of seabuckthorn on ChAT activity in experimental rats

Treated groups (mg/kg po)	ChAT activity (m mol /mg protein/hour)			
	Frontal cortex		Hippocampus	
	Day 7	Day 14	Day 7	Day 14
Vehicle (ACSF) * (N=10)	21.84 ±3.01	23.97 ±4.11	22.05 ±2.92	19.85 ±2.82
Treated with Colchicine * * (N=10)	13.82 ±2.02	10.75 ±1.68	8.92 ±1.72	6.08 ±1.01
Treated with Seabuckthorn * * * (N=10)	20.85 ±3.12	22.70 ±2.91	18.22 ±2.75	19.75 ±2.10
Treated with Colchicine + seabuckthorn * * * * (N=10)	15.58 ±2.33	16.93 ±3.11	14.88 ±3.14	13.87 ±2.91

Comparison:

* vs **	p<0.001	p<0.001	p<0.001	p<0.001
** vs ***	p<0.001	p<0.001	p<0.001	p<0.001
*** vs ****	p<0.01	p<0.05	p<0.05	p<0.01

Table 17.1.3: Effect of seabuckthorn on nor-adrenaline level among experimental rats

Treated groups (mg/kg po)	Nor-adrenaline (μ gm/gm tissue)			
	Frontal cortex		Hippocampus	
	Day 7	Day 14	Day 7	Day 14
Vehicle (ACSF) * (N=10)	0.264 ±0.032	0.271 ±0.041	0.262 ±0.050	0.265 ±0.036
Treated with Colchicine ** (N=10)	0.123 ±0.022	0.106 ±0.028	0.148 ±0.031	0.102 ±0.026
Treated with Seabuckthorn * * * (N=10)	0.271 ±0.048	0.284 ±0.038	0.275 ±0.037	0.288 ±0.041
Treated with Colchicine + Seabuckthorn* * * *(N=10)	0.152 ±0.030	0.164 ±0.034	0.150 ±0.036	0.162 ±0.028

Comparison:

* vs **	p<0.05	p<0.01	p>0.05	p<0.05
** vs ***	p<0.001	p<0.05	p>0.05	p>0.05
*** vs ****	p<0.05	p<0.01	p<0.05	p<0.001

Table 17.1.4: Effect of seabuckthorn on serotonin level among experimental rats

Treated groups (mg/kg po)	Serotonin (μ gm/gm tissue)			
	Frontal cortex		Hippocampus	
	Day 7	Day 14	Day 7	Day 14
Vehicle (ACSF) * (N=10)	0.432 ±0.036	0.442 ±0.020	0.315 ±0.024	0.327 ±0.021
Treated with Colchicine ** (N=10)	0.289 ±0.011	0.221 ±0.014	0.240 ±0.017	0.210 ±0.015
Treated with Seabuckthorn * * * (N=10)	0.489 ±0.032	0.479 ±0.027	0.392 ±0.036	0.385 ±0.030
Treated with Colchicine + Seabuckthorn * * * *(N=10)	0.354 ±0.021	0.362 ±0.024	0.288 ±0.022	0.268 ±0.019

Comparison:

* vs **	$p<0.05$	$p<0.01$	$p>0.05$	$p<0.05$
** vs ***	$p<0.001$	$p<0.05$	$p>0.05$	$p>0.05$
*** vs ****	$p<0.05$	$p<0.01$	$p<0.05$	$p<0.001$

Table 17.1.5: Effect of Seabuckthorn on retention of active avoidance learning among experimental rats

Treated group (mg/kg po)	Retention of active avoidance learning (scores)	
	Day 7	Day 14
Vehicle (ACSF) * (N=10)	7.11 ±1.02	6.82 ±1.06
Treated with Colchicine * * (N=10)	13.03 ±2.13	15.16 ±2.12
Treated with Seabuckthorn * * * (N=10)	6.58 ±1.04	5.21 ±0.92
Treated with Colchicine + Seabuckthorn * * * *(N=10)	10.31 ±1.30	8.94 ±1.04

Comparison:

* vs **	$p<0.001$	$p<0.001$
** vs ***	$p<0.001$	$p<0.001$
*** vs ****	$p<0.001$	$p<0.001$

Methods for Anti-Oxidant Estimation

Catalase estimation was determined by the colorimetric assay of Asruk Sinha (1972). Lipid peroxides was done by the method developed by Hiroshi Ohkawa *et al.,* (1979). Glutathione was done by the method developed by Maria Moran *et al.,* (1979). The results obtained were compared with each group.

17.1.4 RESULTS AND OBSERVATIONS

A neuromodulation activity of seabuckthorn is observed in the present experimental trial. Neurotoxin Colchicine produced marked damage to cholinergic, nor-adrenergic, serotonergic-receptors. The test drug significantly prevented the loss of acetylcholine and ChAT in comparison to colchicine treated animals (Table 17.1, 17.2). Further, the nor-adrenaline and serotonin when measured in different parts of rat brain exerted less damage of the receptors than the only colchicine treated animal group (Table 17.1.3 & 17.1.4). The animals of treated group also showed better learning abilities, as observed from the number of trials reduced significantly to reach to given task in test drug treated animals (Table 17.1.5). In another series of animal, it is noticed that the drug Seabuckthorn has potentiality to check the further loss of acetylcholine. The aged animals when treated with the organic extract of seabuckthorn, the decline in acetylcholine level is significantly lesser than the control group (Table 17.1.6, Fig. 17.1.1). The test drug has shown potential role in the prevention and management of neurodegenerative changes and has beneficial effect in the prevention of early brain aging.

Table 17.1.6: Effect of Seabuckthorn on acetylcholine level among the aged animals

Treated Group		Acetycholine concentration (n mol / gm.)				Comp. Initial vs after 3 weeks
		Initial	After 1 Week	After 2 weeks	After 3 weeks	
Normal control group (N=3)	Frontal cortex	18.45 ±2.14	18.20 ±3.04	16.29 ±3.06	13.92 ±2.44	P<0.05
	Hippo campus	21.72 ±2.28	20.91 ±2.40	17.42 ±2.06	15.11 ±1.92	P<0.01
Treated with test drug (N=3)	Frontal cortex	17.88 ±2.02	18.06 ±3.10	18.72 ±2.53	18.64 ±2.75	P>0.05
	Hippo campus	20.78 ±2.04	20.94 ±3.01	22.75 ±2.11	22.64 ±2.06	P<0.01

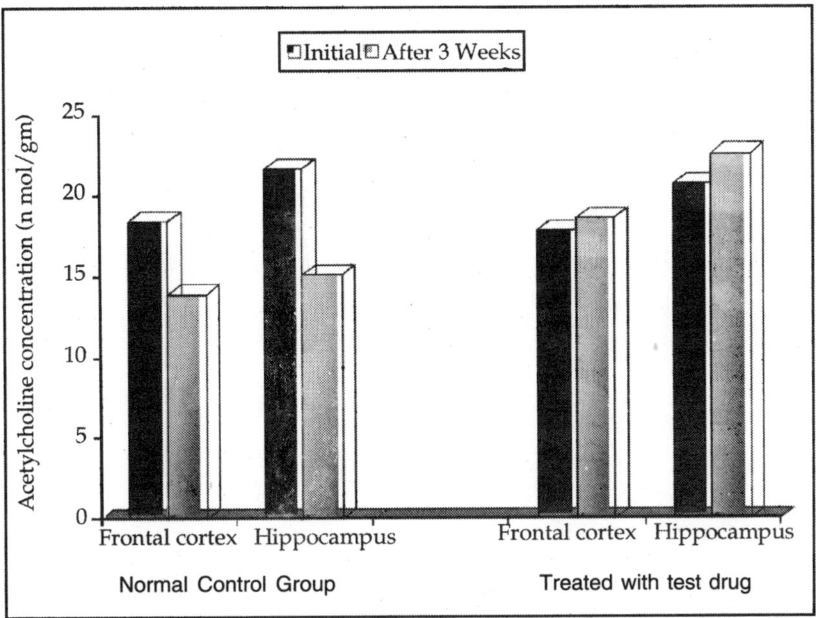
Initial
After 3 Weeks
25
20
15
10
5
0
Acetylcholine concentration (n mol/gm)
Frontal cortex
Hippocampus
Frontal cortex
Hippocampus
Normal Control Group
Treated with test drug

Fig. 17.1.1

□□□

17.2

Modulation of Scholastic Performance of Protein Energy Malnourished (PEM) Children by Seabuckthorn

Abhilasha Sharma, Aruna Agarwal, G. Victor Rajamanickam and G. P. Dubey

CONTENTS

17.2.1 INTRODUCTION

A large majority of children population in the developing countries suffer from undernutrition or malnutrition. Malnutrition is a pathologic condition, which can manifest into over or undernutrition. This condition rises due to deficiency of protein and calorie in diet. Protein Energy Malnutrition (PEM) is one of the most important wide spread nutritional problems among pre-school or school going children of the country. Several workers noticed that nearly two out of three pre-school children in India are malnourished. Growth is influenced by nutrition. The primary cause of malnutrition is inadequate and faulty diet (Beaton Bengoa, 1976; Reddy *et al.* 1993). Conversely PEM may increase susceptibility to severity of infection (Scrimshaw, 1975; Levistsky, 1979; Alleyne *et al.* 1981; Connolly Kvalsiva , 1993). Human brain development depends upon the interaction between the genetic and environment factors like nutrition, stimulation and altitude of the parents. Several finding indicate that behaviours and cognitive functions are impaired by malnutrition. Malnutrition imposed early life not only reduces the growth of the brain but it permanently become smaller in size (Dobbing , 1964; Culley and Uneberges, 1968; Chase *et al.* 1967; Zamenhor *et al.* 1971, Randt and Derby, 1973; Dickerson and Dobbing, 1966). PEM is closely related to the survival, physical and mental development, good health and the overall well-being of the child. Several researchers have found the cerebral cortex, the brain region most closely linked to cognitive and intellectual functioning, exhibiting a reduction in volume and reduction in width after neo-natal malnutrition (West and Kemper, 1976; Morgane *et al.* 1978; Thomas *et al.* 1979, Leuba and Rabinowicx, 1979 a and b; Bedi and Bhide 1988; Dobbing *et al.*, 1971; Cragg, 1972; Saiesi and Saissi, 1973, Noback and Eisenman, 1981).

Sufficient evidence indicates that the malnutrition early in life causes clear delays in psychomotor development in young and malnourished animals (Lynch *et al.* 1975, Oliverio *et al.* 1975, Elias and Samonds; Celedon *et al.* 1980, Reys *et al* 1990, Benefice 1992, Hoorweg & Stanfield 1976, Stoch *et al.* 1982; Galler *et al;* 1984; Galler *et al;* 1985; Galler *et al.* 1987). Various etiological factors like insufficiency of diet, inability to take food properly, lack of digestion, impaired absorption, faulty metabolism of food, etc., have been described causing PEM. PEM impairs mental and cognitive development of the children and also affects Intelligence quotient (IQ), poor school achievement, memory loss, learning disabilities behavioural and emotional problems (Klein *et al.* 1975; Cremen, 1982; Tanuer, 1990). Grantham- Gregor *et al.*, 1992, 2001; Mansell, 1993; Ellis *et al.*, 1975, Horrweg *et al.*, 1976; Evan *et al.*, 1980; Pollitt, 1994). Language scores on development are poorest in most malnutrition status (Winik, 1972; Popkin *et al.*, 1982). Wohl and Goodhart (1968) found infant who are fed protein deficient diet, has retarded physical and mental growth wish anemia, hypo-proteinemia and fatty liver. He has found that kidneys are also damaged resulting in increased excretion of amino-acid in urine. Malnourished children show a deficit in tests of cognitive function or intelligence. Motor skills are generally found to be affected (Hoorweg and Stanfield, 1976; Galler *et al.* 1987; Cravioto and Arri, 1986). PEM children are observed to play with toys for shorter periods and stay closer to their mother (Grantham-McGregor *et al.* 1989) and to be more unresponsive when given a task (Cravito and Arri *et al.* 1986). Malnourished children were often found to have poorer school grades than matched controls (Richardson *et al.* 1973; Pereira *et al.* 1979; Galler *et al.*, 1990). It is reported that school-age children who suffered from early childhood malnutrition have generally been found to have poorer IQ levels, cognitive function, school achievement and greater behavioural problems than matched controls. The disadvantage remains at adolescence (Floud, 1992; Wachs 1995; Grantham-Mc Gregor, 1995; Mahadeviah and Krishna, 1977; Groman, 1995; Benton, 1998; Wainwright, 2000).

Several studies have shown that the maternal malnutrition can affect significantly the intelligence and learning capacity of children. (Donmann *et al.* 1997; Mendez and Adur, 1999; Singh, 2003; Brockman and Rieciuti 1971; Guesry, 1998).

Early malnutrition may produce long term alteration in brain function by altering neurotransmitter metabolism. Most investigators have demonstrated that either pre or post-natal malnutrition causes an increase in brain concentration of monoamines, serotonin and nor-epinephrine (Benesova *et al* 1972, Burns and Brown, 1977; Miller *et al.* 1977, Sobotka *et al.* 1974, Stern *et al.*, 1974, 1975). A decrease in monoamines is also observed by several workers (Deterings *et al.*, 1980 a,b,c; Hisatomi and Nilyama, 1980; Lec and Dubos, 1972; Seidler *et al.*, 1990). In most cases, brain level of these amines are normalized after nutritional rehabilitation (Burns and Brown, 1977; Detering *et al.* 1980 a, b, c). Cortisol levels are also elevated in PEM and are inversely related to growth velocity (Lunn *et al.* 1979). Studies have shown that chronic PEM has also affected the other body organ like liver (Feldman *et al.* 2002), gastrointestinal tract, (Gracey *et al.*, 1977) etc. Malnourished children may have diminished intra-vascular volumes with presence of whole body fluid over load (Patrick *et al.* 1978). Malnutrition cause structural and functional deterioration of the intestinal tract (Brunser *et al.* 1976). Malnutrition affects kidney (Klahr *et al.* 1973). Severe malnutrition suppresses bone marrow, red blood cell and white blood cell production leading to anemia, leucopenia (Rosen, 1975) and lymphocytopenia. Muscle function is impaired by malnutrition because of both a loss of muscle mass and impaired metabolism, (Montgomery, 1962). Severe undernutrition causes diminished anti body production and atrophy of all lymphoid tissues including thymus, tonsils and lymph mass (Good *et al.*, 1976; Dela Fuente, 1991). In ADHD children learning and behavioural problems are more common. The most common clinical feature observed among ADHD case are hyperactivity, perceptual motor impairment, emotional liability, general co-ordination deficit, disorder of attention, impulsivity, disorder of memory, thinking, specific learning disabilities, disorder of speech and hearing, equivocal neurological signs and EEG irregularities (Kaplan and Sadock, 1995). Children with ADHD are at higher risk of anti-social personality as they become adult. (Pliszka, 2000). ADHD appears to be a most important predictor of poor psychosocial functioning including affective anxiety and learning disorder in early adulthood (Rasmussen and Gillberg, 2000).

Children with ADHD were sleepier during the day time and they had a longer reaction time with a deficit in alertness (Lecendrex, 2000). The greatest risk for development of antisocial behaviour by late teenage years is the maintenance of ADHD (Arnold *et al.* 2000). The proper management of ADHD needs medication, diet supplement, behavioural modification etc. The use of tricyclic anti-depressants and noradrenergic specific anti-depressant have also been used. Anti-cholinesterase inhibitors and nortryptyline are effective in reducing the sign and symptoms of ADHD (Biederman and Spencer, 2000; Prince *et al.* 2000; King *et al.* 2000). But the common side effects of stimulants are anorexia, insomnia, abdominal discomfort, bowel upset, loss of appetite (Andress, 1984) increased heart rate and blood pressure.

Keeping the above facts, in view the present study has been planned to investigate the role of seabuckthorn prepared as food supplement in the cognitive development among the PEM children, and children suffering from ADHD. The potential role of seabuckthorn has been evaluated on various neuropsychophysiological parameters applicable in PEM and ADHD cases.

17.2.2 MATERIALS AND METHODS

A short term study has been launched with the view to promote the health status of children showing PEM. A health supplement (Food Supplement) prepared out of the dried fruits of seabuckthorn and given to a group of selected children of PEM and ADHD with the object to assess its beneficial role in improving the protein energy malnutrition and ameliorating the cognitive function of such children. Under this study, 46 children of both sexes show evidence of mild to moderate PEM and 21 children suffering from attention deficit hyper activity disorder (ADHD) have been selected and given a food supplement prepared out of the hydro alcoholic extract of seabuckthorn for a period of three months. The following parameters were applied for the present study:

1. Antheropometric measurement: BMI [weight (kg), height(m)^2]
2. Biochemical parameters: It includes
   - The semi-auto analyzer have been used for assessment of total and differential protein.
   - Serotonin have measured by method developed by Snider *et al.* (1965).
   - Plasma cortisol have determined by method developed by Mattingly (1962).
3. Psychological parameters: various psychological batteries or scales were applied.
   - Tachistoscope instrument is used to measure the attention span.
   - Human maze learning apparatus is used to determine the learning ability of the children, which is based on trial and error principle.
   - Short term memory span is measured by the method developed by Peterson *et al.* (1969), and
   - Long term memory span score is obtained by the scale provided by Chaudhary (1978).
4. Electrophysiological parameters:
   - Multi channel EEG analyzers are used where there is facility to filter the major brain wave.
   - Occipito frontalis muscle action potential is performed by using EMG biofeedback apparatus (Medicaid System 4000).
   - The autonomic over-reactivity is measured by GSR electronic device (Medicaid System 2000).

17.2.3 RESULTS

Seabuckthorn when orally administered in food supplement form to the children of PEM and ADHD is show a significant effect on various psychological symptoms like aggressiveness, hyperactivity, restlessness, distractive behaviour and accident proneness. Hyperactivity also reduced in almost all the cases including ADHD. Other behavioural parameters like restlessness, destructive behaviour and accident proneness also showed significant improvement indicating the effect of seabuckthorn in the modulation of behavioural pattern (Table 17.2.1).

The total protein was found low and a hypoproteinemia was noticed particularly in ADHD cases. After 3 months of seabuckthorn supplement a significant improvement was observed in almost all the groups indicating the improvement in nutritional status under influence of medical health supplement. The children showing evidence of ADHD have shown a significant low values of serum total protein which has increased following 3 months of seabuckthorn therapy (Table 17.2.2 & 17.2.3).

On electrophysiological assessments the brain frequencies like alpha beta and Galvenic skin resistance showed a distinct pattern when performed in PEM and ADHD cases. Almost all the cases of this group of study have revealed a diminished alpha pattern with high beta activity. Similarly, from initial GSR values it is evident, that children of PEM and ADHD have an autonomic over reactivity due to varying reasons which have also reduced following seabuckthorn health supplement therapy, evident from the increase in GSR values. The poor attention span is found to be associated with PEM as well as ADHD cases. The children suffering from ADHD have exhibited a marked deteriorated attention span which has increased significantly after 3 months of Seabuckthorn oral administration. In other two PEM groups also improvement in span of attention has been recorded (Table 17.2.4 & 17.2.5).

A high level of plasma cortisol and circulating 5 HT is also found to be associated with the present group of cases. ADHD cases have exhibited higher values of brain biogenic amines than the mild to moderate PEM cases which have regulated and decreased, following Seabuckthorn therapy (Table 17.2.6).

A memory enhancing property of seabuckthorn has been noticed when the authors conducted the present clinical study in PEM and ADHD subjects. Both the short and long term memory span have enhanced under the novel health supplement, which is found deteriorated before therapy due to malnourishment and various other reasons (Table 17.2.7).

Further, the learning abilities of the children improved following 3 months of therapy which is evident from the task performed on human maze learning by the children under study (Table 17.2.8). On the whole it has been observed that PEM is a major causative factor in deterioration of the cognitive functions particularly, memory, attention and learning abilities of the children. The plant seabuckthorn has exerted beneficial effects in improving both malnourishment and cognitive function as well.

Table 17.2.1: Beneficial role of seabuckthorn on various psychological problems associated with the children of PEM & ADHD

Groups	No. of Cases	Psychological Symptoms in PEM (Improvement In Percent)									
		Aggressiveness		Hyperactivity		Restlessness		Distinctive behavior		Accident Prove	
		BT	AT	BT	AT	BT	AT	BT	AT	BT	AT
Mild PEM + Seabuckthorn (Food Supplement)	25	13 (52.00)	11 (44.00)	14 (56.00)	9 (36.00)	15 (60.00)	10 (40.00)	13 (52.00)	9 (36.00)	11 (44.00)	8 (32.00)
Moderate PEM + Seabuckthorn (Food Supplement)	14	11 (78.57)	9 (64.28)	11 (78.57)	7 (50.00)	9 (64.28)	7 (50.00)	12 (85.71)	8 (57.14)	9 (64.28)	6 (42.85)
ADHD + Seabuckthorn (Food Supplement)	20	19 (95.00)	16 (80.00)	20 (100.00)	17 (85.00)	18 (90.00)	16 (80.00)	17 (85.00)	13 (65.00)	16 (80.00)	12 (60.00)

BT= Before Treatment
AT= After 3 Months Treatment

Table 17.2.2: Improvement in BMI following seabuckthorn supplement among PEM and ADHD children

Groups	No. of Cases	Initial			After 3 Months of Treatment		
		Height (cm)	Weight (kg)	BMI	Height (cm)	Weight (kg)	BMI
Mild PEM + Seabuckthorn (Food Supplement)	25	139.84 ±4.28	32.88 ±5.86	16.54	140.32 ±3.82	35.28 ±3.70	17.85
Moderate PEM + Seabuckthorn (Food Supplement)	14	137.28 ±3.29	29.28 ±4.10	15.62	138.71 ±3.60	32.64 ±2.95	16.90
ADHD + Seabuckthorn (Food Supplement)	20	138.40 ±4.15	27.45 ±3.20	14.39	139.25 ±2.80	31.35 ±2.38	16.04

Table 17.2.3: Improvement in serum total protein, AG ratio following seabuckthorn as food supplement among PEM and ADHD cases

Groups	No. of Cases	Initial			After 3 Months of Treatment		
		Albumin (g/dl)	Globulin (g/dl)	Total Protein (g/dl)	Albumin (g/dl)	Globulin (g/dl)	Total Protein (g/dl)
Mild PEM + Seabuckthorn (Food Supplement)	25	3.02 ±1.06	2.98 ±0.95	5.70 ±1.22	4.32 ±0.85	2.05 ±0.98	6.32 ±1.58
Moderate PEM+ Seabuckthorn (Food Supplement)	14	3.56 ±1.02	1.76 ±0.80	5.10 ±1.05	3.66 ±0.65	1.88 ±0.95	5.58 ±1.60
ADHD + Seabuckthorn (Food Supplement)	20	2.98 ±1.18	1.82 ±0.69	4.78 ±1.40	3.10 ±0.50	1.97 ±0.82	5.10 ±1.68

Table 17.2.4: Effect of seabuckthorn when given as food supplement on brain frequencies and autonomic reactivity among PEM and ADHD cases

Groups	No. of Cases	Alpha frequency (Hz)		Comp. Initial vs after 3 months of therapy	Beta frequency (Hz)	
		Initial	After 3 months of therapy		Initial	After 3 months of therapy
Mild PEM + Seabuckthorn (Food Supplement)	25	7.92 ±1.35	11.25 ±1.85	P<0.001	28.65 ±4.70	22.02 ±3.03
Moderate PEM + Seabuckthorn (Food Supplement)	14	6.25 ±2.62	9.20 ±1.68	P<0.01	26.72 ±5.80	19.62 ±3.52
ADHD + Seabuckthorn (Food Supplement)	20	5.34 ±1.65	8.20 ±1.95	P<0.001	32.75 ±4.98	25.26 ±3.76

Table 17.2.5 : Effect of health supplement prepared out of seabuckthorn on attention span and among PEM and ADHD cases

Groups	No. of Cases	GSR (khoms)		Comp. Initial vs after 3 months of therapy	Attention span (Score out of 30 slides)		Comp. Initial vs after 3 months of therapy
		Initial	After 3 months of therapy		Initial	After 3 months of therapy	
Mild PEM + Seabuckthorn (Food Supplement)	25	158.79 ±18.56	188.85 ±19.85	P<0.01	7.84 ±1.31	8.76 ±1.04	P<0.01
Moderate PEM+ Seabuckthorn (Food Supplement)	14	154.82 ±18.80	204.96 ±17.96	P<0.01	5.72 ±0.97	6.38 ±0.86	P<0.05
ADHD + Seabuckthorn (Food Supplement)	20	146.85 ±19.26	184.86 ±18.25	P<0.001	4.34 ±0.83	6.01 ±0.34	P<0.001

Table 17.2.6: Effect of Seabuckthorn on Serotonin and Plasma cortisol among children showing different degree PEM.

Groups	No. of Cases	Serotonin (µg/ml)		Comp Initial vs after 3 months of therapy	Plasma cortisol (µg/ 100 ml)		Comp Initial vs after 3 months of therapy
		Initial	After 3 months of therapy		Initial	After 3 months of therapy	
Mild PEM + Seabuckthorn (Food Supplement)	25	0.22 ±0.08	0.19 ±0.09	P<0.05	15.84 ±2.64	11.45 ±2.04	P<0.001
Moderate PEM+ Seabuckthorn (Food Supplement)	14	0.26 ±0.08	0.21 ±0.06	P<0.05	17.28 ±3.14	9.02 ±2.19	P<0.001
ADHD + Seabuckthorn (Food Supplement)	20	0.34 ±0.10	0.28 ±0.09	P<0.05	21.35 ±3.45	16.45 ±2.72	P<0.001

17.2.4 DISCUSSION

Protein Energy Malnutrition (PEM) is one of the major causàtive factors for cognitive deficits particularly, learning disabilities and poor attention among the children. Apart from genetic factor, various socio-cultural and ecological settings are responsible for PEM (Rutter *et al.*, 1980; Lango *et al.* 1989).

In a series of controlled studies, it is observed that children's scholastic ability is strongly influenced by protein rich food because it stimulates brain development. In the preset study, the authors have also observed that PEM significantly reduced the child's cognitive development in terms of memory and attention span.

The management of PEM depends on Nutritional status, personal hygiene and nutritional education of the mother. The main goals are to maintain body mass and to encourage physical and mental growth of the PEM children. Several synthetic drugs like, methyl phenidate, amphetamines and tricyclic anti-depressant are used for upliftment of mental health status of PEM children, but continuous use of these substances have shown adverse effects which manifest in term of nervousness, anorexia, increased heart rate, blood pressure, etc.

Table 17.2.7: Effect of seabuckthorn (health supplement) on short term memory span among PEM and ADHD cases.

Groups	No. of Cases	STM (Score)								LTM (Score)					
		Initial			After 3 months of treatment			Comp. Initial vs after 3 months of therapy		Initial		After 3 months of treatment		Comp. Initial vs after 3 months of therapy	
		3 Sec.	9 Sec.	18 Sec.	3 Sec.	9 Sec.	18 Sec.	9 Sec.	18 Sec.	Repetition of word 1 time	Repetition of word 4 time	Repetition of word 1 time	Repetition of word 4 time	Repetition of word 1 time	Repetition of word 4 time
Mild PEM + Seabuckthorn (Food Supplement)	25	5.12 ±2.85	3.04 ±1.62	2.12 ±1.40	6.08 ±1.45	3.24 ±1.82	2.32 ±1.36	P<0.05	P<0.05	2.58 ±1.20	3.23 ±0.65	2.78 ±1.30	3.00 ±0.38	P<0.05	P<0.05
Moderate PEM+ Seabuckthorn (Food Supplement)	14	4.14 ±2.40	2.93 ±0.98	1.93 ±0.86	4.36 ±1.98	3.35 ±0.85	2.21 ±0.64	P<0.05	P<0.05	2.01 ±1.34	2.92 ±1.12	2.45 ±0.95	3.01 ±0.82	P<0.05	P<0.05
ADHD + Seabuckthorn (Food Supplement)	20	3.35 ±1.26	2.45 ±0.96	1.35 ±0.62	3.80 ±1.82	2.95 ±0.98	2.05 ±0.76	P<0.05	P<0.05	1.98 ±0.40	2.32 ±0.24	2.10 ±0.82	2.65 ±0.20	P<0.05	P<0.05

Table 17.2.8: Effect of seabuckthorn (health supplement) on learning ability among PEM and ADHD cases.

Groups	No. of Cases	Human Maze Learning															
		Errors (Score) BT			Error (Score) AT			Comp. Initial [illegible] after 3 months of therapy		Time taken (in Sec) BT			Time taken (in Sec) AT			Comp. Initial vs after 3 months of therapy	
		1st Trial	2nd Trial	5th Trial	1st Trial	3rd Trial	1st Trial	1th Trial	5th Trial	1st Trial	3rd Trial	5th Trial	1st Trial	3rd Trial	5th Trial	1th Trial	5th Trial
Mild PEM + Seabuckthorn (Food Supplement)	25	23.75 ±4.25	18.75 ±2.82	11.78 ±1.96	19.68 ±3.82	15.92 ±2.35	9.82 ±1.87	P<0.05	P<0.05	3.30 ±1.35	2.50 ±1.20	2.38 ±1.37	3.03 ±1.15	2.38 ±1.20	2.13 ±1.28	P<0.05	P<0.05
Moderate PEM + Seabuckthorn (Food Supplement)	14	26.84 ±3.96	22.82 ±4.80	16.68 ±3.75	21.85 ±3.15	19.78 ±2.82	13.52 ±2.75	P<0.05	P<0.05	3.80 ±1.40	2.58 ±1.38	2.42 ±1.08	2.57 ±1.18	2.30 ±1.20	2.05 ±1.32	P<0.05	P<0.05
ADHD + Seabuckthorn (Food Supplement)	20	30.82 ±4.85	26.64 ±3.82	21.85 ±2.95	25.85 ±3.82	22.80 ±2.75	19.65 ±1.68	P<0.05	P<0.05	4.25 ±1.25	4.18 ±1.30	3.58 ±1.20	3.52 ±1.35	3.40 ±1.05	3.21 ±1.26	P<0.05	P<0.05

BT= Before Treatment AT= After 3 Months Treatment

ADHD demonstrates the complex multi-dimensional links between neurobiology, psychology and behaviour (Vance and Luk 2000). The management of PEM induced ADHD are difficult after completing the latency period. Several methods have been advocated to manage the children of ADHD for a time period only. In this present study, ADHD children shown very poor attention span with high beta frequency and low values of GSR indicating inattentiveness, hyperactivity and impulsive behaviour of children. Biochemically, they also exhibited low protein level, high level of plasma cortisol and serotonin level.

In Ayurvedic system of medicine, a group of drugs have been prescribed which come under Medhya-rasayana drugs to ameliorate the cognitive function and overall mental performance. In the present study, seabuckthorn is given as a health supplement to those children who have shown evidence of mild to moderate PEM and ADHD. Hydro-alcoholic extract of seabuckthorn when orally administered to ADHD and PEM children has displayed a gradual improvement in cognitive deficit particularly, in attention span. The psychomotor performance is also improved under oral administration of the drug. Fetal and childhood brain development require plasma and cerebral amino acids at their normal concentration. An adequate amount of vitamins and minerals is also essential for normal brain function (Halliwell and Gutteridge, 1999; Kuhne *et al.*, 1991, Adams and Victor 1993). Seabuckthorn, which is rich source of micro-nutrients like vitamins (Singh *et al.*, 1995), organic acids, minerals, proteins and free amino acids (Chen, 1988; Zhang *et al.*, 1989) exerted beneficial effect in improving the cognitive function of the children. Fruit juice of seabuckthorn has been found useful for week children and pregnant women (Dorjey, 1998). Malnutrition may produce long term alteration in brain function by altering neurotransmitter metabolism. It is reported that pre or post-natal malnutrition are responsible for more concentration of monoamines, serotonin and nor-epinephrine (Burns and Brown, 1977; and Miller *et al.*, 1977). Seabuckthorn when administered to the PEM and ADHD children has exerted beneficial effect to regulate altered level of brain biogenic amines. 5-HT which is rarity among the plant Kingdome is found in seabuckthorn and because of this property, a significant change in various neuro-chemical parameters has been noticed.

Seabuckthorn has capacity to improve overall, mental performance particularly memory and attention span (Gexianoyam *et al.*, 1986; Zhang *et al.*, 1989; Dubey *et al.*, 1990; Chai *et al.*, 1999; Agrawal *et al.*, 2001). The therapeutic efficacy of this plant is exerted beneficial effect in age related deterioration of cognitive functions. (Dubey and Agrawal , 2003, Dubey *et al.*, 2004; Agrawal *et al.*, 2001; Agrawal *et al.*, 2002; Agrawal *et al.*, 2001) It is concluded that the organic extract of seabuckthorn has capacity to improve attention span, impulsiveness, aggression, memory performance and learning capacity of the children.

17.2.5 CONCLUSIONS

Thus, it is concluded that the plant 'seabuckthorn' has multi-dimensional application in the prevention and management of various mental and physical disorders. It is a potent neuromodulatory agent particularly, it enhances of mental competence and also anti-oxidant and immunomodulatory effective measure. Seabuckthorn can be given both to children as well as adults for longer period without any side effect.

❑❑❑

17.3

Seabuckthorn as an Immunomodulatory Agent with Reference to Age-related Immuno-senescence

Aruna Agarwal, Sushma Tiwari, G. Victor Rajamanickam and G. P. Dubey

CONTENTS

17.3.1 INTRODUCTION

There is a diminution in physical, mental, biochemical, enzymatic and hormonal profile of the person which brings about the changes of ageing. Immune system is also affected and shows diminution in its function. Decline in immune efficiency is responsible for impaired clearance of infectious agents, increased frequency of auto-antibody formation and malignancy. Several causative factors like atrophy of thymus and other endocrine glands, deficiency of vitamins, proteins and minerals, physical and mental tension, inhaled and ingested toxic substances are responsible for immunosenescence.

There is an age related decrease in antibody production following primary antigen exposure in humans (Robert, 1974). With advancing age, there is a decrease in the levels of natural iso-antibodies and hetero-antibodies (Patel, 1981). Similarly, there is an increased appearance of auto-antibodies in ageing (Fong 1985; Goodwin *et al.* 1982). Auto-immunity is an interesting aspect of immune dysregulation of old age. Serum levels of immunoglobulins remain normal or even tend to be higher despite decreased production of antibodies against foreign antigens, as a result of production of auto-antibodies against a host of self antigens which include anti-nucleoproteins, anti-IgG anti-thyroglobuline and rheumatoid factor in elderly human subjects (Weksler *et al.*, 1990). Studies have suggested that the presence of auto-antibodies in clinically healthy elderly individuals may be associated with increased mortality (Mathews, 1973). The altered immune response and impaired defence mechanism in the elderly leads to increase mortality and morbidity from a variety of infectious disease like Influenza, pneumococcal, septicaemia and tetanus which occur in persons aged over 50 years. These three conditions are established indications for vaccination in old age (Power, 1992). In fact, the incidence of auto-antibodies of every kind has been reported to be increased in ageing (Schuller *et al.*, 1981).

The importance of immune senescence to the disease of ageing particularly, the infectious diseases in which immunity plays a vital role in resistance and recovery to infection is essential to study as regards the prevention and management of clinical conditions of aged is concerned. The maintenance of normal cognitive function is the most important aim of health professionals as loss of memory, depression and various systemic functions rapidly declines due to ageing process (Venkoba Rao 1996) and due to this reason, the elderly population is the heavy user of medical and personal care services. The World Health Organization has clearly recognized the utility of Ayurvedic system of medicine to fulfill the world wide health need of present scientific era. To study the concept of resistance against disease (immunity) and the idea that such resistance can be modified by using several pharmacologic and non-pharmacologic approaches is of clinical use. In Ayurvedic system of medicine several natural medicinal plants have been mentioned which are being used as a potent agent to modify the higher brain dysfunctions and also to enhance the body immunity among the human beings. Keeping the facts in view, it was decided to validate the therapeutic potential of seabuckthorn (*H. rhamnoides*) for the improvement and maintenance of mental functions as well as to enhance body immunity against various infections among the elderly people.

17.3.2 MATERIALS AND METHODS

Eighty-eight elderly subjects of both sex groups with an age range of 58-65 years at the time of commencement of the study provided the data. Out of 88 cases, 54 elderly individuals were given the hydroalcoholic extract of *H. rhamnoides* and remaining 34 cases were kept on placebo treatment to serve as control. In this series, only those elderly subjects were selected. They had major complaints such as loss of memory with poor attention and, frequent cough and cold due to reduced body immunity. Persons under investigation who showed any metabolic or persistent endocrine abnormality and having any major psychiatric symptoms were excluded from the series. The elderly subjects who did not follow the treatment schedule or developed any complication during the course of trial were discarded from the series.

Apart from detailed geriatric enquiries a basal recording of various psychophysiological and certain biochemical assessments like IgG, IgM, HB % and serum protein levels were measured among all the subjects under investigation. Mini mental state examination developed by (Folstein *et al.*, 1975) was used to determine the mental state of the elderly subjects. It contains eleven items and maximum score awarded is 30. It helps to assess the different cognitive functions including short-term and long-term memory, language, calculation, writing and drawing abilities. The subjects who scored less than 20 scores were considered to have cognitive impairment. Quantitative estimation of serum immunoglobulins was done by the method of single radical immune-diffusion technique (Fahey, 1968). After basal recording of the parameters, the hydroalcoholic extract of dry fruits of *H. rhamnoides* in the dose of 800 mg in a divided dose (i.e. twice in a day) was given orally continuously for six months. Each test drug capsule contained 400 mg organic extract of *H. rhamnoides* which was prepared by adding an additional additive to make each capsule of 500 mg. Like test formulation capsules, placebo capsules were also prepared by using barley powder and was administered upto same duration in a separate group. The data obtained were statistically analyzed on student's paired test.

17.3.3 RESULTS AND OBSERVATIONS

It is observed that specific clinical features such as frequent cough and cold, general weakness including early fatigue and weight loss considerably improved following 6 months administration of hydroalcoholic extract of *H. rhamnoides*. Poor cognitive functions including deteriorated attention span and memory span is found to be associated with ageing process. The scores obtained out of the Mini mental state examination increased significantly in drug treated group, indicating the better mental performance of the elderly people (Table 17.3.1). It is observed that the hydroalcoholic extract of *H. rhamnoides* has better influence in the regulation of deteriorated mental functions particularly, the memory and attention performance.

When the effect of *H. rhamnoides* was evaluated on certain biochemical parameters haemoglobin and serum protein showed mild increase in the levels in test drug treated group, though the changes were statistically not significant (Table 17.3.2). The value of IgG and IgM which were initially found significantly towards lower side of normal range, after six months of treatment with *H. rhamnoides* a considerable increase was observed (Table 17.3.3). The result indicates the overall improvement in mental state under influence of *H. rhamnoides* and the immunomodulatory property of the drug.

Table 17.3.1: Effect of *H. rhamnoides* on declined mental state of the elderly people

Groups	Sex	No. of cases	Mini mental state examination (Score)			Attention span (Score)		
			Initial	After 6 months	Initial vs 6 months	Initial	After 6 months	Initial vs 6 months
Placebo Treated	Male	18	16.72 ±1.82	16.03 ±2.01	P>0.05	14.20 ±2.65	14.01 ±2.39	P>0.05
	Female	16	15.14 ±2.02	15.02 ±2.12	P>0.05	13.31 ±3.45	12.72 ±1.46	P>0.05
Drug Treated	Male	31	16.25 ±3.11	18.23 ±3.02	P<0.001	14.45 ±2.38	16.43 ±3.03	P<0.001
	Female	23	15.10 ±2.82	16.55 ±2.04	P<0.02	13.12 ±2.60	15.23 ±3.72	P<0.001

Table 17.3.2: Effect of *H. rhamnoides* on haemoglobin and serum protein among the elderly people

Groups	Sex	No. of cases	Hemoglobin (gm%)			Serum protein (mg/dl)		
			Initial	After 6 months	Initial vs 6 months	Initial	After 6 months	Initial vs 6 months
Placebo Treated	Male	18	12.35 ±1.22	12.11 ±1.31	P>0.05	5.82 ±0.72	5.72 ±0.81	P>0.05
	Female	16	11.50 ±1.25	11.45 ±1.30	P>0.05	5.63 ±0.75	5.41 ±0.78	P>0.05
Drug Treated	Male	31	12.38 ±1.24	13.02 ±1.38	P>0.05	5.83 ±0.91	6.27 ±0.95	P>0.02
	Female	23	11.46 ±1.02	12.06 ±1.28	P>0.05	5.80 ±0.87	5.98 ±0.77	P>0.02

Table 17.3.3: Effect of *H. rhamnoides* on immunoglobulin level among the elderly people

Groups	Sex	No. of cases	Serum IgG (mg/dl)			Serum IgM (mg/dl)		
			Initial	After 6 months	Initial vs months	Initial	After 6 months	Initial vs 6 months
Placebo Treated	Male	18	986.48 ±98.72	955.30 ±81.75	P>0.05	1.52 ±0.33	1.32 ±0.75	P>0.05
	Female	16	874.95 ±91.88	836.72 ±85.34	P>0.05	1.38 ±0.43	1.36 ±0.62	P>0.05
Drug Treated	Male	31	992.42 ±101.70	1272.28 ±113.39	P<0.001	1.43 ±0.55	1.88 ±0.62	P<0.01
	Female	23	925.31 ±93.65	1143.66 ±102.74	P<0.001	1.41 ±0.92	1.75 ±0.56	P<0.02

17.3.4 DISCUSSION

The ageing population of the world at present is the largest in the history of mankind and shows increasing trend. India is also a part of the global scenario with its elderly population rising fast. Over this quarter century, expectancy of life at 60 and 70 years has shown significant rise from 13.8 and

8.9 years, respectively, to 16.2 and 10.6 years. Early cross-sectional studies have indicated a peak of mental ability at age 24 years with a decline thereafter at the age 30 years and continue up to old age. Further those workers have demonstrated that perceptual motor skills decline with increasing age (Schaie *et al.*, 1974). Studies on learning and memory, demonstrate decrements in performance with increasing age. Studies on memory (Poon 1985) indicate that older subjects have more difficulty than younger subjects with short-term memory tasks.

Immunosenescence is the effect of age on the immune system and defence mechanism. A physician can assume among the elderly patients that individual has defective defence mechanism and is on higher morbidity and mortality. Immunosenescence greatly affects the outcome of many illnesses that have an age-related incidence such as related incidence like emphysema, post operative infection, urinary tract infection, diabetes, bowel inflammation, etc. The increased susceptibility of the elderly to infections, neoplastic disease and vascular injury may be a consequence of immune senescence (Miller, 1991). It has been predicted that if the pathobiology of ageing was related to the loss of immune competence with age, the survival of individuals with an impaired immune response will be lesser in comparison to the individuals whose immune competence is maintained.

The plant *H. rhamnoides* is being used for enhancing immunity and body resistance by the local people of Himalayan range since last many years. Several studies have been carried out proving the therapeutic potentials of *H. rhamnoides* (Zhang *et al.*, 1987; Chai *et al.*, 1999). Those workers have successfully studied the beneficial role of *H. rhamnoides* in the management of atherosclerosis and cardiovascular disorders. In one of the International Symposiums held at China, the beneficial effects of various compounds of Hippophae rhamnoides was discussed and in a series of reports the anti-viral, anti-bacterial, anti-inflammatory, wound healing, anti-anxiety properties were proved. In the present series of clinical trial, the hydroalcoholic extract of *H. rhamnoides* have exerted efficacious effects on altered mental functions and immune senescence among the elderly. A considerable improvement in memory and attention span including early fatigue and general weakness has indicated that this drug has potentiality to facilitate the deteriorated mental function of the elderly and it has capacity to modify the altered immunoglobulin levels. According to Dubey *et al.* (1990), the hydroalcoholic extract of *H. rhamnoides* has the capacity to reduce the neuropsychological reactivity and thus, it enhances the tolerance capacity towards the stress.

17.3.5 CONCLUSIONS

The present study indicates that along with the advancing age anxiety and stress, precipitate a number of medical problems related to mental function and general body immunity. The *H. rhamnoides* drug has shown immunomodulatory as well as adoptogenic property which is helpful in the management of immunosenescence and thus, it has a significant effect in reducing the frequent infections particularly the cold and cough and thus, is able to protect the elderly people from future occurrence of pneumonia, tuberculosis, etc. In brief, it has been concluded that the test drug improves the overall mental performance by correcting the memory span and also modulates the immune system among the elderly population. The formulation has definite potential to prevent the further cognitive decline of the elderly. The immunomodulatory property of the drug is evident from the measurement of IgG and IgM levels which have shown optimum increase after six months of test drug treatment. It indicates that the drug has capacity to modify the impaired immunity reflected in terms of immunoglobulin measurements and clinically assessed by reduced incidence of recurrent infections and frequent cold and cough among the elderly. Hence, *H. rhamnoides* may be proposed as an effective remedial measure for the prevention and management of various neuro-psychophysiological deteriorations among the elderly persons. The drug *H. rhamnoides* did not show any neurologic, metabolic, allergic hemopoetic or endocrinal side effects even after its continuous administration.

□□□

17.4

Parkinson Disease and Role of Seabuckthorn

Aruna Agarwal, Sushma Tiwari, G. Victor Rajamanickam, and G.P. Dubey

CONTENTS

17.4.1 INTRODUCTION

Parkinson Disease (PD) is a progressive neurodegenerative disorder associated with dopaminergic nigrostriatal neurons. It is reported that 60 - 80 % dopaminergic neurons are lost before the motor signs of PD emerge. It is one of the common neurologic disorders affecting approximately 1% of individuals older than 60 years. The incidence has been estimated to be 4.5-21 cases per 1,00,000 population per year. PD is about 1.5 times more common in men than in women. The average age of onset is approximately 60 years. No standard criteria exist for the diagnosis of PD. Onset of PD is asymmetric with the common initial finding being asymmetric resting tremour in an upper extremity. Apart from this rigidity, brady kinesia, postural instability are also associated with PD. Depression, emotional changes, difficulty with swallowing and chewing, speech changes, urinary problem or constipation, skin problems, sleep problem, dementia or other cognitive problems, muscle cramps and dystonia, pain, fatigue and loss of energy and sexual dysfunction are also presenting clinical features of PD (Stern *et al.*, 1991; Hughes, 2004). PD is believed to be due to a combination of genetic and environmental factors (Brice, 2005). The environmental factor from the development of PD is use of pesticides, living in a rural environment, consumption of well water, exposure to herbicides and proximity to industrial plants.

17.4.2 EXPERIMENTAL STUDY

17.4.2.1 Materials and Methods

Six Albino - rats of Charles Foster strain with an average body weight of 224.95±24.90 gm were taken for the experiment. The age of the rats varied from 6-8 weeks. The rats were housed in polypropylene cages at room temperature. All animals were kept in one room and no other species being housed with them. The room was well ventilated with fresh air and normal temperature. The animals had free access to palliated feed of standard composition. Aqua-guard water was provided ad-libitum. Animals were marked on their back with 0.5 % picric acid. The animals were distributed randomly into different groups.

Name of the drug	Prescribed effective dose for human consumption	Experimental dose /kg/ day average Indian weight 50 kg
Seabuckthorn (Part used fruits leaves)	Hydro-alcoholic extracts of the drug 1200 mg/ day	24 mg/kg

Following test were performed to assess the beneficial role of test drug on P. D.

Experimental design : The study was divided into four groups :

Group-I : This group served as normal control series and rats were kept on normal laboratory condition.

Group-II: In this group the neurotoxin methyl-4-phenyl-1, 2, 3, 6-tetra-hydropyridine (MPTP) was given in the dose of 3 mg/day intra-peritoneally for 5 days continuously to all the 6 rats.

Group-III: The rats were given test drug containing the hydro-alcoholic extract of Seabuckthorn (fruits & leaves) 24 mg/kg per day for 14 days. After administration of the drug upto 14th days the rats were sacrificed after 2 hours.

Group-IV: The animals of this group were given the test drug for 14th days along with MPTP doses (5 days).

Biochemical analysis: Dopamine estimation was carried out following method developed by Chang (1964). Nor-adrenaline and 5-HT estimations were done by fluorometric method of Speeg (1974). The results obtained were statistically analyzed and compared with each other groups.

Brain Dissection

Rats were decapitated by cervical dislocation after anesthetization with ether. Brain of each animal was quickly removed and placed on a glass plate over ice and dissected following method described by the Glowinski and Iverson (1966).

17.4.2.2 Results

MPTP treatment to experimental animals produced marked reduction in dopamine, nor-epinephrine and serotonin levels when measured in total brain and mid brain part of rats. These animals when treated with test drug containing hydro-alcoholic extract of fruits and leaves of seabuckthorn in a prescribed dose revealed significantly rise in the levels of the neurochemical assessments under investigation suggesting the anti-Parkinsonism and anti-oxidant properties of the test drug (Tables 17.4.1-3).

17.4.3 CLINICAL STUDY

17.4.3.1 Material and Methods

In order to determine the beneficial role of the test drug 20 aged suffering from Parkinson's disease were selected and treated with test drug for a period of 3 months. 11 cases were treated with the plant of seabuckthorn (hydro-alcoholic extract of fruit and leaves) and 9 cases were kept on placebo therapy.

The cases which showed positive evidence of major complaints like tremors, rigidity, bradykinesia, postural instability with decreased level of dopamine were selected. Nor-epinephrine was also measured among Parkinson's disease cases. Circulating dopamine was estimated by method (Chang, 1964) and nor-epinephrine by Griffiths *et al.*, (1970). The result obtained in placebo and treated group were compared with each other.

Table 17.4.1: Beneficial role of seabuckthorn on dopamine level in MPTP treated rats

Groups	Dopamine (µgm/gm)			
	Total brain		Mid brain	
	7 days	14 days	7 days	14 days
Normal control * (N=6)	0.796 ±0.101	0.804 ±0.098	1.308 ±0.102	1.288 ±0.117
Treated with MPTP ** (N=6)	0.334 ±0.094	0.206 ±0.084	0.152 ±0.072	0.110 ±0.085
Treated with Seabuckthorn * * * (N=6)	0.870 ±0.102	0.896 ±0.068	1.276 ±0.110	1.289 ±0.103
Treated with MPTP **** + Seabuckthorn (N=6)	0.512 ±0.068	0.527 ±0.079	0.568 ±0.099	0.986 ±0.087

Comparison : *Vs **, **Vs ***, ***Vs **** at P<0.001

Table 17.4.2: Beneficial role of seabuckthorn on nor-epinephrine level in MPTP treated rats

Groups	Nor - epinephrine µg/gm			
	Total brain		Mid brain	
	7 days	14 days	7 days	14 days
Normal control * (N=6)	0.365 ±0.088	0.379 ±0.062	0.624 ±0.098	0.638 ±0.101
Treated with MPTP ** (N=6)	0.152 ±0.045	0.124 ±0.035	0.278 ±0.055	0.203 ±0.071
Treated with Seabuckthorn *** (N=6)	0.386 ±0.085	0.398 ±0.092	0.638 ±0.076	0.652 ±0.094
Treated with MPTP **** + Seabuckthorn(N=6)	0.208 ±0.075	0.228 ±0.092	0.374 ±0.085	0.399 ±0.086

Comparison : *Vs **, **Vs ***, ***Vs **** at P<0.001

Table 17.4.3: Beneficial role of seabuckthorn on serotonin level following MPTP treatment in experimental rats

Groups	Serotonin (μg/gm)			
	Total brain		Mid brain	
	7 days	14 days	7 days	14 days
Normal control * (N=6)	0.448 ±0.094	0.451 ±0.089	0.622 ±0.092	0.634 ±0.098
Treated with MPTP ** (N=6)	0.293 ±0.077	0.210 ±0.091	0.272 ±0.058	0.245 ±0.066
Treated with Seabuckthorn *** (N=6)	0.461 ±0.088	0.487 ±0.065	0.634 ±0.101	0.659 ±0.107
Treated with MPTP **** + Seabuckthorn (N=6)	0.329 ±0.062	0.346 ±0.091	0.358 ±0.055	0.372 ±0.088

Comparison : *Vs **, **Vs ***, ***Vs **** at P<0.001

17.4.3.2 Results

In a preliminary clinical study, 20 cases of Parkinson's disease showing evidence of tremours, rigidity, bradykinesia and postural instability were selected. The symptoms like depression, sleep problem, cognitive problems, muscle cramp, dystonia, etc., were also found to be associated more or less with Parkinson's disease cases. When these cases were treated with test drug improvement in clinical symptomatology and some of the associated complaints were noticed. Though, the degree of improvement varied individual to individual, good number of cases reported improvement in tremours, rigidity of muscles, bradykinesia, postural instability, sleep pattern, memory, attention span, depression and anxiety (Tables 17.4.4-5). The neuro-chemical assessments like dopamine and nor-epinephrine were modulated under drug treatment. Both the factors' neuro-chemical levels were significantly low before therapy. The placebo group of cases did not show any significant change during trial period (Tables 17.4.6-7).

17.4.4 DISCUSSION

MPTP crosses the blood brain barrier and it oxidized to MPP^+ by the enzyme MAO. MPP+ accumulates in mitochondria and infers. the function of complex respiratory chain. One of the most important hypotheses is that the increased dopamine turnover due to levodepa administration could increase oxidative damage and accelerates loss of dopamine neurons. In the present series of study, dopamine level when measured in different parts of rat brain showed a significant fall in the level following MPTP treatment. The test drug therapy upto 14 days checked the loss of dopamine following MPTP administration in comparison to group where only MPTP was given. A mild level of dopamine also enhanced in test drug treated group indicating the dopamine modulatory property of test drug. Some effects, like feeling of well being, improved sleep pattern, etc., may be due to the synergistic effect of test drug, which has shown anti-oxidant, anti-anxiety and immunomodulatory properties of the test drug in several studies.

PD is associated with increased dopamine turnover, decreased protective mechanisms of glutathione, increased iron (a peroxidation molecule) and evidence of increased lipid peroxidation. The test drug has shown significant rise in dopamine level along with its anti-oxidant effects indicating anti-Parkinsonism property.

Dopamine is a clinical messenger responsible for transmitting signals between the substantia nigra and the next 'relay station' of the brain to produce smooth, purposeful movement. Loss of dopamine results in abnormal nerve firing within the brain causing impaired movement (Girault and Greengard, 2004; Walts, 1997). Parkinson's disease patients also have a loss of nerve ending that produces nor-epinephrine. The loss of nor-epinephrine is responsible for several non motor feature observed like fatigue and abnormalities of blood pressure regulation in PD cases. In the present clinical trial elevation in dopamine and nor-epinephrine levels indicates that the drug acts on dopaminergic neurons and prevents the damage to dopaminergic pathway and thus, modulates the dopamine transmitting activity.

Research as suggested that mitochondria may play a role in the development of the Parkinson's disease. Mitochondria are the energy-producing components of the cell and are major source of free radicals/ molecules that damage membranes, proteins, DNA and other parts of the cell. This damage is often referred to as oxidative stress. Oxidative stress-related changes, including free radicals damage to DNA, proteins and fats have been detected in brain of Parkinson's disease patients. Mitochondrial dysfunction, oxidative stress, inflammation or over-stimulation of cells and many other cellular processes may contribute to Parkinson's disease. Several studies have shown that a fruit, seed and leaves of seabuckthorn contain anti-oxidant and immunomodulatory properties. The fruits and leaves of Seabuckthorn contain β-carotene, vitamin E, flavonoides, vitamin C, catechins, elagic and ferulic acid of potententently antioxidant agents (Rosch *et al.*, 2003; 2004; Eccleston *et al.*, 2002; Gao *et al.*, 2000; Heinonen *et al.*, 1998; Prior *et al.*, 1998; Rice-Evanc *et al.*, 1994). Seabuckthorn also protects the mitochondrial damage which is generated by oxidative stress through its antioxidant and immunomodulatory properties (Goel *et al.*, 2005; Narayanans *et al.*, 2004).

Table 17.4.4: Effect of test drug on clinical symptoms among Parkinson's disease cases

Treated group		**Symptoms in Parkinson's disease**							
		Tremors		**Rigidity**		**Bradybinesia**		**Postural instability**	
		Before therapy	**After therapy**	**Before therapy**	**After therapy**	**Before therapy**	**After therapy**	**Before therapy**	**After therapy**
Parkinson's disease	Placebo treated (N=9)	9 (100%)	9* (100%)	7 (77.78%)	7* (77.78%)	8 (88.89%)	8* (88.89%)	9 (100%)	9* (100%)
	Drug treated (N=11)	11 (100%)	7# (63.63%)	8 (72.72%)	6# (75%)	11 (100%)	7# (63.63%)	8 (72.72%)	6# (75%)

Table 17.4.5: Effect of test drug on associated symptoms among Parkinson's disease cases

Treated group		**Associated symptoms with Parkinson's disease**									
		Depression		**Sleep problems**		**Cognitive problems**		**Muscle cramp and dystonia**		**Fatigue**	
		Before therapy	After therapy	Before therapy	After therapy	Before therapy	After therapy	Before therapy	After therapy	Before therapy	After therapy
Test drug treated	Placebo treated (N=9)	9 (100%)	9* (100%)	9 (100%)	9* (100%)	8 (88.89%)	8* (88.89%)	8 (88.89%)	8* (88.89%)	9 (100%)	9* (100%)
	Drug treated (N=11)	10 (90.90%)	6# (60%)	11 (100%)	7# (63.63%)	11 (100%)	9# (81.81%)	11 (100%)	8# (72.72%)	11 (100%)	9# (81.81%)

* No improvement after therapy ; # Improvement after therapy

Table 17.4.6: Dopamine enhancing effect of seabuckthorn among Parkinson's disease cases

Treated group		Dopamine (µgm/ml)		Initial vs . after 3 months
		Initial	After 3 months	
Parkinson's disease	Placebo treated (N=9)	0.884 ±0.039	0.869 ±0.048	P>0.05
	Drug treated (N=11)	0.659 ±0.210	1.72 ±0.104	P<0.001

Normal Range: 1-3 µg/ml

Table 17.4.7: Changes in nor-epinephrine level following seabuckthorn treatment in Parkinson's disease cases

Treated group		Nor-epinephrine (µg/ml)		Initial vs. after 3 months
		Initial	After 3 months	
Parkinson's disease	Placebo treated (N=9)	5.72 ±1.09	5.33 ±1.23	P>0.05
	Drug treated (N=11)	5.20 ±2.01	7.64 ±2.11	P<0.05

Normal Range: 8-12 ng/ml

Parkinson's disease may affect memory, social judgment, language, reasoning or other mental skill. The Drug donepezil may reduce the behavioural symptoms in Parkinson's disease cases but long term effects causes high blood pressure, nausea, vomiting, etc. Several recent studies have indicated that Seabuckthorn contains biologically active substance which has potential role in the prevention and management of age related deterioration of cognitive function. It has capacity to improve over-all mental performance particularly, memory and attention span (Gexianoyam *et al.*, 1986; Zhang *et al.*, 1987; Dubey *et al.*, 1990; Agrawal *et al.*, 2001).

Parkinson's disease develops aching muscle and joints because of the rigidity and abnormal postures often associated with the disease. People with Parkinson's disease may develop pain due to compression of nerve roots or dystonia related muscle spasms. Dopaminergic drugs, opiates, antidepressants and other types of drugs may be used to treat this type of pain. These agents produce a variety of side effects. Several evidences are available that hydro-alcoholic extract of fruits and leaves of seabuckthorn support the immune system, muscle nourishment, strengthens cell walls, and have a significant anti-inflammatory activity, pain reduction property and stimulate tissue regulation (Yu let *et al.*, 1993; Zhone Fei *et al.*, 1989; Zhang *et al.*, 1987; Cheng *et al.*, 2003).

Depression is the most common problem appears in the course of the disease (Zesiewicz 1999; Hughes *et al.*, 2004). Parkinson's disease related depression or use of anti depressant medication also may cause decreased sex drive and other problems (Agrawal *et al.*, 2000). It is concluded that Seabuckthorn has potential role in the management of neuro-psychiatric disorders associated with aged.

The most common drug for Parkinson's disease is dopamine precursor-substance such as levodopa that crosses the blood brain barrier and then changed in to dopamine. Levodopa is very successful in reducing the tremors and other symptoms of Parkinson's disease during the early stages of the disease. Levodopa usually helps most with bradykinesia and rigidity. The most common initial side-effects of synthetic dopamine replacement therapy are nausea, vomiting, low blood pressure and restlessness. Long term use of levodopa some time causes hallucinations and psychosis (Montastruc *et al.*, 1994; Murer *et al.*, 1998).

Ferrous iron is the essential co-factor for L-dopa biosynthesis (Birkmayer and Birkmayer, 1986) and seabuckthorn is the rich source of folic acid and has potent antioxidant activity (Rosch *et al.*, 2003). Therefore, the test drug enhances the capacity of biosynthesis of dopamine as observed in the present experimental study. Further, the plant has capacity to facilitate blood circulation and increased plasma catecholamine concentration (Zhang *et al.*, 2001).

Dopamine agonists' drugs like bromocriptine, pergolide, apomorphine, pramiperole, mimic supplement the dopamine in the brain (Walts, 1997) and their adverse reaction is similar to L-dopa including drowsiness, sudden sleep onset, confusion, edema and vomiting. In the present experimental study, Seabuckthorn has prevented the loss of dopaminergic, nor-adrenergic and serotonergic receptors among MPTP treated animals.

17.4.5 CONCLUSIONS

It is concluded that since the plant seabuckthorn has been proven as one of the potent therapeutic agents showing antioxidant, anti-inflammatory, immunomodulatory including neuro-chemical modulation particularly, dopamine. Thus, it can be proposed as one of the suitable remedial measures for the management of Parkinson's disease. The drug is safe and can be given for a long period.

❑❑❑

17.5

Antioxidant Effect of Seabuckthorn

Aruna Agarwal, Sushma Tiwari, G. Victor Rajamanickam and G.P. Dubey

CONTENTS

17.5.1 INTRODUCTION

Oxidative stress plays an important role in neuro-degenerative diseases particularly Parkinson's and Alzheimer's diseases. These disorders make the elderly to lead a difficult routine life. They cannot move freely. There are known anti-oxidants like Schisanhe (Sal) and Salviaholic acid A (Sal A), a Chinese synthetic new compound which has shown a significant potential role in the protection of cerebral cells the injuries induced by oxidative stress. It is necessary to find out some plant based safe and effective anti-oxidants to prevent or to slow down the process of diseases related to neuro-degeneration. The standard extract of seabuckthorn has been widely employed for its significant benefits in neuro-degenerative disorders. Although anti-oxidative actions has been studied in different trials but the mechanisms involved in its pharmacological activity is not established properly. Parkinson's and Alzheimer's diseases are frequently associated with oxidative stress. In the present series of study, the lipid peroxidation (LPO), the activity of the anti-oxidant enzyme catalase (CAT) and glutathione (GSH) were evaluated in different parts of the brain including liver and kidney following seabuckthorn treatment. A significant decrease of LPO in hippocampus was observed. Such change indicates the anti-oxidant property of the seabuckthorn. The present finding suggests the possible role of seabuckthorn extract in the prevention and management of diseases involving free radicals and oxidative damage.

Table 17.5.1: Effect of seabuckthorn on lipid peroxidase activity following colchicine neuro-toxin among different parts of experimental rats

Groups	Catalase (millimoles of H_2O_2 consumed/min/mg protein)								
	Cortex			Hippocampus			Hypothalamus		
	Control *	7 days **	14 days ***	Control *	7 days **	14 days ***	Control *	7 days **	14 days ***
Colchicine Treated (N=10)	0.284 ±0.068	0.614 ±0.051	0.684 ±0.072	0.412 ±0.047	0.648 ±0.063	0.701 ±0.075	2.103 ±0.054	3.704 ±0.066	3.941 ±0.059
Colchicine + Test Drug Treated (N=10)	--	0.451 ±0.071	0.447 ±0.054	--	0.561 ±0.062	0.482 ±0.058	--	3.011 ±0.051	2.584 ±0.061

Comparison:	* vs **	* vs ***	* vs **	* vs ***	* vs **	* vs ***
Colchicine Treated	P<0.001	P<0.001	P<0.001	P<0.001	P<0.001	P<0.001
Colchicine +Test drug Treated	P<0.05	P<0.01	P<0.001	P<0.001	P<0.001	P<0.001

Table 17.5.2: Anti oxidant effect of test drug determined in terms of catalase activity

Groups	Catalase (millimoles of H_2O_2 consumed/min/mg protein)					
	Liver			Kidney		
	Control *	7 days **	14 days ***	Control *	7 days **	14 days ***
Colchicine Treated (N=10)	0.194 ±0.039	0.841 ±0.048	0.875 ±0.082	0.206 ±0.069	0.784 ±0.068	0.806 ±0.056
Colchicine + Test Drug treated (N=10)	--	0.665 ±0.032	0.483 ±0.026	--	0.629 ±0.041	0.582 ±0.061

Comparison:	* vs **	* vs ***	* vs **	* vs ***
Colchicine Treated	P<0.001	P<0.001	P<0.001	P<0.001
Colchicine +Test drug Treated	P<0.001	P<0.001	P<0.001	P<0.001

Table 17.5.3: Effect of seabuckthorn on glutathione level in colchicine treatment albino rats

Groups	Glutathione (μmoles of GSH oxidized/mg protein)								
	Cortex			Hippocampus			Hypothalamus		
	Control *	7 days **	14 days ***	Control *	7 days **	14 days ***	Control *	7 days **	14 days ***
Colchicine treated (N=10)	2.360 ±0.106	1.023 ±0.102	1.284 ±0.112	3.061 ±0.122	1.608 ±0.104	1.225 ±0.098	21.106 ±2.036	11.76 ±1.262	9.450 ±1.020
Colchicine + Test Drug treated (N=10)	--	1.942 ±0.088	2.011 ±0.101	--	2.081 ±0.201	2.892 ±0.104	--	16.110 ±2.092	13..81 ±2.061

Comparison:	* vs **	* vs ***	* vs **	* vs ***	* vs **	* vs ***
Colchicine Treated	P<0.001	P<0.001	P<0.001	P<0.001	P<0.05	P>0.05
Colchicine +Test drug Treated	P<0.001	P<0.001	P<0.001	P<0.001	P<0.05	P<0.01

Table 17.5.4: Glutathione modulation under seabuckthorn treatment among colchicine treated rats

Groups	Glutathione (μmoles of GSH oxidized/mg protein)					
	Liver			Kidney		
	Control *	7 days **	14 days ***	Control *	7 days **	14 days ***
Colchicine treated (N=10)	2.006 ±0.123	1.101 ±0.302	1.006 ±0.028	1.986 ±0.505	1.202 ±0.112	0.984 ±0.036
Colchicine + Test Drug treated (N=10)	--	1.684 ±0.201	1.987 ±0.048	--	1.002 ±0.054	1.201 ±0.082

Comparison:	* vs **	* vs ***	* vs **	* vs ***	
Colchicine Treated		P<0.001	P<0.001	P<0.001	P<0.001
Colchicine +Test drug Treated	P<0.001	P<0.001	P<0.001	P<0.001	

The evidence for oxidative stress in Alzheimer's disease is experimentally observed. Lipid peroxidase has been implicated in the pathogenesis of Alzheimer's disease. To determine the anti-oxidant effects of seabuckthorn, we have analyzed certain oxidative stress markers like lipid peroxidase in terms of Malodialdehyde (MDA) in association with certain anti-oxidant parameters like glutathione (GSH), lipid preoxidase and catalase. Much emphasis has been given on oxidative stress and free radical scavenging and anti-oxidant system of the disease. Therefore, effect of drug have been assessed on glutathione, lipid peroxidase and catalase activity following colchicine treated.

17.5.2 CLINICAL STUDY

17.5.2.1 Material and Methods

Under the field survey programme, 95 people with age range of 65-75 yrs of both sexes were selected. Out of 95 cases, 41 were normal and remaining 54 showed evidence of Senile Dementia of Alzheimer's Type (SDAT). Among them, a sustained deterioration of memory along with three of the five complaints such as poor orientation, poor judgement, problem solving difficulties, trouble in the functioning of community affairs, inability to function independently in home, during hobbies and personal care were noticed to be present. Only those people were taken into consideration to diagnose SDAT cases. Among the aged subjects only those who exhibited more than three above mentioned complaints were kept in SDAT group and were assessed on

Table 17.5.5: Changes in lipid peroxidase activity under seabuckthorn treatment in different parts of brain in colchicine treated rats

Groups	LPO activity (μmoles of MDA formed/mg protein)								
	Cortex			Hippocampus			Hypothalamus		
	Control *	7 days **	14 days ***	Control *	7 days **	14 days ***	Control *	7 days **	14 days ***
Colchicine treated (N=10)	0.389 ±0.028	0.586 ±0.053	0.682 ±0.071	0.322 ±0.030	0.560 ±0.042	0.571 ±0.073	0.352 ±0.051	0.606 ±0.082	0.659 ±0.068
Colchicine + Test Drug treated (N=10)	--	0.509 ±0.051	0.492 ±0.022	--	0.514 ±0.045	0.406 ±0.091	--	0.589 ±0.056	0.628 ±0.031

Comparison	* vs **	* vs ***	* vs **	* vs ***	* vs **	* vs ***
Colchicine Treated	P<0.001	P<0.001	P<0.001	P<0.001	P<0.001	P<0.001
Colchicine +Test drug Treated	P<0.001	P<0.001	P<0.001	P<0.001	P<0.001	P<0.001

Table 17.5.6: Changes in lipid peroxidase activity in liver and kidney under seabuckthorn treatment following colchicine in experimental rats

Groups	LPO activity (micromoles of MDA formed/mg protein)					
	Liver			Kidney		
	Control *	7 days **	14 days ***	Control *	7 days **	14 days ***
Colchicine treated (N=10)	0.356 ±0.028	0.626 ±0.092	0.654 ±0.075	0.301 ±0.024	0.490 ±0.086	0.486 ±0.039
Colchicine + Test Drug treated (N=10)	--	0.592 ±0.032	0.486 ±0.046	--	0.406 ±0.032	0.356 ±0.024

Comparison:	* vs **	* vs ***	* vs **	* vs ***	
Colchicine Treated		P<0.001	P<0.001	P<0.001	P<0.001
Colchicine +Test drug Treated	P<0.001	P<0.001	P<0.001	P<0.001	

various neuropsychophysiological parameters. Dementia rating scale-2 (Jurica, 2001) has been utilized in the present study. Based on this scale the score obtained for diagnosed SDAT cases in the present study is as follows :

Complaints	Normal control	SDAT
Attention	79.55	32.22
Initiation/ preservation	23.55	10.04
Construction	21.37	8.93
Conceptualization	2.55	1.45
Memory	21.18	9.79
Total score	10.91	5.87

After diagnosis of SDAT, various electropsychophysiological tests like Mini-mental State Examination (Folstein *et al.*, 1975) EEG, EMG, Memory, Attention span and circulating 5-HT and nor-adrenaline were measured before therapy and after therapy in both normal as well as SDAT cases. Parameters were repeated at the end of 2 months of therapy. Study continued upto 6 months. After baseline recording of the parameters out of 54 aged in the group of SDAT, 29 cases were treated with test drug containing the hydroalcoholic extract of dried fruits and dried leaves (equal quantity) of seabuckthorn and 25 aged cases of SDAT were treated with placebo to establish the beneficial role of test drug. Similarly, in the normal control group 20 aged were given placebo and 21 were treated with test drug. The data obtained were statistically analysed and compared.

Preparation of test drug		Dose
Hydroalcoholic extract of leaves of seabuckthorn	-	450 mg
Hydroalcoholic extract of fruits of seabuckthorn	-	450 mg

Both extracts were mixed together in equal quantity and the capsules were prepared with the addition of additive to make final capsule as 500 mg and were administered to the selected aged twice in a day for a period of 6 months.

17.5.2.2 Results

The diagnosed cases of SDAT were treated with the hydro-alcoholic extract of seabuckthorn have shown beneficial effects in the prevention and management of neuro-degeneration particularly the drug which has exhibited effects in modulation of cognitive functions when assessed on various neuropsychophysiological parameters.

Table 17.5.7: Improvement in various complaints associated with SDAT cases following oral administration

Treated groups		Complaints (In percent)									
		Aggression		Sleep disturbance		Communication difficulties		Depression		Psychotic features	
		Before treatment	After treatment	Before treatment	After treatment	Before treatment	After treatment	Before treatment	After treatment	Before treatment	After treatment
Normal ageds	Placebo treated (N=20)	8 (40%)	8✖ (40%)	17 (85%)	18✖ (90%)	–	–	10 (50%)	10✖ (50%)	2 (10%)	2✖ (10%)
	Test drug treated (N=21)	14 (66.67%)	11✡ (78.57%)	16 (76.19%)	19✡ (87.5%)	–	–	16 (76.19%)	11✡ (68.75%)	4 (19.04%)	3✡ (75%)
Aged SDAT cases	Placebo treated (N=25)	24 (96%)	24* (96%)	25 (100%)	24* (90%)	25 (100%)	25* (100%)	25 (100%)	25* (100%)	17 (68%)	18* (72%)
	Test drug treated (N=29)	26 (89.65%)	20# (76.92%)	28 (96.55%)	21# (75%)	29 (100%)	17# (58.62%)	29 (100%)	20# (68.96%)	18 (62.06%)	12# (66.66%)

B= Percent of symptoms present before therapy
A✖= No improvement in percent A✡= Improvement in percent A*= No improvement in percent ***A#= Improvement in percent***

The cognitive screening tests of SDAT cases were done. Mini Mental State Examination which assess the orientation, concentration, mental flexibility (i.e., working memory) short term memory and constructional praxis reported a mild improvement of scores in SDAT aged whereas in normal aged a further deterioration was noticed (Table 17.5.8).

Fluctuations in affected were also reduced under influence of test drug in both normal as well as SDAT ageds (Table 17.5.9).

Table 17.5.8 : Effect of seabuckthorn on Mini-mental state examination among aged cases of Senile Dementia of Alzheimer's Type

Treated groups		Mini-mental state examination (Score)				Initial vs 6 months treatment
		Initial	After 2 months	After 4 months	After 6 months	
Normal ageds	Placebo treated (N=20)	18.82 ±2.01	18.74 ±2.11	17.78 ±3.02	17.76 ±2.14	P>0.05
	Test drug treated (N=21)	17.84 ±1.76	18.28 ±2.04	18.65 ±1.92	19.02 ±1.88	P>0.05
Aged SDAT cases	Placebo treated (N=25)	12.72 ±1.01	10.64 ±1.14	10.67 ±1.06	9.85 ±1.22	P<0.05
	Test drug treated (N=29)	11.22 ±1.00	10.64 ±1.24	11.89 ±2.01	12.28 ±2.06	P<0.05

The test drug has shown better memory performance among SDAT cases in comparison to placebo treated group. Though improvement is not so marked rather it is observed that there is no further decline in the memory scores (Table 17.5.10-11).

Anxiety and depression score were found significantly high in normal as well as cases of SDAT at initial level. After 6 months of test drug treatment a marked reduction in scores were obtained. The placebo series did not show any such change (Table 17.5.12-13).

The drug seabuckthorn has shown neurochemical modulatory effects which are associated with SDAT like Acetycholine, ChAT, nor-adrenaline, serotonin, dopamine, GABA etc. In the present series of study a significant low values of nor-adrenaline and 5-HT were recorded in SDAT cases which modified under influence of test drug. The drug has potentiality to act on various brain receptors and it prevents the mental capability of the aged (Table 17.5.14-15).

Table 17.5.9 : Effect of seabuckthorn on attention status of the aged cases of Senile Dementia of Alzheimer's Type

Treated groups		Fluctuation of attention (per minute)				Initial vs. 6 months treatment
		Initial	After 2 months	After 4 months	After 6 months	
Normal ageds	Placebo treated (N=20)	22.84 ±3.02	25.24 ±2.41	26.02 ±3.71	25.88 ±3.45	P>0.05
	Test drug treated (N=21)	24.34 ±3.16	19.41 ±2.84	18.26 ±2.72	16.89 ±2.06	P<0.001
Aged SDAT cases	Placebo treated (N=25)	39.72 ±4.12	38.11 ±5.26	41.71 ±4.84	40.82 ±3.92	P>0.05
	Test drug treated (N=29)	38.79 ±5.11	33.24 ±4.28	30.45 ±3.06	27.04 ±3.26	P<0.001

17.5.3 DISCUSSION

Several clinical trials found that acetyl-L-carnitine supplementation delays the progression of Alzheimer's disease (Pettegrew *et al* 1995), improves memory (Salvioli *et al* 1994, Rai *et al* 1990, Sano *et al* 1992) and enhances overall performance in some people with Alzheimer's disease (Cucinotta *et al* 1988, Bonavita 1986).

2-dimethylaminoethanol (DMAE) may increase the levels of the brain neuro-transmitter acetylcholine. In one preliminary trial, people with senile dementia when given DMAE supplement, showed positive behaviour changes (Ferris *et al.,* 1977). Emerging evidence suggests a possible benefit of DHEA supplementation in people with Alzheimer's disease. Most of the studies have indicated that people with Alzheimer's disease to have lower blood DHEA levels than the normal control (Hillen *et al.,* 2000; Nasman *et al.,* 1991, Sunderland *et al.,* 1989, Yanase *et al.,* 1996).

Table 17.5.10: Effect of seabuckthorn on short term memory span among aged SDAT cases population

Treated groups		Short term memory (Score)												Iinitial vs 6 months treatment	
		Initial			After 2 months			After 4 months			After 6 months				
		3 sec.	9 sec.	18 sec.	3 sec.	9 sec.	18 sec.	3 sec.	9 sec.	18 sec.	3 sec.	9 sec.	18 sec.		
Normal aged	Placebo treated (N=20)	5.42 ±0.85	5.06 ±1.01	4.96 ±0.76	5.28 ±0.90	5.10 ±1.00	4.82 ±0.75	5.17 ±0.68	5.12 ±1.11	4.78 ±0.78	4.97 ±0.5	4.98 ±0.77	4.92 ±0.82	P>0.05	P>0.05
	Test drug treated (N=21)	5.11 ±1.02	4.96 ±0.75	4.72 ±0.82	5.27 ±1.02	4.98 ±0.75	4.88 ±0.65	5.36 ±0.92	5.10 ±0.98	4.98 ±0.78	5.52 ±1.03	5.21 ±1.02	5.01 ±0.94	P>0.05	P>0.05
Aged SDAT cases	Placebo treated (N=25)	2.84 ±0.40	2.35 ±0.39	1.85 ±0.55	2.71 ±0.76	2.14 ±0.65	1.68 ±0.51	2.14 ±0.56	2.06 ±0.48	1.45 ±0.71	2.12 ±0.60	1.84 ±0.35	1.36 ±0.54	P<0.001	P<0.01
	Test drug treated (N=29)	2.88 ±0.45	2.25 ±0.38	1.90 ±0.45	2.94 ±0.62	2.56 ±0.68	2.08 ±0.57	2.98 ±0.62	2.80 ±0.48	2.17 ±0.36	2.95 ±0.60	2.88 ±0.62	2.25 ±0.48	P>0.05	P<0.05

Table 17.5.11: Effect of seabuckthorn on long term memory span among aged Senile Dementia of Alzheimer's Type

Treated groups		Long term memory (Score)								Initial vs 6 months treatment	
		Initial		After 2 months		After 4 months		After 6 months			
		Repetition of words 1 time	Repetition of words 4 time	Repetition of words 1 time	Repetition of words 4 time	Repetition of words 1 time	Repetition of words 4 time	Repetition of words 1 time	Repetition of words 4 time	Repetition of words 1 time	Repetition of words 4time
Normal ageds	Placebo treated (N=20)	2.91 ±0.45	2.64 ±0.58	2.84 ±0.76	2.45 ±0.86	2.81 ±0.92	2.31 ±0.77	2.54 ±0.82	2.44 ±0.48	P>0.05	P>0.05
	Test drug treated (N=21)	2.85 ±0.52	2.56 ±0.82	2.90 ±0.71	2.66 ±0.68	2.92 ±0.82	2.76 ±0.55	3.06 ±0.91	2.75 ±0.76	P>0.05	P>0.05
Aged SDAT cases	Placebo treated (N=25)	1.22 ±0.54	0.92 ±0.22	1.04 ±0.71	0.88 ±0.12	1.06 ±0.42	0.91 ±0.16	1.02 ±0.22	0.84 ±0.21	P>0.05	P>0.05
	Test drug treated (N=29)	1.31 ±0.45	0.84 ±0.31	1.45 ±0.20	1.04 ±0.34	1.51 ±0.35	1.18 ±0.37	1.62 ±0.40	1.07 ±0.36	P<0.05	P<0.05

Table 17.5.12: Effect of seabuckthorn on anxiety level among the aged cases of Senile Dementia of Alzheimer's Type

Treated groups		Anxiety Level (Score)				Initial vs. 6 months treatment
		Initial	After 2 months	After 4 months	After 6 months	
Normal aggeds	Placebo treated (N=20)	61.45 ±2.80	61.75 ±3.04	61.85 ±2.92	62.11 ±2.75	P>0.05
	Test drug treated (N=21)	62.45 ±3.00	59.45 ±3.11	59.24 ±3.46	58.45 ±3.61	P<0.01
Aged SDAT cases	Placebo treated (N=25)	65.98 ±3.72	65.41 ±3.04	64.82 ±3.11	64.97 ±3.17	P>0.05
	Test drug treated (N=29)	66.75 ±3.41	63.12 ±4.22	62.49 ±3.12	61.22 ±3.24	P<0.001

Table 17.5.13: Effect of seabuckthorn on depression level among the aged cases of Senile Dementia of Alzheimer's Type

Treated groups		Depression level (Score)				Comparison initial vs. 6 months treatment
		Initial	After 2 months	After 4 months	After 6 months	
Normal ageds	Placebo treated (N=20)	8.94 ±1.22	9.14 ±1.31	9.20 ±1.42	9.15 ±1.22	P>0.05
	Test drug treated (N=21)	9.02 ±1.34	7.76 ±1.20	8.25 ±1.02	7.25 ±1.06	P<0.001
Aged SDAT cases	Placebo treated (N=25)	12.11 ±1.20	12.25 ±1.22	12.45 ±1.32	12.16 ±1.45	P<0.01
	Test drug treated (N=29)	12.33 ±1.22	10.85 ±1.28	10.45 ±2.10	10.02 ±2.14	P<0.001

Table 17.5.14: Effect of seabuckthorn on nor-adrenaline level among aged cases suffering from Senile Dementia of Alzheimer's Type

Treated groups		Nor-adrenaline (μg/ml)				Initial vs. 6 months treatment
		Initial	After 2 months	After 4 months	After 6 months	
Normal ageds	Placebo treated (N=20)	5.44 ±1.10	6.27 ±1.97	5.85 ±1.20	6.21 ±2.05	P>0.05
	Test drug treated (N=21)	5.01 ±0.68	4.96 ±0.92	5.06 ±0.85	5.80 ±0.81	P>0.05
Aged SDAT cases	Placebo treated (N=25)	2.62 ±0.32	2.66 ±0.45	2.25 ±0.48	2.14 ±0.76	P<0.01
	Test drug treated (N=29)	2.35 ±0.91	2.84 ±0.88	3.26 ±0.75	3.91 ±0.82	P<0.001

Table 17.5.15: Effect of seabuckthorn on circulating 5-HT level among the aged cases of Senile Dementia of Alzheimer's Type

Treated groups		5-HT (μg/ml)				Initial vs. 6 months treatment
		Initial	After 2 months	After 4 months	After 6 months	
Normal ageds	Placebo treated (N=20)	5.08 ±1.02	5.41 ±0.80	6.05 ±1.44	5.25 ±1.01	P>0.05
	Test drug treated (N=21)	6.11 ±1.02	5.93 ±1.01	5.05 ±0.94	5.12 ±0.85	P>0.05
Aged SDAT cases	Placebo treated (N=25)	1.20 ±0.14	1.16 ±0.22	1.38 ±0.25	1.65 ±0.31	P<0.001
	Test drug treated (N=29)	1.34 ±0.26	1.84 ±0.34	2.30 ±0.20	2.88 ±0.21	P<0.001

In a preliminary study, people who used antioxidant supplement (Vitamin C or Vitamin E) had a lower risk of Alzheimer's disease, compared with people who did not take antioxidants (Morris *et al.*, 1998). Large amount of supplement of Vitamin E may slow the progression of Alzheimer's disease

(Sano *et al.*, 1997). Fruit of this plant is quite rich in Vitamin C (300-1,600 mg/100 gm), which is 4-100 times higher than many of fruits and vegetables (Singh *et al.*, 1995; Rousi and Aulin, 1977; Plekhanova, 1988; Wahlberg and Jeppsson, 1990; 1992; Yao *et al.*, 1992). There is only one word to attribute. It is nothing but Seabuckthorn is an antioxidant. The fruits, seed and leaves contain an impressive array of anti-oxidant compound. The leaves are an equally rich source of important anti-oxidants including β carotene, vitamin E, flavonoids, catechins, elagic acid, ferulic acid, (Rosch *et al.*, 2003, 2004 ; Eccleston *et al.*, 2002; Gao *et al.*, 2000; Heinonen *et al.*, 1998; Prior *et al.*, 1998; Rice - Evanc *et al.*, 1994, 1997).

Vitamin B is involved in nerve transmission in parts of the brain called cholinergic neurons that deteriorate in Alzheimer's disease (Eder *et al.*, 1976, 1980). The activity of vitamin B1 - dependent enzymes has been found to be lower in the brains of people with Alzheimer's disease (Gibson *et al.*, 1988). Studies suggested that vitamin B1 supplementation could slow the progression of Alzheimer's disease (Meador et al 1993; Blass *et al.*, 1988). Fruit oil is very rich in vitamin E (203 mg) (Fe *et al.*, 1993; Huang and Xiao, 1993; Jablczynska *et al.*, 1994) and β-carotene (249 mg) (Bernath and Foldesi, 1992; Wolf and Wegert, 1993) Vitamin E concentration can be up to 160 mg/100 gm of berries (Zhang *et al.*, 1989; Ma and Cui, 1989; Wahlberg and Jeppsson, 1990, 1992).

Some researchers have found an association between Alzheimer's disease and deficiency of vitamin B12 and folic acid (Clarke *et al.*, 1998; Snowdon *et al.*, 2000). In a study of elderly Canadian, those with low level of folate were more likely to have dementia of all types including Alzheimer's disease, than those with higher levels of folate (Ebly *et al.*, 1998). Seabuckthorn is also the rich source of folic acid and has a high potent anti-oxidant capability (Rosch *et al.*, 2003, Zhao *et al.*, 1997). Alcoholic extract of leaves and fruits are used as anti-oxidant and immunomodulatory property (Geetha *et al.*, 2002).

In one of the reports, two people with a hereditary form of Alzheimer's disease received daily coenzyme Q10, iron (sodium citrate) and vitamin B6. Mental status improved in both patient, and one became almost normal after six months (Imagawa *et al.*, 1992). In the present preliminary clinical study, the aged cases of SDAT who received the test drug seabuckthorn which contained vitamin B1, B2, B6, iron and folic acid, after a treatment for a period of 6 months, their mental performance were improved and better quality of life was reported by most of the subject.

Leaves of seabuckthorn are equally important in medicinal properties, as they carry some unique flavonoïdes (Guseva *et al.*, 1995; Sasov *et al.*, 1996;

Shipilina *et al.*, 1995). Seabuckthorn is also rich in flavonoid (vitamin P) and contains appreciable amounts of water soluble and fat soluble vitamins (Zhang *et al.*, 1989; Solonenko and Shishkina, 1989; Schapiro, 1989). There are at least 24 chemical elements present in Seabuckthorn juice e.g., nitrogen, phosphorous, iron, manganese, boron, calcium, aluminium, silicon and others (Wolf and Wegert, 1993; Zhang *et al.*, 1989, Tong *et al.*, 1989).

Research in the late 1950's and early 1960's reported that 5-Hydroxytryptamine isolated from Seabuckthorn bark inhibited tumour growth (Sokoloff *et al.*, 1961). The immunomodulatory property of Seabuckthorn is well established in several studies. The ripen fruits of Seabuckthorn contains malic acid, oxalic acid, phospholipid, vitamin C, A, B complex, E and K and other nutrients like fat, protein, organic acids and flavonoides. The peel of steam and fruit contains 5-HT which is a rare occurrence in plant kingdom. Several recent studies have indicated that Seabuckthorn contains biologically active substance, which enhances immunity and reduces the cardiovascular over-activity (Dubey *et al.*, 2002; Gupta *et al.*, 1990; Agrawal *et al.*, 2001; Dubey *et al.*, 2004; Dubey *et al.*, 2005; Dubey *et al.*, 2005). The therapeutic efficacy of this plant has exerted beneficial effect in age related deterioration of cognitive functions (Dubey Agrawal, 2003; Dubey *et al.*, 2004; Agrawal *et al.*, 2001; Agrawal *et al.*, 2002). Several studies have shown the anti-inflammatory and anti-anxiety properties. It has also capacity to improve overall mental performance particularly memory and attention span (Gexianoyam *et al.*, 1986, Zhang *et al.*, 1987; Dubey *et al.*, 1990; Chai *et al.*, 1999; Agrawal *et al.*, 2001; Agrawal *et al.*, 2004). Seabuckthorn has potentiality to enhance the adaptability by reducing adverse reactivity towards any kind of stress of emotional reaction (Dubey *et al.*, 2001). Seabuckthorn is the plant beneficial in the prevention and management of cardiovascular and respiratory disorders and also used for enhancing immunity and general body resistance (Dubey *et al.*, 1990).

The Compound extraction of seabuckthorn strengthens non-specific immunity (Zhang *et al.*, 1989) and the seed oil of the plant has capacity to restore under inhibited state of immune function. Seabuckthorn oils have unique anti-aging properties and stimulate tissue regulation (Yu Let *et al.*, 1993; Zhone Fei *et al.*, 1989, Zhang *et al.*, 1987; Cheng *et al.*, 2003). An alcoholic extract of seabuckthorn which mainly contains flavonoids, protect the bone marrow damage due to radiation and also enhance non-specific immunity (Agrawal and Goel, 2002). Seabuckthorn oil protects the liver damaging from toxic chemicals, as revealed in laboratory studies (Gao *et al.*, 2003; Cheng *et al.*, 1990). It has also protected the function of integrity of mitochondria from radiation-induced oxidative stress (Goel *et al.*, 2005; Narayanans *et al.*,

2004; Bray, 1999). The total flavones of seabuckthorn given for 8 weeks to 35 hypertensive patients have prevented supine isometric exercise induced increase in heart rate, blood pressure, and plasma catecholamine concentration (Zhang *et al.*, 2001).

17.5.4 CONCLUSIONS

Dementia generally gets worse with advancing age but the development of symptoms can often be slowed with a proper management with drugs or other methods. General drugs which prevent the breakdown of acetylcholine in the brain are called "Cholinesterase inhibitors" and they are supposed to act by increasing the levels of acetylcholine in the synopses. These agents may improve the memory as well as slowing down the changes in mood of demented cases. In the present study, novel test drug has prevented the acetylcholine loss as noted in experimental studies improved the mental status of the aged subjects. The memory and attention improved and no further deterioration was noticed. Another series of experimental trial have further suggested that the process of decline of the acetylcholine level due to age is slowed down under the influence of test drug.

References

Brice, A. "Genetics of Parkinson's disease. 2005. LRRK2 on the rise (Scientific Commentary)". *Brain* 128 (12): 2760–2762.

Adams, R.D., and Victor, M. 1993. Disease of the nervous system due to nutritional deficiency, In: Adams, R.D., and Victor M., (eds) Principle of Neurology: (5th ed) Publishers Mc. Grow-Hill International 850-876.

Agrawal, A.; Dixit S.P., and Dubey GP. 2000. Role of a herbal formulation in the management of age related neuropsychophysiological deterioration among the elderly. *Indian Journal of Gerontology*, Vol 14, 1 &2, 15-23.

Agrawal, A.; Khuntia, B.B., and Dubey G.P. , 2001. Seabuckthorn modilates the mental health status and immunosenescence among the elderly-A double blind placebo control study. *Indian J. Gerontology*, vol & pp.

Agrawal, A, Dixit, S.P.; Dubey, G.P., and Medhya Rasayana. 2002. Property of seabuckthorn (*Hippophae rhamnoides*) in age related cognitive deficits, Ayurvedic conference on Rasayana, Rashtriya Ayurveda Vidyapeeth (National Academy of Ayurveda), Ministry of Health and Family Welfare, Govt. of India., Dhanwantari Bhawan , Road No. 66, Punjabi Bagh (West), New Delhi.

Agrawal, A; Tewari, S. and Dubey, G.P. 2001. Role of seabuckthorn (*Hippophae rhamnoides*) in the management of age related memory disorder. Proceeding of International workshop on Seabuckthorn (V. Singh and P.K. Khosla (eds) pp 225-230, New Delhi.

Agrawal, A; Vasanthi, H.R.; Attrey, D.P., Dubey, G.P. 2004. An ayurvedic drug therapeutic strategy in the prevention and management of neuropsychiatric problems among elderly. In : Seminar on Seabuckthorn , A resource for Environment, Health and Economy. Organised by : Directorate of Life Sciences, Defence Research and Development Organisation, Chanakya Bhawan New Delhi, March 12-13., 2004.

Agrawal, P.K., and Goel, H.C. 2002. Protective effect of RH-3 with special reference to radiation induced micronuclei in mouse bone marrow, *Indian Journal of Experimental Biology*, May; 40 (5): 525-530.

Albert, M.S., and Drachman, D.A. 2000. Alzheimer's disease: what is it, how many people have it, and why do we need to know? *Neurology*, 55, 166–168.

Alleyne, G.A.O.; Hay, R.W.N; Picou, D.I; Stanfield J.P, and Whitehead R.G. 1981. In: Protein Energy Malnutrition. The ELBS and Edward Arnold Ltd. London, ; 1-3.

Andress, G. 1984. Contemporary drug abuse In: Harshberger S.E. (eds.) Medicinal Pharmacology, C.V. Mosby Co. Toranto, 348-49.

Arnold, L .E; Pinkham S.M., and Votalato, N. 2000. Does Zinc Moderate essential fatty acid and amphetamine treatment of ADHD. *J. Child Adolese Psychopharmacol Summer;* 10, 111-17 (2000).

Sinha, Asru K. 1972. Colorimetric Assay of Catalase. *Analytical Biochemistry* 47, 389-394.

Ballard, C.; Bannister, C and Solis M. 1996. The prevalence, associations and symptoms of depression amongst dementia sufferers. *J. Affect Disord* 36(3-4), 135-44, (1996).

Beaton, G.H., and Bengoa JM. 1976. WHO Nutrition in preventive Medicine, *WHO*, 2-3, Geneva.

Bedi, K.S. and Bhide, P.G. 1988. Effects of environmental diversity on brain morphology, *Early. Hum. Dev.* 17:107-143.

Benefice, E. 1992. Physical activity and anthropometric and functional characteristics of mildly malnourished Senegalese children. *Ann. Trop. Paediatr.* 12, 55-66.

Benesova, O., Frankova, S., Tikal, K., Benes, V. and Kunz, K. 1972. The Effect of pyrithioxine (Encephobol Merck) on behavior, learning and biochemical variables of brain in rats malnourished in early life. I. Individual and social behavior of rats in relation to brain monoamines and liver tryptophan-Pyrrolase activity. *Act. Nerv. Super.* (Praha) 14, 172-173.

Benton, D. 1998. Breakfast, blood glucose and cognition Am J. Clin. Nutr. 67 (suppl) 7725-7785.

Bernath, J. and Foldesi, D. 1992. Seabuckthorn (*Hippophae rhamnoides* L.): a promising new medicinal and food corp. *J. Herbs, Spices & Medicinal Plants* 1(1/2), 27-35.

Beveridge, T.T.S.C.; Li, B.D. Oomah and A. Smith. 1999. Sea buckthorn products: manufacture and composition. *J. Agric. Food Chem.*, 47, 3480-3488(1999).

Biederman. J, and Spencer T. 2000. Non-stimulant treatments for ADHD. *Eur Child Adolesc. Psychiatry* 9, 1151-9.

Blass, JP., Gleason, P., and Brush D, 1988. Thiamine and Alzheimer's disease. A pilot study. *Arch. Neurol.* 45 833–5.

Bonavita, E. 1986. Study of the efficacy and tolerability of L-acetylcarnitine therapy in the senile brain. *Int. J. Clin. Pharmacol. Ther. Toxicol,* 24 511–6.

Bray, T.M. 1999. Antioxidants and oxidative stress in health and disease. Introduction. *Proc. Soc. Expl. Biol. Med.* 222, 195.

Birkmayer and J. G. D. Birkmayer. "Iron, a new aid in the treatment of Parkinson patients". *Journal of Neural Transmission* 67 (3–4): 287–292.

Brockman. L, and Rieciuti H. 1971. Severe protein calorie deficiency and cognitive development. *Dev. Psychol.* 4, 312-319.

Brunser, O; Castillo. C, and Araya M. 1976. Fine structure of the small intestinal mucosa in infantile marasmic malnutrition. *Gastroenterology,* 70, 495-507.

Burns, E.M. and Brown, K.B. 1977. Prenatal malnutrition:effects on brain norepinephrine content. *Brain Res. Bull.* 2, 313-316.

Buschke, H.; Kulansky, G. and Katz M. 1999. Screening for dementia with the Memory Impairment Screen. *Neurology* 52, 231–238.

Ellis, C.E. and Hill, D.E. 1975. Growth, Intelligence and school performance in children with cystic fibrosis who have had an episode of malnutrition during Infancy. *J. Pediat,* 87, 565 (1975).

Celedon, J.M. and de Andrace, I. 1979. psychomotor development during treatment of severely marasmic infants. *Early Hum. Dev.* 3, 267-275.

Celedon, J.M., Csaszor D., Middleton, J. and deAndraca I. 1980. The effect of treatment on mental and psychomotor development of marasmic infants according to age of admission. *J. Ment. Defic. Res.* 24, 27-35.

Chai Quiyan (1999).The experimental studies on the cardiovascular pharmacology of Seabuckthorn extract form *Hippophae rhamnoides* In: Proceeding of International workshop on Seabuckthorn, Beijing, China.

Chang, C.C. 1964. *Int J. Neuropharmac.* 3, 643-649.

Chase, H.P., Dorsey, J. and Mckhann, G.M. 1967. The effect of malnutrition on the synthesis of a myelin lipid. *Pediatrics,* 40, 551-559.

Chatterjee, A., Strauss, M.E., Smyth, K.A., Whitehouse, P.J. 1992. Personality changes in Alzheimer's disease. *Arch. Neurol.,* 49, 486-491.

Chaudhary, O.P. 1978. Manual for long term memory. Published by Manovigyanic Parikchhan Sansthan, U.H.B., 28 Sanjay Nagar colony, Chowkaghat, Varanasi.

Chen, T. 1988. Studies of the biochemical composition of Hippophae and its quality assessment in Gansu Province. *Hippophae*, 1, 19-26 (1988).

Cheng, J. 2003. Inhibitory effects of total flavones of *Hippophae rhamnoides* on thrombosis in mouse femoral artery and in vitro platelet aggregation, *Life Sciences;* 72(20), 2263-2271.

Cheng, T. *et al* 1990. Acute toxicity of flesh oil of *Hippophae rhamnoides* and its protection against experimental hepatic injury, Journal *of Traditional Chinese Medicine* 15(1), 45-47., 64.

Clarke, R., Smith, D., Jobst and K.A., et al. 1998. Folate, vitamin B12, and serum total homocysteine levels in confirmed Alzheimer disease. *Arch. Neurol.*, 55:1449–55.

Connolly, K.J., and Kvalsvig, J.D. 1993. Infection, Nutrition and cognitive performance in children. *Parasilology*, 107 suppl: 5187-200.

Cooper, J.K.; Mungas, D.; Verma, M, and Weiler, P.G. 1991. Psychotic symptoms in Alzheimer's disease. *Int. J. Ger. Psychiatry*, 6, 721-726.

Cragg, B.G. 1972. The development of Cortical synapses during starvation in the rat. *Brain* 95:143-150.

Cravioto, J. and Arrieta, R. 1986. Nutrition, Mental development and learning. In: Human Growth, Faulkner, F. and Tanner JM (eds) Vol 3, PP. 501-536. Plenum Publishing, New York, NY and London, UK.

Cremen, H.D., 1982. Influence of Malnutrition on mental development and behaviors. *Bibl. Nutr. Dieta.* (31): 32-9.

Crook, T., Petrie, W., Wells, C., and Massari D.C. 1992. Effects of phosphatidylserine in Alzheimer's disease. *Psychopharmacol Bull.* 28:61–6.

Cucinotta, D. et al. 1988. Multicenter clinical placebo-controlled study with acetyl-L-carnitine (LAC) in the treatment of mildly demented elderly patients. *Drug Development Res.* 14:213–6.

Culley, W.J. and Uneberger, R.O. 1968. Effect of under nutrition on the size and composition of the rat brain. *J. Nutr.* 96, 375-381.

Cummings, J.L. and Benson, D.F. 1992. *Dementia: A clinical approach.* Stoneham, MA: Butterworth-Heineman.

Dammann, O. and Leviton, A. 1997. The role of peri-natal brain damage in development disabilities: an epidemiological perpetive. *Mental retardation and development disorders Research Review*, 3, 13-21.

Delafuente JE., 1991. Nutrient and immune response. *Rheum. Dis Clin. North AM.* 17, 203-211.

Detering, N., Collins R.M., Jr Hawkins; R.L., Ozand P.T. and Karahasan 1980b. A. comparative effects of ethanol and malnutrition on the development of catecholamine neurons: changes in neurotransmitter levels. *J. Neurochem.* 34, 1587-1593 (1980b).

Detering, N., Collins, R.M.; Jr. Hawkins R.L.; Ozand P.T. and Karahasan, A. 1980a. Comparative effects of ethanol and malnutrition on the development of catecholamine neurons: changes in noeepinephrine turnover. *J. Neurochem.* 34, 1788-1791.

Detering, N., Edwards, E., Ozand, P and Karahasan, A. 1980c. Comparative effects of ethanol and malnutrition on the development of catecholamine neurons: changes in specific activities of enzymes. *J. Neurochem.* 34: 297-304.

Dickerson, J.W. and Dobbing, J. 1966. Some peculiarities of cerebellar growth in pigs. *Proc. Soc. Med.* 59, 1088.

Dobbing, J. 1964. The influence of early nutrition on the development and myelination of the brain proc. *Roy. Soc. Biol.* 159, 503.

Dobbing, J., Hopewell, J.W. and Lynch, A. 1971. Vulnerability of developing brain VII permanent deficit of neurons in cerebral and cerebellam cortex following early mild undernutrition. *Exp. Neurol.* 32. 439.

Dorjey, T. 1998. Seabuckthron – A wonder plant in Ladakh. Paper presented in symposium on "Role of women in eco-development in Ladakh" held at Divisional forest office, Leh, Ladakh, J&K, India. 1st -4th Oct.

Dubey, G.P., Agrawal A, and Dixit, SP., 2001. Neuro-physiological adaptation following stress under effect of Seabuckthorn. In. Proceeding of International Workshop on Seabuckthorn, eds. Singh, V, Khosla, PK, pp-199-204, 18-21 Feb 2001.

Agrawal A, Khuntia BB, Dubey GP, Seabuckthorn modulates the mental health status and its effect on immunosenescence among the elderly- A double blind placebo control study. In. Proceeding of International Workshop on Seabuckthorn, Eds. Singh, V, Khosla PK, pp-214-220, 18-21 (Feb 2001).

Dubey, G.P., Agrawal, A, Dixit, S.P., 2002. Role of seabuckthorn (*Hippophae rhamnoides*) in the maintenance of cardiovascular homeostasis following cold stress. *Journal of natural remides.* Vol 3 / 136-40.

Dubey, G.P., Agrawal, A; Gupta, B.S. and Udupa, K.N. 1990. Neurophysiological adaptation following cold stress under the influence of Badriphal (*Hippophae rhamnoides* L.) *Pharmacophychoecologia,* 3, 59-63.

Dubey, G.P.; Agrawal, A.; Wilson, E.; Vyas, N. and Rajamanickam, G.V. 2005. Novel markers of atherothrombotic risk – therapeutic benefits. In Second International Conference of The International Society for Ayurveda & Health and International seminar on Chronic Diseases and its Management by Complementary and Alternative Medicine (CAM), organised by Deptt. of Medicinal Chemistry, IMS, BHU & University of Connecticut , School of Medicine, USA, 23rd -25th Sept. 2005.

Dubey, G.P., Agrawal, A. 2003. Memory enhancing potentials of seabuckthorn (*Hippophae rhamnoides*) - A placebo controlled study. In: Congress of the International Seabuckthorn Association - ISA. Organised by : International Seabuckthorn Association - ISA German Seabuckthorn Society - Sanddorn E.V. Humboldt-University, Berlin, Technical University, Berline Sept 14-18, 2003.

Dubey, G.P., Attreya D.P., Agrawal, A . 2004. Multi-dimensional therapeutic potential of Seabuckthorn (*Hippophae rhamnoides*) with special reference to memory disorder. In seminar on Seabuckthorn, a resource for environmental, health and economy directorate of life sciences, defence research and development organization Chanakya Bhawan, New Delhi.

Dubey, G.P., Singh, R.G., Agrawal, A., and Rajamanickam, G.V. 2005b. Assessment of atherothrombotic profile for the targeting of Ayurvedic formulation treatment in the prevention of coronary event. In 38th Annual Conference, Indian Pharmacological Society at Institute of Pharmacology, Madras Medical College, Chennai, and Tamilnadu, 28th to 30th March 2005.

Ebly, E.M., Schaefer, J.P., Campbell, N.R., and Hogan, D.B. 1998. Folate status, vascular disease and cognition in elderly Canadians. *Age Ageing*, 27, 485–91.

Eccleston, C; Baoru, Y; Tahvonen, R; Kallio, H; Rimbach, G.H. and Minihane, A.M. 2002. Effects of an antioxidant rich juice (sea buckthorn) on risk factors for coronary heart disease in humans. *Journal of Nutritional Biochemistry*, 13,346-54. (2002)

Eder, L.; Dunant, Y., and Loctin F. 1980. Thiamine and cholinergic transmission in the electric organ of Torpedo. *J Neurochem*. 35, 1278–96.

Elias, M.F. and Samonds, K.W. 1979. Protein and calorie malnutrition during deprivation and rehabilitation. *Am. J. Clin. Nutr.* 230, 355-366.

Emerich, D.F. and Walsh, T.J., 1990. Cholinergic cell loss and cognitive impairments following intraventricular or intradentate injection of colchicines, *Brain Res.* 517, 157-167. (1990)

Evans, D.E., Bowie, M.D., and Hansen, J.D.L. et al. 1980. Intellectual development and nutrition. *Journal of pediatrics*, 97, 358-363.

Evans, D.A.; Funkenstein, H.H., and Albert, M.S. 1989. Prevalence of Alzheimer's disease in a community population of older persons. *Journal of the American Medical Association, 262*, 2551-2556.

Fahey, J.L. and Makelvey, E.M. 1968. Quantitative determination of serum immunoglobulin in antibody agar plate. J. Immu, 94, 84.

Fe, Q.; Yang Q.Y., Yang G.D., Shi, J, Chang, C. and Chen, H.J. 1993. Determination of á-tocopherol content in Seabuckthorn oil by reverse – phase high performance liquid chromatography. *Journal of xi a Medical University*, 14 (2), 181-183.

Feldman, M., Friedman, and LS, Sleisenger, M.H. 2002. Gastrointestinal and Liver Disease; Pathophysiology Diagnosis and Management. 7th ed.; 1:273-274 (2002).

Ferris, S.H.; Sathananthan, G; Gershon S, et al. 1977. Senile dementia. Treatment with Deanol. *J. Am. Geriatr. Soc*; 25, 241.

Ferris, S.H., and Kluger, A. 1996. Commentary on age-associated memory impairment, age- related cognitive decline and mild cognitive impairment. *Aging Neuropsychology, and Cognition*, 3, 148-153.

Floud R. 1992. Anthropometric measure of nutritional status in industrialized societies : Europe and North America since 1750. *In* : Osmani, S.R. (ed). Nutrition and poverty. Oxford: Oxford University Press, 218-42.

Folstein, M.F., Folstein, S.E., and McHugh, PR. 1975. "Mini-Mental State": a practical method for grading the cognitive state of patients for the clinician. *J. Psychiatr. Res* 12 (3), 189-98.

Fong, S. 1985. Origin and age associated changes in the expression of physiologic autoantibody. *Gerontology*, 31, 236.

Friedland R. 2000. American Academy of Neurology's 52nd Annual Meeting in San Diego, CA, April 29–May 6, 2000.

Galler, J.R.; Ramsey, F. and Solimano, G. 1985. A Follow up study of the effects of early malnutrition on subsequent development. II. Fine motor skills in adolescence Pedi.

Galler, J.R., Ramsey, F.; Morley, D.S., Archer, E. and Salt, P. 1990. The long lerm effect of early Kwashiorkor compared with marasmus. IV. Performance on the national high School entrance examination. *Pediatr. Res*. 28, 235-239.

Galler, J.R., Ramsey, F.; Salt P and Archer, E. 1987. Long term effects of early Kwashiorkor compared with marasmus. III. Fine motor skills. *J. Pediatr. Gastroenterol. Nutr*. 6 : 855-859.

Galler, J.R., Ramsey, F.; Solimano, G.; Kucharski, L.T., and Harrison, R. 1984. The influence of early malnutrition on subsequent behavioral development. IV. Soft neurological pediatr. Res. 18, 826-832.

Gao, S.; Hendrie, H.C., and Hall, K.S., et al. 1998. The relationships between age, sex, and the incidence of dementia and Alzheimer's disease. A meta-analysis. Arch. Gen. Psychiatry.; 55, 809–815.

Gao, Z.L. et al. 2003. Effect of sea buckthorn on liver fibrosis: a clinical study, World *Journal of Gastroenterology*, 9 (7), 1615-1617.

Gao, X.; Ohlander, M.; Jeppsson, N.; Bjork, L. and Trajkovski, V. 2000. Changes in antioxidant effects and their relationship to phytonutrients in fruits of seabuckthorn (*Hippophae rhamnoides* L.) during maturation. *Journal of Agricultural Food Chemistry*, 48, 1485-90.

Geetha, S., Sai Ram, M., Singh, V., Ilavazhagan G. and C.Sawhney. R., 2002. Anti-oxidant and immunomodulatory properties of seabuckthorn (*Hippophae rhamnoides*). An in vitro study. *J. Ethnopharmacol.*, 79, 373-378.

Gexianoyam, pl. elaborate, 1986. Preliminary research on the chemical composition of Seabuckthorn. *Chinese herbs,* 17, 42-44.

Gibson, G.E., Sheu, K.F., Blass, J.P., pl. elaborate.1988. Reduced activities of thiamine-dependent enzymes in the brains and peripheral tissues of patients with Alzheimer's disease. *Arch. Neurol.* 45:836–40.

Girault, J.A. and Greengard, P. 2004. The neurobiology of dopamine signaling. *Arch. Neurol.* 61(5), 641-4.

Glowinski, J. and Iverson, L.L. 1966. Regional studies of catecholamines in the rat brain. I the disposition of (3H) dopamine and (3H) DOPA in various regions of rat brain. *J. Neurochem.* 13, 655-699.

Goel, H.C., Gupta, S., Garg, A.P. and Bala, M. 2005. Protection of mitochondrial system by *Hippophae rhamnoids* L. against radiation-induced oxidative damage in mice. *Journal of Pharmacy and Pharmacology,* 57, 135-43.

Good, R.A., Fernandes, E.J., Yunis WC pl. elaborate. 1976. Nutritional deficiency immunological function and disease. *A.M. J. Pathol,* 84, 599-614.

Goodwin, J.S., Searles, R.P. and Tung, K.S.K. 1982. Immunological responses of a healthy elderly population. *Clin. Exp. Imminol.* 48, 403.

Gracey, M., Gullity, G.J., Suharjone S. 1977. The stomach in malnutrition. *Archives of Diseases in Childhood,* 52, 325-327.

Grant W.B. 1997. Dietary links to Alzheimer's disease. *Alzheimer Dis. Rev.* 2, 42–55.

Grantham MC Gregor SM. , 2001. Ani CC Undernutrition and mental development. *Nestle Nutr Workshop ser clin perform programme;* (5): 1-14 (2001).

Grantham, M.C., Gregor, S.M., Fernald, L.C., 1992. Nutritional deficiencies and subsequent effects on mental and behavioural development in children. *Southest Asian J. Trop. Med. Public Health,* 28, Suppl 2, 58-68.

Grantham, M.C. and Gregor, S. 1995. A review of studies of the effect of severe malnutrition on mental development *Journal of nutrition* 125, 22325-85.

Grantham, M.C., Gregor, S.M., Schofield, W. and Haggard, D. 1989b. Maternal child interaction in survivors of severe malnutrition who received psychosocial stimulation. *Eur. J. Clin. Nutr.* 43, 45-52.

Griffiths, J.C., Leung, F.Y.T., Mc Donad, J.J. 1970. Flurometric determination of plasma catecholamines: Normal human epinephrine and norepinephrine levels, *Clin. Chem. Ecta.* 30, 395.

Groman, K.S. 1995. Malnutrition and cognitive development evidence from experimental/ quasi experimental studies among the mild-to-moderately malnourished. *J. Nutr.* 125 (suppl-8), 22395-22445.

Guesry, P. 1998. The role of nutrition in brain development. *Prev. Med,* 27, 189-194.

Gupta, B.S., Agrawal, A., Dubey, G.P., Udupa, K.N., 1990. Anti-ageing effect of badrifal (*Hippophae rhamnoides*) *Pharmaco-psychoecologia,* 3,53-57.

Gurevich, S.K. 1956. The application of seabuckthorn oil in ophthalmology. *Vesttn Oftalmologii,* 2, 30-33.

Guseva, A.R. and Nestuk, M.N. 1995. Kmetadike opredelenia flavonolovykh veshestva v rasteniakh. *J. Biokhem,* 18 (4), 480 -485.

Haba, K.; Ogawa, N.; Kawata, N. and Mori, A. 1988. A method for parallel determination of choline acetyltransferase and muscarinic cholinergic receptors; application in aged-rat brain. *Neurochem. Res.* 13, 951-955.

Halliwell, B. and Gutteridge, J.M.C. 1999. Oxidative stress and disorders of the nervous system: general principles. In : Halliwell, B. and Gutteridge, J.M.C. (eds) Freeradicals in biology and medicine, (3rd ed), Oxford Science Publications, Oxford, 721-731.

Hebert, L.E., Scherr, P.A., Beckett, L.A., pl. elaborate. 1995. Age-specific incidence of Alzheimer's disease in a community population. *JAMA,* 273, 1354–1359.

Heinonen, M.; Meyer, A. and Frankel, E. 1998. Antioxidant activity of berry phenolics on human low-density lipoprotein and liposome oxidation. *J. Agric. Food Chem.* 46, 4107-4112.

Heiss, W.D., Kessler, J. and Mielke, R, pl. elaborate. 1994. Long-term effects of phosphatidylserine, pyritinol and cognitive training in Alzheimer's disease. A neuropsychological, EEG, and PET investigation. *Dementia,* 5, 88–98.

Hillen, T.; Lun, A. and Reischies, F.M., pl. elaborate. 2000. DHEA-S plasma levels and incidence of Alzheimer's disease. *Biol. Psychiatry,* 47, 161–3.

Hiroshi, O.; Nobuko O. and Kunin Y. 1979. Assay for lipid peroxides in animal tissues by thiobarbituric acid reactions. *Analytical Biochemistry,* 95, 351-358.

Hisatomi, K. and Niiyama. Y. 1980. Effects of postnatal under nutrition on the catecholamine and serotonin contents of suckling rat brain *J. Nutr. Sci. Vitaminol* (Tokyo), 26,279-292.

Hoorweg, J. and Stanfield, J.P. 1976. The effects of protein energy malnutrition in early childhood on intellectual and motor abilities in later childhood and adolescence. *Dev. Med. Child Neurol.,* 18,130-105.

Huang, Y.Q. and Xiao, Y.Y. 1993. Determination of á-tocopherol and its isomers in seabuckthorn oils by high performance liquid chromatographic method. *Hippophae,* 6(3), 35-40.

Imagawa, M.; Naruse, S.; Tsuji, S.; pl. elaborate. 1992. Coenzyme Q10, iron, and vitamin B6 in genetically-confirmed Alzheimer's disease. *Lancet.* 340, 671.

Jablczynska, R.; Krawazyk, U. and Minkowski, K.; 1994. Fatty acids and tocopherols in some varieties of *Hippophae rhamnoides. Acta poloniae pharmaceutica–drug research Jl* (3), 267–269.

Jurica, P.J., Leittern, C.L. and Mattis, S. 2001. Dementia Rating Scale-2, Professional Manual; Psychological Assessment Resources, Inc, Florida.

Kalmijn, S.; Lauher, L.J.; Ott, A.; pl. elaborate. 1997. Dietary fat intake and the risk of incident dementia in the Rotterdam study. *Ann. Neurol.* 42, 776–82.

Kapalan, H.I., and Sadock, B.J. 1995. Attention Deficit Hyperactivity Disorder Comprehensive Textbook of Psychiatry Vol II, Williams and Wilkins: Baltimore, 2295-2310.

Kaszniak, A.W., and Christenson, G.D. 1994. Differential diagnosis of dementia and depression. In Storandt M. and VandenBos G.R. (eds.), *Neuropsychological Assessment of Dementia and Depression in Older Adults: A Clinician's Guide* (pp. 81-118). Washington, DC: American Psychological Association.

King, J.A.; Kelly, T.A., and Deiville, Y. 2000. Adult level of testosterone after catecholamine innervations in an animal model of ADHD. *Neuropsychobiology*, 42, 163-68.

Klahr, S., and Alleyne, G.A.O., 1973. Effect of chronic protein-calorie malnutrition on the kidney. *Kidney International*, 3, 129-141.

Klein, P.S., Forbes, G.R., and Nadir PR. 1975. Effect of starvation in infancy on subsequent learning abilities. *Journal of Pediatrics*, 87, 18-15.

Kuhne, T.; Bubl, R. and Baungartner, R. 1991. Maternal vegan diet causing a serious infantile neurological disorder due to vitamin B12 deficiency. *European Journal of Pediatric*, 150, 205-208.

La Rue, A. 1992. Aging and neuropsychological assessment. New York: Plenum.

Lango, R.H., P Fisher D, Molinari L, Kandu S, Lipp A, and Duc G. 1989. Significance of prenatal and postnatal factor in the development of age preterm infants Sat 5 to 7 year. *Development Medicine and child neurology*, 31: 440-456. (1989).

Larrabee, G.J.; Largen, J.W. and Levin, H.S. 1985. Sensitivity of age-decline resistant ("Hold") WAIS subtests to Alzheimer's disease. *Journal of Clinical and Experimental Neuropsychology, 7*, 497-504.

Lec., C.J. and Dubos, R. 1972. Lasting biological effects of early environmental influences. 8 effects of neonatal infection. Perinatal malnutrition and crowding on crowding on catecholamine metabolism of Brain. *J. Exp. Med.* 136, 1031-1042.

Lecendrex, M.; Konafal, E.; Bouvard, M.; Falissard, B., and Simeoni, M.C. 2000. Sleep and alertness in children with ADHD. *J. Child Psychol. Psychiatry*, 41:803-12.

Lethem, R. and Orrell, M. 1997. Antioxidants and dementia. *Lancet*, 349, 1189–90.

Leuba, G. and Rabinowicz, T. 1979a. Long-term effects of postnatal undernutrition and Maternal malnutrition on mouse cerebral cortex. I Cellular densities, cortical volume and total number of cells. *Exp. Brain Res.*, 37, 283-298.

Leuba, G. and Rabinowicz,T. 1979b. Long-term effect of postnatal undernutrition and maternal malnutrition on mouse cerebral cortex .II. Evaluation of dendritic branching and spines in the visual region exp. *Brain Res.*, 37, 299-308.

Levistsky, D.A. and Strupp, B.J. 1979. Malnutrition and the brain, changing concepts, changing concerns. *J. Nutr.* (In press).

Lezak, M. 1995. Neuropsychological assessment (3rd ed.). New York: Oxford.

Lunn, P.G.; Whitehead, R.G.; Cole, T.J. 1979. The relationship between hormonal balance and growth in malnourished children and rats. *British Journal of Nutrition*, 41, 73-84.

Lynch, A.; Smart, J.L. and Dobbing, J. 1975. motor co-ordination and cerebellar size in adult rats undernourished is early life. *Brain Res.*, 83, 249-259.

Stern M. 1991. The epidemiology of Parkinson's disease. *Archives of Neurology*, 48 (9), 903–907.

Ma, Z. and Y, Cui. 1989. Studies on the fruit character and biochemical composition of some forms within the Chinese seabuckthorn (*Hippophae rhamnoides* ssp. *sinensis*) in Shanxi, China. *Proc. Int. Symp. Seabuckthorn* (*H. rhamnoides* L.), Xian, China. p. 106-112.

Mahadeviah; Rukmini, K., and Swany 1997. Role of pediatrician in learning disability in children. *Journal of Practical pediatrics*, 5 (1).

Mansell, J. 1993. Services for people with learning disabilities and challenging behaviour or mental health needs. HMSO, London (1993).

Maria, S. Moran; Joseph, W.; Depierse and Bengt, Mannervik. 1979. Levels of Glutathione, Gluathione Reductase and Gluathione-S- Transferase activities in Rat lung and Liver, *Biochemica et Biophysica Acta*.

Mathews, J.D. 1973. Association of auto-antibodies with smoking cardiovascular morbidity and death in the Busselton population, *Lancet*, 2, 754.

Mattingly, D. 1962. A single flurometric for the study of tree II-hydroxycortio steroid in human plasma, *J. Clin. Phathol.*, 15, 374.

Meador, K.; Loring, D.; Nichols, M. 1993. Preliminary findings of high-dose thiamine in dementia of Alzheimer's type. *J. Geriatr. Psychiatry Neurol.*, 6, 222–9.

Mendez, M.A., Adaur, L.S| 1999. Severity and timing of stunting in the first 2 yrs. of life affect performance on cognitive tests in rate childhood. *J. Nutr*, 129, 1555-1562.

Miller, R.A. 1991. Accumulation of hyporesponsive calcium extruding memory T-cells as a key feature of age dependent immune dysfunction. *Clin. Immunol. Immunopathol.*, 58(3), 305-17.

Miller, M.; Leahy, J.P.; Stern, W.C.; Morgane, P.J. and Resnick, O. 1977. Tryptophan availability. Relation to elevated brain serotonin in developmentally protein-malnourished rats. *Exp. Neurol.*, 57, 142-157.

Miyanaga, K.; Ynoemura, K.; Takagi, T. 1995. Clinical effects of DHA in demented patients. *J. Clin. Ther. Med.*, 11, 881-901.

Montastruc, J.L., Rascol, O.; Senard, J.M. 1994. A randomized controlled study comparing bromocriptine to which levodopa was later added, with levodopa alone in previously untreated patients with Parkinson's disease: a five year follow up. *J. Neurol. Neurosurg. Psychiatry*, 57, 1034-38.

Montgomery, R.D. 1962. Muscle morphology in infantile protein Malnutrition. *Journal of Clinical Pathology*, 15, 511-521.

Morgane, P.J.; Kemper, T.; Stern, W.; Hawrylewicz, E. and Resnick, O. 1978. The effect of protein malnutrition on the developing central Nervous system in the rat. *Neurasci, Blobehav. Rev.*, 2, 137-230.

Morris, M.C., Beckett, L.A.; Scherr, P.A. 1998. Vitamin E and vitamin C supplement use and risk of incident Alzheimer disease. *Alzheimer Dis. Assoc. Disord.*, 12, 121–6.

Murer, M.G.; Dziewczapolski, G. and Menalled, L.B. 1998. Chronic levodopa is not toxic for remaining dopamine neurons, but instead promotes their recovery, in rats with moderate nigrostriatal lesions. *Ann. Neurol.* 43, 561-575.

Narayanan, S.; Ruma, D.; Gitika, B.; Sharma, S.K.; Pauline, T.; Sai Ram, M.; Illavazhagan, G.; Sawhney, R.C.; Kumar, D. and Banerjee, P.K. 2004. Antioxidant activities of seabuckthorn (*Hippopahe rhamnoides*) during hypoxia induced oxidative stress in glial cells, 278(1-2), 9-14.

Nasman, B.; Olsson, T.; Backstrom, T. 1991. Serum dehydroepiandrosterone sulfate in Alzheimer's disease and in multi-infarct dementia. *Biol. Psychiatry*, 30, 684–90.

Noback, C.R. and Eisenman, L.M. 1981. Ssome effects of proteincalorio. Undernutrition on the developing central nervous system of the rat. *Anat. Rec.*, 201, 67-73.

Oliverio, A.; Castellano, C. and Allegra, S.P. 1975. Eeffects of genetic and nutritional factors on post-natal reflex and behavioral development in the mouse. *Exp. Aging Res.*, 1, 41-56.

Patel, P.J. 1981. Ageing and antimicrobial immunity. *J. of Experimental Medicien*, 154, 821.

Patrick, J.; Reeds, P.J.; Jackson, A.A. 1978. Total body water in malnutrition: the possible role of energy intakes. *British Journal of Nutrition*, 39, 417-424.

Pereira, S.M.; Sundaraj, R. and Begum, A. 1979. Physical growth and neuro-integrative performance of survivors of protein energy malnutrition. *Br. J. Nutr.*, 42, 165-171.

Peterson, L.R.S. and Peterson, M.J. 1969. Short-term retention of individual verbal items. *J. Exp. Psycho.*, 58, 193-198.

Pettegrew, J.W., Klunk WE, Panchalingam K, 1995. Clinical and neurochemical effects of acetyl-L-carnitine in Alzheimer's disease. *Neurobiol Aging* 1995; 16:1–4.

Plekhanova, M.N. Seabuckthorn. 77 p. (in Russian).

Pliszka, S.R. 2000. Pattern of psychiatric comorbidity with attention deficit/hyperactivity disorder. *Child Adolesc. Psychiatry. Clin. N. Am.*, 9, 626-29.

Pollitt, E. 1994. Poverty and child development: relevance of research in developing to the United States. *Child Development*, 65, 283-295.

Poon, L.W. 1985. Differences in human memory with ageing In: Nature, causes and clinical implications. In: Handbook of the psychology of Ageing (J.E. Birren and K.W. Schaie, Eds. 2nd Ed.) New York, Van Nostrand Reinhold, P. 427.

Popkin, B. and Lim –Ybanez, M. 1982. Nutrition and School achievement. *Soc. Sci. Med.*, 16, 53-61.

Power, D.C. 1992. Immunological principles and emerging strategies of vaccination for the elderly., *J. Am. Geriatr. Soc.*, 40, 81-9.

Priest, N.D. 1993. Satellite symposium on Alzheimer's disease and dietary aluminum. *Proc. Nutr. Soc.*, 52, 231–40.

Prince, J.B. Wilens, T.E.; Biederman, J. 2000. A controlled study of NT in Children and adolescent with ADHD. *J Child Adolesc. Psychopharma Col.*, 10, 193-04.

Prior, R.; Cao, G.; Martin, A.; Sofic, E.; McEwen, J.; O'Brien, C.; Lischner, N.; Ehlenfeldt, M.; Kalt, W.; Krewer, G.; Mainland, C. 1998. Antioxidant capacity as influenced by total phenolic and anthocyanin content, maturity, and variety of *Vaccinium* species. *J. Agric. Food Chem.*, 46, 2686-2693.

Rai, G.; Wright, G.; Scott, L. 1990. Double-blind, placebo controlled study of acetyl-l-carnitine in patients with Alzheimer's dementia. *Curr. Med. Res. Opin.* 11, 638–47.

Randt, C.T. and Derby, B.M. 1973. Behavioral and brain correlates in early life nutritional deprivation, *Arch. Neurol.*, 28, 167-172.

Reddy, Vinodinl, Prahlad Rao, N.; Gowrinath, Sastry J. and Kashinat K. 1993. In: Nutrition Trends in India. National Institute of Nutrition: Hyderabad, 31-32.

Rediess, S., and Caine, E.D. 1996 Aging, cognition, and DSM-IV. *Aging Neuropsychology and Cognition, 3*, 105-117.

Reys, M.R.; Valdecanas, C.M.; Reyes, O.L. and Reyes, T.M. 1990. The effects of malnutrition on the motor, perceptual and cognitive function of Filipino children *Int. Disabil Stud.* 12, 131-136.

Rice-Evans, A. and Miller, N. 1994. Total antioxidant status in plasma and body fluids. *Meth. Enzym.*, 234, 279-293.

Rice-Evans, A.; Miller, N. and Paganga, G. 1997. Antioxidant properties of phenolic compounds. *Trends in Plant Science*, 2 152-159.

Richardson, S.A.; Birch, H.G. and Hertzig, M.E. 1973. School performance of children who were severely mal-nourished in infancy. *Am. J. Ment. Defic.*, 77, 623-632.

Robert Thompson, I.C. 1974. Agein Immune response and mortality, *Lancet*, 2, 368.

Rosch, D.; Bergmann, M.; Knorr, D. and Kroh, L.W. 2003. Structure-antioxidant ef.ciency relationships of phenolic compounds and their contribution to the antioxidant activity of seabuckthorn juice. *Journal of Agricultural and Food Chemistry* 51, 4233-9.

Rousi, A. and Aulin. H. 1977. Ascorbic acid content in relation to ripeness in fruit of six *Hippophae rhamnoides* clones from Pyhäranta, SW Finland. *Ann. Agric. Fenn.*, 16, 80-87.

Rutter, M.L.; Madge, N. and Tizard, J. 1980. Intelligence. In : Rutter, M.L. (ed) Scientific Foundation of Development Psychiatry. Heinemann Medical Books, London.

Saissi, F. and Saissi, B. 1973. Differential effects of protein calorie restriction and subsequent repletion on neuronal and nonneuronal components of cerebral cortex in newborn rats. *J. Nutr.*, 103, 1625-1633.

Salvioli, G., and Neri, M. 1994. L-acetylcarnitine treatment of mental decline in the elderly. *Drugs Exp. Clin. Res.*, 20, 169–76.

Sano, M.; Bell, K.; Cote, L.; 1992. Double-blind parallel design pilot study of acetyl levocarnitine in patients with Alzheimer's disease. *Arch. Neuro.*, 49, 1137–41.

Sano, M.; Ernesto, C.; Thomas, R.G. 1997. A controlled trial of selegiline, alpha-tocopherol, or both as treatment for Alzheimer's disease. *N. Engl. J. Med.*, 336, 1216–22.

Sasov, S.A.; Gerasimova, G.K. and Toteeva, N.N. 1996. Biologically active natural complex of poliphenol compounds. In: 2nd Inter. Symp. of the chemistry of Natural compounds. Eskiseher – Turky p 24.

Schaie, K.W., Labouvie and Vief, G. 1974. Generationalversus ontogenetic components of change in adult cognitive behaviour : A fourteen year sequential study. *Develop Psychol.*, 10, 35.

Schapiro, D.C. 1989. Biochemical studies on some hopeful forms and species of seabuckthorn in USSR. *Proc. Int. Symp. Sea-buckthorn* (*H. rhamnoides* L.), Xian, China. p. 64-66.

Schmidt, R.; Hayn, M.; Reinhart, B.; 1998. Plasma antioxidants and cognitive performance in middle-aged and older adults: results of the Austrian Stroke Prevention Study. *J. Am. Geriatr. Soc.*, 46, 1407–10.

Schuller, E.; Allinquant, B. and Reboul, J. 1981. Immunologic studies in human aging, associated increase in anti-RNA and anti DNA antibodies. *J. Clin. Lab. Immunol.* 6, 107.

Scrimshaw, NS. : 1975. Nutrition and infection. *Prog. Food Nutr. Sci.* 1:393-420.

Seidler, F.J., Bell, J.M. and Slotkin, T.A. 1990. Undernutrition and overnutrition in the neonatal rat : long-term effects on noradrenergic pathways in brain regions. *Pediatr. Res.*, 27, 191-197.

Sen, A.P. and Bhattacharya, S.K. 1991. Effect of selective muscarinic receptor agonists and natagonists on active-avoidance learning acquisition in rats. *Indian J. Exp. Biol.*, 29, 136-139.

Shipilina, L.D.; Vichkanova, S.A. and Fateeva, T.V . 1995. In: Pisha, Ecol, Chelovek. Moskwa, P: 7.

Singh, M . 2003. Nutrition, Brain and Environment: How to have smarter Babies. Vol 40, PP 213-220.

Singh, V.; Singh, B. and Awasthi . C.P.1995. Distribution taxonomy and nutritional value of Seabuckthorn (*Hippophae* L.) in dry temperate Himalayas in: *Proc. Int. Workshop Seabuckthorn*, Beijing, China p. 52-59.

Smith, M.A.; Petot G.J. and Perry, G. 1997. Diet and oxidative stress: a novel synthesis of epidemiological data on Alzheimer's disease. *Alzheimer Dis. Rev*, 2, 58–9.

Smith, G.E.; Petersen, R.C.; Parisi, J.E.; Ivnik, R.J.; Kokmem, E.; Tangalos, E.G.; and Waring, S. 1996. Definition, course, and outcome of mild cognitive impairment. *Aging, Neuropsychology, and Cognition*, 3, 141-147.

Snider, S.H.; Axelrad, J. and Zewig, M. 1965. A sensitive and specific fluorescence assay for tissue serotonin, biochem. *Pharmacol.*, 14, 831.

Snowdon, D.A.; Tully, C.L.; Smith, C.D. 2000. Serum folate and the severity of atrophy of the neocortex in Alzheimer disease: findings from the Nun study. *Am. J. Clin. Nutr.*, 71, 993–8.

Sobotka, T.J.; Cook, M.P. and Brodie, R.E. 1974. Neonatal malnutrition: neurochemical, hormonal and behavioral manifestations. *Brain Res.*, 65, 443-458.

Sokoloff, B.K.; Funaoka, M.; Fujisawa, C.C. Saelhof, E. Raniguchi L.B. and Miller. C. 1961. An oncostatic factor present in the bark of *Hippophae rhamnoides*. *Growth*, 25(5), 401-409.

Solonenko L.P. and Shishkina. E.E. 1989. Electrophoretic and amino acid analysis of proteins in the seabuckthorn fruit. *Proc. Int. Symp. Seabuckthorn* (*H. rhamnoides* L.), Xian, China. p. 91-95.

Speeg Jr., K.V. 1974. A sensitive and specific fluorometric method for the determination of acetylcholine. In: Choline and Acetylcholine: Handbook of Chemical Assay Methods. Edited by Hanin, I. Raven Press, New York, pp. 129-133.

Stern, W.C.; Forbes, W.B.; Resnick, O. and Morgane, P.J. 1974. Seizure susceptibility and brain amine levels following protein malnutrition during development in the rat. *Brain Res.*, 79, 375-384.

Stern, W.C.; Miller, M.; Forbes, W.B.; Morgane, P.J. and Resnsick, O . 1975. Ontogeny of the levels of biogenic amines in Various parts of the brain and in peripheral tissues in normal and protein malnourished rats. *Exp. Neurol.*, 49, 314-326.

Stoch, M.B.; Smythe, P.M.; Moodie, A.D. 1982. Psychosocial outcome and CT findings after gross undernourishment during infancy: a 20 year developmental study. *Developmental Medicine and Child Neurology*, 24, 419-436.

Sunderland, T.; Merril, C.R.; Harrington, M.G. 1989. Reduced plasma dehydroepiandrosterone concentrations in Alzheimer's disease. *Lancet*, 2:570.

Hughes, T.A.; Ross, H.F.; Mindham, R.H.S.; Spokes. E.G.S. 2004. Mortality in Parkinson's disease and its association with dementia and depression. *Acta Neurologica Scandinavica*, 110 (2), 118.

Tanuer, J.M. 1990. Foetus into man: physical growth from conception to maturity, N.A: Harvard University Press.

Thomas, Y.M.; Bedi KS.; Davies, C.A. and Dobbing, J. 1979. A Stereological analysis of the neuronal and synaptic content of the frontal and cerebellar cortex of weanling rats undernourished from birth. *Farly Human Dev.*, 3, 109-126.

Tong, J.; Zhang, C.; Zhao, Z.; Yang Y. and Tian, K. The determination of the physical-chemical constants and sixteen mineral elements in raw seabuckthorn juice. *Proc. Int. Symp. Sea-buckthorn* (*H. rhamnoides* L.), Xian, China. p. 132-137.

Vance, A.L. and Luk, E.S. 2000. Attention deficit hyperactivity disorder, current progress and controversies. *Aust. NZJ Psychiatry*, 34, 179-30.

Venkoba Rao, A. 1996. Depression and Bipolar affective disorder: *In*. Psychiatry for the Developing World by "Tantam D.; Appleby, L. and Duncan, A. (eds), Churchill, Living Stone, New Delhi, pp. 185-220.

Wachs, T. 1995. *J. Nutr*, 125 (Suppl), 2245-54.

Wahlberg, K. and Jeppsson. N. (1990). Development of cultivars and growing techniques for sea-buckthorn, black chokeberry, *Lonicera* and *Sorbus*. Sverigges Lantbruksuniversitedt Balsgård-Avdelningenför Hortikulturell Växförädling Verksamhetsberättelse 1988-1989 p. 80-93. (in Swedish, summary in English).

Wahlberg, K. and Jeppsson. N. 1992. Development of cultivars and growing techniques for seabuckthorn, black chokeberry, honeysuckle and rowan. Sverigges Lantbruksuniversitedt Balsgard-Avdelningenför Hortikulturell Växförädling Verksamhetsberättelse 1990-1991 p. 86-100. (in Swedish, summary in English).

Wainwright, P . 2000. Nutrition and behaviour. The role of n-3 fatty acids in cognitive function. *Brit. J. Nutr.*, 83, 337-339.

Waltts, R.L. 1997. The role of dopamine agonists in early Parkinson's disease. *Neurology*, 49: S34-48.

Weksler, M.E.; Tusuda, T.; Kim, Y.T. and Ssiskind, G.W. 1990 .Immunobiology of ageing and cancer. *Detect. Preven.* 14, 609-11.

West, C.D. and Kemper, T.L . 1976. The effect of a low protein diet on the anatomical development of the rat brain. *Brain res.*, 107, 221-237.

Winik, M. 1972. Nutrition an development. New York. Wiley.

Wohl, M.G. and Goodhart, R.S. 1968. (eds) Mordern Nutrition in Health and Disease : Leasfebiger Philadelphia.

Wolf, D. and Wegert. F. 1993. Experience gained in the cultivation, harvesting and utilization of sea-buckthorn. *In*: Cultivation and Utilization of Wild Fruit Crops. Bernhard Thalacker Verlag GmbH & Co., pp. 23-29 (in German).

Yanase, T.; Fukahori, M.; Taniguchi, S. 1996. Serum dehydroepiandrosterone (DHEA) and DHEA-sulfate (DHEA-S) in Alzheimer's disease and in cerebrovascular dementia. *Endocr. J.*, 43, 119–23.

Yao, Y.; Tigerstedt P.M.A. and Joy. P. 1992. Variation of vitamin C concentration and character correlation between and within natural sea-buckthorn (*Hippophae rhamnoides* L.) populations. *Acta Agric. Scand.*, 42, 12-17.

Yu, Let. 1993. Effects of *Hippophae rhamnoides* juice on immunologic and antitumor functions, *Acta Nutrimenta. Sinica.*, 15 (3): 280-283.

Zamenhor, S.; Vanmarthens, E.; and Gauel, L. 1971. DNA (cell number) and protein in neuonatla rat brain : alteration by timing of maternal dietary protein restriction. *J. Nutr.*, 101, 1265-1270.

Zesiewicz, T.A.; Gold, M.; Chari, G. 1999. Current issues in depression in Parkinson's disease. *Am. J. Geriatr. Psychiatry*, 7, 110-8.

Zhang Maoshun, 1987. Treatment of ischemic heart diseases with flavonoids of Hippophae rhamnoides, *Chinese Journal of Cardiology*, 15(2), 97-99.

Zhang, W. J.; Yan, J.; Duo, B.; Ren and J.; Guo. 1989. Preliminary study of biochemical constitutes of berry of sea buckthorn growing in Shanxi Province and their changing trend. *Proceeding of International Symposium on Sea Buckthorn* (*H. rhamnoides.* L.), Xian, China.

Zhang, X.; Zhang, M.; Gao, Z.; Wang, J. and Wang, Z. 2001. Effect of total flavones of Hippophae rhamnoids L. on sympathetic activity in hypertension. *Hua Xi Yi Ke Da Xue Xue Bao* 32, 547-550.

Zhao, Yuzhen and Fuheng. W. 1997. Seabuckthorn flavonoids and their medical value. *Hippophae*, 10, 39-41.

Zhone, Fei. 1989. Study on the immunopharmacology of the components extracted from *Hippophae rhamnoides* L. In : *Proceeding of International symposium of Seabuckthorn*, p. 368-370, Xi' an, China.

□□□

18

Conservation of Genetic Resources of Seabuckthorn

Sandhya Gupta and Sanjai K. Dwivedi

CONTENTS

Plant Genetic Resources (PGR) include plant species that not only provide food, feed fodder, fibre and medicine but also are sources of shelter, energy and other uses that support the livelihood. Thus PGR encompasses not only the cultivated species and their wild relatives but other species occurring in the wild state as well. The number of cultivated types has narrowed as a result of crop improvement efforts for productivity and quality. The related wild and weedy species are ruthlessly being cut for fodder, fuel and timber because these species have no other immediate profitable use. The need to conserve the genetic resources is far greater in the present time than ever before. The genetic variability has to be properly conserved for the use in improvement programme to cater to the national and international needs.

18.1 STRATEGIES

Conservation of PGR can be achieved either *in situ* or *ex situ*. Both the approaches are complementary. That means both static (seed banks, culture collection etc.) and dynamic (conservation of populations in evolving natural environments) ways are needed to cater the need of conservation of the germplasm. There are two basic strategies for germplasm conservation:

(i) *In situ* conservation: Conservation of a species in its original habitat/ agro-ecosystem. Natural habitat ensures the maintenance and recovery of viable populations of species in their surroundings. In case of cultivated species, *in situ* would refer to on-farm cultivation, where cultivars have acquired and sustained their unique features. These include biosphere reserve, on-farm and home gardens. The *in situ* conservation, which involves protection or keeping aside the ecosystems or habitat harbouring biodiversity from human intervention, is good for forestry species, wild relatives of crop species, forage crop species, medicinal and aromatic plants and the species that are under various levels of threat or about to extinct. The approaches which are being followed to meet the objectives of in situ conservation are as below:

- Ecosystem approach
- Habitat approach

(ii) *Ex situ* conservation: It refers to the germplasm conservation attempted outside the natural habitat. Most of the genetic resources for food and agriculture have been conserved following ex situ approaches. For example, seed genebank, *in vitro* genebank, cryo genebank, pollen genebank, DNA bank, field genebank and botanical garden. Suitable ex situ approaches may be adopted for cost effective conservation to meet the desired objectives. Thus, *ex situ* conservation approach may be classified as follows:

- *Plant conservation:* Botanical garden, herbal garden, arboreta etc.
- *Seed conservation:* Storage of orthodox (desiccation tolerant) seed at low temperature)
- *In vitro conservation*: Maintenance of aseptic cultures of cell, tissue and organ
- *Cryopreservation:* Preservation of desiccation sensitive seed, organ and cultures using cryogenics
- *DNA conservation:* genomics, DNA, DNA libraries, gene construct

Species producing orthodox seed (i.e. seed propagated species) are conserved in the genebank in cold storage modules at various temperatures for active and base collections. Seeds of most of the food crops such as cereals, pulses etc. can be dessicated to reduce their moisture content, thus are conserved in the seed genebanks. Most of the temperate fruit species are perennial and produce orthodox seeds, such as apple, cherry and grapes, which can also be stored in seed genebank. However, these plants are cross-pollinated and produce heterozygous progeny. Thus, germplasm conservation through seed is not useful as far as conservation of true-to-type genotype is concerned. On the basis of temperature regime the storage is classified into medium and long-term storage.

Medium-term storage at 4°C: active collection

Long term storage at -20°C: base collection

Cryopreservation at -196°C: base collection

Species which are propagated through vegetative means of propagation (i.e. vegetatively propagated species) are conserved either in field genebank or in the clonal or in vitro repository. Some of these species, for example, almond, walnut, pecan nut etc. produce recalcitrant or intermediate seeds, which cannot be dried to the low moisture level required for proper seed storage and, thus, are not amenable to storage in seed genebank.

Field genebank: active collection

In vitro repository at 10-25°C: active collection

Cryopreservation at -196°C: base collection

Conservation of PGR in India: Conservation programmes are being implemented in India at three levels, namely, genotypes, species and ecosystems. *In situ* conservation of wild flora through protection of habitats and ecosystems is being implemented by the Ministry of Environment and Forest. Fourteen Biosphere reserves have been established (Table 1) in different bio-geographical zones, to protect representative ecosystem.

Table 18.1: Biosphere reserves established in India

Name of Site	Year of notification	Location	Bio-geographic zones
Nilgiri	1986	Tamil Nadu, Kerala and Karnataka	Western Ghats
Nanda Devi	1988	Uttranchal	West Himalayas
Nokrek	1988	Meghalaya	East Himalayas
Manas	1989	Assam	East Himalayas
Sunderbans	1989	West Bengal	Gangatic Delta
Gulf of Mannar	1989	Tamil Nadu	Coasts
Great Nicobar	1989	Andaman and Nicobar Island	Andaman and Nicobar Island
Similipal	1994	Orissa	Deccan Peninsula
Dibru-Saikhowa	1997	Assam	East Himalayas
Dehang Debang	1998	Arunachal Pradesh	East Himalayas
Pachmarhi	1999	Madhya Pradesh	Semi-Arid-Gujrat-Rajputana
Kanchanjunga	2000	Sikkim	East Himalayas
Agasthyamali	2001	Kerala	
Achanakmar-Amarkantak	2005	Madhyapradesh-Chattisgarh	Chattisgarh

Source: MoEF (2005-2006)

Ex situ conservation requires collection and systematic storage of seed/ propagules outside the natural habitat of species for short-medium and long-term. *Ex situ* conservation of genetic variability of cultivated plants and their wild relatives is the sole responsibility of the National Bureau of Plant Genetic Resources (NBPGR) that operates under the Indian Council of Agricultural Research (ICAR). NBPGR, as the nodal agency for *ex situ* conservation of PGR for food and agriculture, is maintaining active and base collections of various crop species in a network of genebanks in the country. Besides seed storage there are adequate infrastructure facilities for *in vitro* conservation and cryopreservation of the germplasm.

18.2 RECENT DEVELOPMENTS CONCERNING PLANT GENETIC RESOURCES (PGR)

Earlier, the PGR used to be considered as 'common heritage of humankind' and access to PGR was fairly liberal in and across the countries. But now the trends has been changed and issues concerning access to PGR for sustainable use, fair and equitable sharing of benefits from their commercial use came up, following related developments at the international and national level.

Convention of Biological Diversity: The Convention of Biological Diversity (CBD) came into force in 1993. The CBD reaffirmed the sovereign rights of countries over their biological resources, and provided the authority to the National Governments to determine access to genetic resources occurring in their territories, subject to National legislations, prior informed consent and mutually agreed terms.

International Treaty on Plant genetic Resources for Food and Agriculture: The International Treaty on Plant genetic Resources for Food and Agriculture (ITPGRFA) was adopted in 2001. It is a legally binding International treaty, considering needs of farmers and plant breeders, the conservation of PGR, their sustainable use, and the fair and equitable sharing of benefits accrued from their use.

Protection of Plant Varieties and Farmers' Rights Act: The Protection of Plant Varieties and Farmers' Rights Act (PPVFRA), 2001 of India provides for protection of plant varieties and farmers rights. As per the Act, a new variety shall be registered at PPVFR Authority if it conforms to the criteria of novelty, distintiveness, uniformity and stability. The registration is to be valid for 9 years in case of trees and vines and 6 years in case of other crops and may be reviewed for 9 more years. There is a provision for the establishment of a National Gene Fund to meet the requirements of the amount to be paid by way of benefit sharing and compensation, expenditure for supporting the conservation and sustainable use of genetic resources including *in situ* and *ex situ* collections and for strengthening the capability.

Biological Diversity Act: The Biological Diversity Act (BDA), 2002 of India, is a broad policy statement identifying the strategies and the actions that need to be taken for access of genetic resources, their conservation, sustainable use, and fair and equitable sharing of benefits as per the provisions of the CBD. The Act envisages establishing a National Biodiversity Authority, State Biodiversity Boards and local level Biodiversity Management Committees, to provide for regulated access to bio-resources and the traditional knowledge associated with them.

18.3 CHARACTERIZATION OF SEABUCKTHORN

The seabuckthorns (*Hippophae* L.) are deciduous shrubs in the genus *Hippophae*, family Elaeagnaceae. It is also referred to as sandthorn or seaberry. There has been a rapidly growing recognition of seabuckthorn berries for their consumer product potential, exceptional nutrient content and antioxidant qualities, giving them commercial status as a novel superfruit. The genus *Hippophae* is divided into two sections, Hippophae and Gyantsenses. Each has three important species in commercial terms. Section Hippophae has *H. salicifolia, H. rhamnoides* (with 8 subspecies) and *H. goniocarpa* (with 2 subspecies), and section Gyantsenses includes *H. gyantsensis, H. neurocarpa* (with 2 subspecies) and *H. tibetana.* The genetic variability available within the genus is considerable (Avdeev, 1983). The 6 species and 12 subspecies are native over a wide area of Europe and Asia, including China, Mongolia, India, Nepal, Pakistan, Russia, Great Britain, France, Denmark, Netherlands, Germany, Poland, Finland, Sweden and Norway. *Hippophae* spp. grows in the forest areas of Mountains from Hindu Kush - Himalayan region, adjoining areas of China and parts of Europe. In India, four species namely, *H. rhamnoides* L., *H. rhamnoides* var. *Turkestanica* Rousi., *H. salicifolia* D. Don, *H. tibetana* Schltr. occur wild in the Himalayan region.

Availability of extensive genetic variability and its scientific utilization are prerequisites for any crop improvement programme. Variability in this genus is huge, and its conservation and scientific exploration for crop improvement are necessary (Lian and Chen 1993; Lian et al., 1995, 1998), for which is required a comprehensive set of morphological descriptors that accommodate all possible variations. Dwivedi *et al. (2007)* have extensively surveyed and evaluated 200 fruit samples of seabuckthorn (H. *rhamaoides)* in India and have conserved 19 promising genotypes The dendrogram developed by Mathew et al (2008) showed that the variability of seabuckthorn in India is huge and the country has the potential to emerge as a leader in the improvement and commercial exploitation of this crop. Previous efforts to characterize *Hippophae* germplasm using morphological characters had made use of 30 quantitative and 7 qualitative characters. A semi-strict tree of the 117 equally parsimonious trees with the length of 140 steps failed to distinguish these 3 species, whereas a semi-strict tree of 14 equally parsimonious trees with a length of 122 steps clustered *H. rhamnoides* and *H. tibetana* together, and *salicifolia* in a different but closer cluster (Hyvonen, 2003). Reports by Rousi (1971) and Huang (1995) point to a closer association between *H. rhamnoides* and *H. salicifolia* than between *H. rhamnoides* and *H. tibetana.*

The results obtained through use of the present descriptor list match that observation. Further, these values for genetic distances are confirmed by RAPD (Bartish *et al.,* 1999) and isozyme (Yao and Tigerstedt, 1993) analyses.

18.4 CONSERVATION

Seabuckthorn, despite being a common and widespread species, is deserving of conservation measures. Its distribution pattern has been described as "highly fragmented," i.e. it tends to occur as isolated patches, and these are often genetically distinctive. A recent analysis of Sea Buckthorn genetic resources noted, "The high nutrient and medicinal values of the fruit have led to uncontrolled exploitation and even destruction of Sea Buckthorn resources in some parts of its natural distribution. Thus, protection and preservation of the valuable germplasm have become urgent." In Hungary, wild Sea Buckthorn is rarely observed, and the plant is protected as an endangered species. Such protection needs to be extended, especially to the Asian Sea Buckthorns, which occupy small distinct ranges. As a fruit crop, Sea Buckthorn requires considerable development, and the wild plants constitute an invaluable resource for selecting superior agricultural plants (Small et al., 2002). In India, propagation of seabuckthorn through hard wood stem cutting has been developed (Dwivedi *et al. 2001)* and also through soft wood cutting (Dwivedi *et al.,2007),* which may play important role in conservation of valuable germplasm using asexual methos of propagation.

18.5 CONCLUSIONS

Conservation of plant biodiversity is of great significance. We must acknowledge the fact that conservation of PGR has come a long-way in protecting the bioresources, a national heritage, for present and future use. Seabuckthorn has great potential for sustainable environmental protection and commercial exploitation in cold desert areas of Himalayas. This plant is a deciduous shrub, widely distributed in temperate and sub tropical zones of Asia. It withstands extreme temperature ranges from -40°C to +40°C and adapts very well in cold deserts. This plant also has an outstanding ability to conserve water and moisture, apart from fixing atmospheric Nitrogen. This plant grows, flourishes and dies naturally in the valleys and mountain areas and thereby highly fertilizes the soil. It has an extraordinary economic benefits and it would be commercially well exploited. Each and every part of the plants is well utilized. The fruit is the richest source of vitamin C and is also used as raw material for producing food, cosmetic, medicine etc. The fruit pulp and seeds contain high quality medicinal oil. The fruit skin (after extracting pulp) is utilized for making tea. The leaves from male plant are also commercially used to make tea. The whole plant is an excellent source of fuel wood and fodder. The Seabuckthron plant offers an amazing opportunity to mountain farmers for sustainable livelihood by protecting their fragile ecosystem by soil and moisture conservation. It also has tremendous value addition properties. This plant is the best option to mitigate the problems of fragile ecosystem, economic sustainability of marginal farmers in these mountain areas. Plants can tolerate extremely harsh winters and poor soils.

Future Thrust

- There is a need to adopt complementary conservation strategies involving both *in situ* and *ex situ* approaches. The safe conservation of PGR would be achieved through protection of biodiversity rich areas, on-farm with participation of farmers and local communities, and by strengthening of infrastructure, scientific capabilities and improved coordination between various national institutes involved in conservation of PGR.
- Proper scientific technologies have not been developed for seabuckthorn propagation and plantation. This crop has to be researched extensively for its scientific management, tissue culture techniques, plant protection etc.
- Suitable products from *Hippophae* species can raise the economic status of the local people. The main limitations are the development of an infrastructure for a processing market. The processors are not likely to get interested unless there is a market and conversely, there is not likely to be a market developed unless there are processed products in place.

References

Anderson, M.M. and Wahlberg, K. 1994. The breeding potential of sea buckthorn (*Hippophae L.*). *New Plantsman* Dec. pp 207-217.

Avdeev, V.T. 1983. New Taxonomy for the genus Hippophae L. Ahborti Akademijai Fanhoi RSS Tocikiston, Su'Bai Fanhoi Biologi 4:11-17.

Bartish, I.V.; Jeppsson, N.; Bartish, G.I.; Lu, R. and Nybom, H. 2002. Inter-and intra-specific genetic variation in Hippophae investigation by RAPD markers. In: N. Jeppsson. Genetic variation and fruit quality in sebuckthorn and chokeberry. Doctoral thesis, part III, Swedish University of Agricultural Sciences, Alnarp, Sweden. pp 1-20.

Bernáth, J. and Földesi, D. 1992. Sea buckthorn (Hippophae rhamnoides L.): a promosing new medicinal and food crop. *J. Herbs Spices Med. Plants*, 1: 27-35.

Dwivedi SK, Singh R & Ahmed Z 2007 a. Identification and evaluation of biodiversity of Seabuckthorn in cold arid regions of Ladakh *J. of Agric Sc.* 77(7): 427-431.

Dwivedi, S.K.,Singh, Ranjit and Ahmed, Z. 2006 a. The Seabuckthorn. Field Research Laboratory, DRDO, Leh, Ladakh (J&K) India.

Gao, W.; Ramanatha Rao, V. and Zhou, Ming-De. (eds) 1999. Proceedings of National Workshop on Conservation and Utiilization of Plant Genetic Resources, 25-27 October 1999, Beijing, China.

Huang, Q. 1995. A review of seabuckthorn breeding in China. Proceedings of the International workshop on seabuckthorn, Beijing, China, 12-17 December 1995. China Science and Technology Presss, Beijing, China. pp 111-117.

Hyvonen, J. 2003. Phylogeney of seabuckthorn. In: Singh, V. (ed). Seabuckthorn (Hippophae L.)-A multipurpose wonder plant (Vol. 1). Indus Publishing Company, New Delhi. India. pp. 47-63.

Kam, Barbara and Nora Bryan 2003. The Prairie Winterscape: Creative Gardening for the Forgotten Season. Fifth House Ltd., 108-110. ISBN I-894856-08-2.

Li, T.S.C. 2002. Product development of sea buckthorn. p. 393–398. In: J. Janick and A. Whipkey (eds.), Trends in new crops and new uses. ASHS Press, Alexandria, VA.

Li, T.S.C. and Schroeder, W.R. 1996. Seabuckthorn (*Hippophae rhamnoides* L.): A multipurpose plant. HortTechnology 6(4): 370-378.

Lian Y.S. and Chen X.L. 1993. A study on the germplasm resources of genus Hippophae L. In: Proceedings of International Symposium on Seabuckthorn (Hippophae L.), Novosibirsk, Sioberia, Russia, 23-27 October 1993. International Centre for Research and Training on Seabuckthorn. pp.157-161.

Lian Y.S., Chen X.L. and Lian H. 1998. Systematic classification of the genus Hippophae L. Seabuckthorn Research 1:13-23.

Lian Y.S., Chen X.L. and Sun K. 1995. New discoveries of the genus *Hippophae* L. In: Proceedings of International workshop on Seabuckthorn, Beijing, China, 12-17 December 1995, China Science and Technology Press, Beijing, China, pp 60-66.

Mathew, D.; Parimelazhagan, T.; Gomez, Saji and Ahmed, Z. 2008. Characterization of Seabuckthorn (*Hippophae* spp.) genetic resources in India using morphological descriptors 149: 22-26.

MoEF 2006. Annual Report 2005-2006. Ministry of Environment and Forest, Government of India, New Delhi, India.

Naithani, H.B. 2004. *Hippophae* Linn. (Seabuckthorn) in India: a review. *Indian Forester*, 130(9): 1045-1056.

Pietilä, M. 1998. Buckthorn, the plant for tomorrow's healthy living. Joonas International news forum for professional berry growers, Joensuu, Finland.

Rousi, S. 1971. The genus *Hippophae* L. A taxonomic study. *Annales Botanical* Fennici, 8: 177-227.

21. Small, E.; Catling, P.M. and Li, T.S.C. 2002. Blossoming Treasures of Biodiversity: 5 Sea Buckthorn (*Hippophae rhamnoides*) - an ancient crop with modern virtues. *Biodiversity* 3 (2): 25-27.

22. Yao, Y and Tigerstedt, P.M.A. 1993. Isozyme studies of genetic diversity and evaluation in Hippophae. *Genetic Resources and Crop Evaluation*, 40: 153-164.

23. Zeb A. 2004. Important therapeutic uses of sea buckthorn (*Hippophae*): a review. *J Biol Sci* 4:687-693.

24. Zeb A. 2006. Anti-carcinogenic potential of lipids from *Hippophae*: - evidence from the recent literature, *Asian Pacific J Cancer Prev* 7:32-5.

□□□

19

Exploring Seabuckthorn Genome Using Molecular Tools

R. Ghangal, S. Raghuvanshi and P.C. Sharma

CONTENTS

Due to its high nutritional, medicinal and environmental importance, seabuckthorn has recently become an important subject of plant science research. The plant is also being explored as a source of candidate genes involved in the management of abiotic stresses like drought, cold, etc. Efforts have been made to study the basic biology of the plant, its geographical distribution and to develop improved seabuckthorn cultivars and products. The knowledge of the extent and distribution of genetic diversity of a species provides a better scope for selection in plant breeding. A repertoire of currently available molecular tools helps in exploring plant genomes and transcriptomes. The molecular data generated using these tools have been useful in augmenting the pace and performance of conventional plant breeding programmes. The botanical, taxonomical, distributional, nutritional and medicinal importance of seabuckthorn has been discussed at appropriate places in this book. In this chapter, we attempt to review the application of molecular markers in understanding intra- and inter-specific genetic diversity and molecular tagging in *Hippophae*. Our research on gene expression profiling in this important plant species is also briefly described.

Although many studies have been conducted on seabuckthorn's origin, evolution, geographical distribution and phylogenetic relationships (Yu et al. 1993, Chen and Lian 1994), the taxonomical classification of this genus and the relationship among its taxa still remains controversial (Avdeyev 1983, Lian 1988, Lian and Chen 1992, 1996, Lian et al. 1997, Hyvonen 1996, Bartish et al. 1999a, 2000a, 2002, Bartish et al. 2000b, Lian et al. 2000). Primitive taxonomy was presented by Rousi (1971) who divided genus *Hippophae* into three subspecies namely *H. rhamnoides*, *H. salicifolia* and *H. tibetana*. *H. rhamnoides* was further divided into nine subspecies. Lian (1988), Lian and Chen (1992, 1996) and Lian et al. (1997) conducted extensive systemic studies based on fruit morphology and established two sections i.e. sect. *Hippophae* and sect. *Gyantsensis*. Surviving in harsh environment leads to some changes in a species and accumulate more genetic variation in order to adapt itself to various pressures (Li et al. 1999). In due course of evolution, *Hippophae* gained extensive genetic variation and differentiated into various kinds of geographical populations. Though extensively studied, circumscription of species in this genus remains unresolved in particular when subspecies have been promoted to species level and when some species have been designated as subspecies by different authors. Precisely, the subspecies of *H. rhamnoides* have been accepted differently as these are not always easily distinguishable. Balsgard Department of Horticulture and Plant Breeding, Sweden maintains the world's most diverse *Hippophae* gene bank (Bartish et al. 2000a). Such a germplasm collection should ideally represent the global genetic diversity available for seabuckthorn in the natural environment.

As mentioned above, a clear morphological identification of different varieties is not possible in seabuckthorn and therefore molecular marker based identification has become important tool to study genetic diversity. DNA polymorphism based approaches have been frequently exploited to probe the genetic variation of species and to understand the genetic relationships among and within species or populations. In seabuckthorn also, different genetic marker systems have been used to access phylogenetic relationships and diversity. DNA-based molecular markers have acted as versatile tools and have found their own position in various fields like taxonomy, physiology, embryology, genetic engineering, etc. The discovery of PCR (polymerase chain reaction) was a landmark in this effort and proved to be a unique process that brought about a new class of DNA profiling markers. A brief idea of these molecular markers and their application in seabuckthorn genome analysis has been presented in the following text.

19.1 GERMPLASM CHARACTERIZATION USING MOLECULAR MARKERS

1. Isozymes and Protein Profiling

Yao and Tigerstedt (1993), provided information on the genetic variation, differentiation and evolution in *Hippophae* by analyzing seed samples of 25 populations from China, Finland and Russia. In the 11 population of *H. rhamnoides* ssp. *sinensis*, they could only reveal 6 allozyme loci and out of six loci investigated, only four were proposed as good genetic markers for identifying species and subspecies. This indicates a low level of genetic diversity at the population level and provided only limited information on the population genetics of *H. rhamnoides* ssp. *sinensis*. Moreover, as expected, there was higher variation in the study done by Sun et al. (2006) (RAPD analysis of the natural population of *H. rhamnoides* ssp. *sinensis*) than that revealed by Yao and Tigerstedt (1993) because RAPD detects much higher genetic diversity than allozyme data. Ahmad and Kamal (2002) also used total seed protein and performed the SDS-PAGE to inspect the genetic diversity in four genotypes of *H. rhamnoides* ssp. *turkestanica*, a natural seabuckthorn population found growing in Pakistan. They also conducted morphological studies of these genotypes to correlate the results obtained from the total seed protein profile. Based on both morpho-molecular analysis Ahmad and Kamal (2002) concluded that their existed immense genetic diversity in the local seabuckthorn population of *H. rhamnoides* ssp. *turkestanica*.

2. Random Amplified Polymorphic DNA (RAPD)

The RAPD marker system (Welsh and McClelland 1990, Williams et al. 1990) has been widely used to estimate the genetic diversity in various plant species. RAPD marker system have several advantages over other DNA markers due to its low cost, speed, ease and use of small amount of DNA/ plant material. In seabuckthorn, RAPD markers have been used for diverse studies in the field of taxonomy, distribution and geographic origin, sex determination and population genetics.

RAPDs are DNA fragments amplified by PCR using short synthetic primers (generally 10 bp) of random sequence. These oligonucleotides serve as both forward and reverse primer, and are usually able to amplify fragments from many genomic sites simultaneously. Amplified fragments, usually within 0.5–5 kb size range, are separated by agarose gel electrophoresis, and polymorphisms are detected after ethidium bromide staining as the presence or absence of bands of particular size. These polymorphisms are considered to be primarily due to variation in the primer annealing sites, but they can also be generated by length differences in the amplified sequence between primer annealing sites.

The main advantage of RAPDs is that they are quick and easy to assay. Because PCR is involved, only low quantities of template DNA are required, usually 5–50 ng per reaction. Since random primers are commercially available, no sequence data for primer construction are needed. Moreover, RAPDs have a very high genomic abundance and are randomly distributed throughout the genome. This technique also has several advantages over isozymes and other DNA markers, including ease, speed and low cost. RAPDs therefore have been widely used for estimating genetic diversity and relatedness in plant populations (Heum et al.1994, Ge et al. 1999). However, the RAPDs sometimes suffer with the problem of reproducibility as variation in experimental parameters like concentration of template, primer DNA and $MgCl_2$, and PCR profile can change the resulting RAPD fingerprints. Therefore, it is mandatory to define experimental conditions very precisely and to stick to these conditions to ensure reproducible patterns.

Bartish et al. (2000b) investigated the genetic diversity of all the taxa representing the genus *Hippophae*. They obtained 219 polymorphic RAPD markers and found 16 fixed RAPD markers. The fixed markers represent PCR products that are either present or absent in all individual plants from the same population. These can be regarded as taxon specific markers. Earlier Bartish et al. (1999a) got four fixed alleles out of the 156 polymorphic RAPD loci in subspecies of *H. rhamnoides*. Loci with fixed alleles between species and subspecies are especially valuable for phylogenetic reconstruction based on character state difference such as parsimony or maximum likelihood analysis. The highest diversities were observed in *H. goniocarpa* population and *H. tibetana* while *H. salicifolia* was the least diverse among the species. Principal co-ordinate (PCO) analysis also revealed that *H. salicifolia* is clearly differentiated from other taxa. AMOVA (Analysis of Molecular Variance) analysis indicated that 48.6% of the total variability resides between species and 51.4% within species whereas 59.6% of the variability was found between subsp. and 40.4% within subspecies (Bartish et al. 2000b). A study by Bartish *et al.* (2000b) showed good correspondence with the taxonomy suggested by Lian and Chen (1993) who grouped *H. goniocarpa, H. gyantsensis, H. neurocarpa* and *H. tibetana* in the section Gyantsensis and *H. rhamnoides, H. salicifolia* and *H. sinensis* in the section *Hippophae*. *H. rhamnoides* consists of 6 subspecies (*carpatica, caucasia, fluviatilis, mongolica, rhamnoides* and *turkestanica*). *H. sinensis* consists of two subspecies, *sinensis* and *yunnanensis* which were earlier included in *H. rhamnoides* by Rousi (1971) and Lian et al. (1998). One more contradiction with Rousi (1971) is regarding the *H. gyantsensis,* which Rousi regarded as a subspecies of *H. rhamnoides,* has been found to be distinct species in the study conducted by Lian and Chen (1993) and Bartish *et al.* (2000a). On morphological basis Hyvonen (1996) suggested that *H. gyantsensis* and *H. rhamnoides* ssp. *turkestanica* are closely related but RAPD study by Bartish *et al.* (2000b) does not support this conclusion. Lian and Chen (1993) and Lian *et al.* (1995) suggested that *H. goniocarpa* ssp. *goniocarpa* and ssp. *litangensis* have originated from interspecific hybridization between *H. neurocarpa* ssp. *neurocarpa* and *H. rhamnoides* ssp. *sinensis* and between *H. neurocarpa* ssp. *stellatopilosa* and *H. rhamnoides* ssp. *yunnanensis,* respectively and fixed marker data of Bartish *et al.* (2000b) supports this hybridization theory for both taxa. The partitioning of molecular variance within and among populations is directly influenced by the overall diversity of the plant sample. Bartish *et al.* (1999a) accounted 85% variability of the total variance within ssp. *rhamnoides* but when the analysis was extended to the entire genus *Hippophae* (Bartish *et al* 2000) the variability came down to 40.4%. Bartish *et al.* (1999a) found 89.7% and 55.2% polymorphism in ssp. *rhamnoides* at the subspecies and population levels, respectively and interestingly genetic diversity within population of ssp. *sinensis* (Sun et al. 2006) was comparable to ssp. *rhamnoides*. However, Bartish et al. (1999a) regarded this variation within population somewhat lower than expected and attributed the lower value to the fragmented distribution and isolation between island populations. While Sun et al. (2006) attributed this lower value to the successional stage as ssp. *sinensis* mainly occupies area of early successional stages and such species generally tend to have relatively lower levels of variability within population.

Moreover, Lian *et al.* (2000) found ssp. *sinensis* exhibiting higher degree of vegetative reproduction through root suckers. Therefore, high level of clonal reproduction may also be responsible for the prevalence of relatively lower genetic diversity.

H. rhamnoides ssp. *wolongensis* and *H. rhamnoides* ssp. *sinensis* share close relationship as they cluster together showing high genetic identity (Sheng *et al.* 2006). Unlike Bartish *et al.* (2002), Sun *et al.* (2002) who placed *H. tibetana* and *H. salicifolia* at relative isolated positions in cluster analysis, Sheng *et al.* (2006) found them to be closely related although they obviously differed in morphology. Some degree of genetic diversity and genetic differentiation was found in samples collected at different times for *H. rhamnoides* ssp. *sinensis*, thus reflecting a gradual differentiation with time (Sheng *et al.* 2006). This result is consistent with the theory of Lian *et al.* (1998) who indicated that the evolutionary process in *Hippophae* was a process of gradual development.

Bartish *et al.* (2000a) conducted a study on 55 plants of *H. rhamnoides* L. including subspecies *carpatica, caucasica, fluviatilis, mongolica* and *rhamnoides* and intraspecific hybrids, and reported that variation was considerably higher within subspecies than between subspecies. Analysis of molecular variance in ssp. *rhamnoides* showed that 85% of variation was within population and 15% among population. Similarly Singh *et al.* (2006) also reported high variability within *H. rhamnoides* population of 17 morphotypes from Ladakh, India.

3. *Amplified Fragment Length Polymorphism (AFLP)*

AFLPs are DNA fragments (80–500 bp) obtained from digestion with restriction enzymes, followed by ligation of oligonucleotide adapters to the digestion products and selective amplification by the PCR (Vos *et al.* 1995). AFLPs therefore involve both the reproducibility of RFLP and versatility of PCR. The PCR primers consist of a core sequence (part of the adapter), and a restriction enzyme specific sequence and 1–5 selective nucleotides (the higher the number of selective nucleotides, the lower the number of bands obtained per profile). The AFLP banding profiles are the result of variations in the restriction sites or in the intervening region. Both RAPD and AFLP are useful molecular marker techniques. However, to distinguish closely related genotypes such as different clones within a given cultivar or divergence due to small mutational events producing minimal genetic differences, AFLP marker system shows higher multiplex ratios (Cervera et al 1998, Cabrita *et al* 2001).

Ruan and Li (2005) attempted AFLP technique to study the 15 Seabuckthorn cultivated varieties of Russia, Mongolia and China viz. *Hippophae rhamnoides* ssp. *mongolica, H. rhamnoides* ssp. *sinensis* and hybrid varieties. For selective amplification, 8 primer combinations were used. A total of 731 AFLP bands were generated out of which 645 were polymorphic thus representing 88% polymorphism. Thus, AFLP markers have been demonstrated to be useful for fingerprinting and detecting genetic relationship among seabuckthorn cultivars of diverse origin.

4. *Inter Simple Sequence Repeats (ISSRs)*

Since 1994, a molecular marker technique called Inter Simple Sequence Repeat (ISSR) is being used in areas of genetic diversity, phylogenetic studies, gene tagging, genome mapping and to detect polymorphism (Zietkiewicz *et al.* 1994). ISSR is a simple and quick method which combines high polymorphism of microsatellites, also known as Simple Sequence Repeats (SSRs), to the ease and universality of RAPD (Reddy *et al.* 2002). When compared to RAPD, it can be highly variable within a species and have the advantage of long primers allowing for more stringency and higher reproducibility of polymorphic bands. The main advantage of ISSRs is that no sequence data for primer construction are needed. Because the analytical procedures include PCR, only low quantities of template DNA are required (5–50 ng per reaction). Furthermore, ISSRs are randomly distributed throughout the genome.

Tian et al. (2004a) studied the genetic diversity of 11 natural populations of *H. rhamnoides* from north-eastern and north-western China by ISSR analysis. Later, Tian *et al* (2004b) used the same technique to compare the genetic variation in natural population of three subspecies of *H. rhamnoides* i.e. ssp. *sinensis*, ssp. *yunnanensis* and ssp. *gyantsensis*. Tian *et al.* (2004a) obtained a total of 207 polymorphic bands from eight primers used, while Tian *et al.* (2004b) got 288 polymorphic bands using the same number of primers. Tian *et al.* (2004b) observed that genetic diversities of 15 populations were not significantly different thus indicating no obvious difference in mean genetic diversity of population from different subspecies. As the coefficient of gene differentiation (Gst) for ssp. *yunnanensis* and ssp. *sinensis* population was 0.2790 and 0.4184, respectively, implying that ssp. *sinensis* has greater genetic diversity among population than ssp. *yunnanensis*. Gst of 15 population indicated that 14.5% of the total genetic variation existed among populations, while 85.5% existed within population (Tian et al. 2004b). On the other hand, Tian et al. (2004a) found Gst value as 0.0679 among 11 populations indicating that 6.79% of the total genetic diversity existed among the population and 93.21% existed within population. These studies suggested that genetic diversity of seabuckthorn mainly lies within population and conservation of genetic resource can be sought by preserving more individuals than populations (Tian *et al.* 2004a, 2004b).

Based on genetic distance among population studied by Tian et al. (2004b), UPGMA dendogram showed each population clustered together in distinct clusters. The finding of Rousi (1971) that ssp. *yunnanensis* and ssp. *sinensis* are closely related was later supported by Tian et al. (2004b) on the basis of ISSR analysis. The dendogram also showed two clades, where ssp. *sinensis* and ssp. *yunnanensis* constituted one monophyletic clade and second clade constituted ssp. *gyantsensis*, which is the last clade at the base of the cladogram. Therefore, the species status of *H. gyantsensis* proposed by Lian (1988) and suggested by RAPD data (Bartish et al. 2000a) can be justified by ISSR analysis.

5. Microsatellites or Simple Sequence Repeats (SSRs)

Microsatellites or simple sequence repeats (SSRs) are tandemly repeated motifs of 1-6 bases ubiquitous in all prokaryotic and eukaryotic genomes. Variations in repeat number at a given locus generate multiple allelic forms within a population, revealing high polymorphism at microsatellite loci. They are present in both coding and noncoding regions and are usually characterized by a high degree of length polymorphism. The origin of such polymorphism is still under debate though it appears most likely to be due to slippage and other errors during DNA replication (Tautz 1989). SSRs have been developed for many plant species. However, the high cost of initial isolation and characterization hinder their application in many plant species. If nucleotide sequences of the flanking regions of the microsatellite are known, specific primers (generally 20–25 bp) can be designed to amplify the microsatellite by PCR. Microsatellites and their flanking sequences can be identified by constructing a small-insert genomic library, screening the library with a synthetically labelled oligonucleotide repeat for identification of fragments carrying the repeat sequences and subsequently sequencing those fragments. Alternatively, microsatellites may be identified by screening sequence databases for microsatellite sequence motifs from which adjacent primers may then be designed (Wu and Tanksley 1993, Temnykh et al. 2000). In addition, primers may be used that have already been found functional for closely related species. The strengths of microsatellites include the codominance of alleles, their high genomic abundance in eukaryotes and their random distribution throughout the genome, with preferential occurrence in low-copy regions (Morgante et al. 2002). Surprisingly, despite their proven superiority and importance in genome analysis, no attempts has been made till date to develop SSR markers for seabuckthorn.. One of the main drawbacks of microsatellites is that high development costs are involved if adequate primer sequences for the species of interest are unavailable, making them difficult to apply to unstudied groups. Moreover, unavailability of nucleotide database also hampers exploitation of this marker system in a plant species like seabuckthorn. Seeing the importance and utility of this plant our laboratory has started the work to develop the SSR marker in seabuckthorn. Since, microsatellite markers developed for a species can be cross amplified to some extent in related species, it is worth exploring to test microsatellites from related species. However, unavailability of microsatellite markers in other *Hippophae* species forces isolation of microsatellite markers from the genome of a seabuckthorn species.

We screened the seabuckthorn EST sequences developed in our laboratory using a microsatellite search tool, MISA (Thiel et al. 2003), for the presence of SSRs. As expected, a small proportion of ESTs showed the presence of microsatellites. We plan to use these genic microsatellites for screening seakbuckthorn genomes in near future. Since no genomic microsatellites are available till date for seabuckthorn, these EST-SSRs may prove very useful for genome analysis.

6. Internal Transcribed Spacer (ITS) region

Several authors have regarded nuclear DNA as a valuable information in phylogenetic studies of plants and particularly the ITS (Internal Transcribed Spacer) regions of nuclear ribosomal DNA (nrDNA), to resolve the phylogenetic relationships at different taxonomic levels. ITS region evolving relatively at rapid rates helps in resolving the phylogenetic relationship at intra specific level. Sun et al. (2002) analysed the nucleotide sequence of ITS region of nuclear ribosomal DNA from 15 taxa of *Hippophae* (containing two putative hybrids *H. goniocarpa* and *H. litangensis*) and three out groups to reconstruct the phylogeny within the genus. The complete ITS region comprised of ITS 1, 5.8 S, ITS 2, and this region was PCR amplified and sequenced. Phylogeny based on ITS region analysis highly supported the monophyly of the genus *Hippophae* (100% bootstrap value). *H. tibetana* occupies basal position within the genus in strict consensus tree. The remaining taxa form a clade (72% bootstrap value) with two well defined subclade, one of which includes all subspecies of *H. rhamnoides* (97% boot strap value). The other subclade (91% bootstrap value) included two putative hybrids (*H. goniocarpa* and *H. litangensis*), two subspecies of *H. neurocarpa* (subsp. *neurocarpa* and subsp. *stellatopilosa*), *H. gyantsensis* and *H. salicifolia*. Though, when the two putative hybrid sp. were excluded, the strict consensus tree displayed similar topology but the bootstrap values increased considerably and the phylogenetic relationship among the *H. neurocarpa*, *H. gyantsensis* and *H. salicifolia* was well resolved.

Table 19.1: Study of Genetic Diversity in Seabuckthorn using different Molecular Markers

S. No	Authors	Marker System	Sample Population	Observation
1	Yao & Tigerstedt (1993)	Isozyme analysis	*H. rhamnoides* ssp. *sinensis*, *H. tibetana*, *H. neurocarpa*	Six isozyme loci revealed low level of genetic diversity in 11 population of *H. rhamnoides ssp. sinensis*
2	Persson & Nybom (1998)	RAPD	*H. rhamnoides* varieties Leikora X PollmixI BHi 10224 X 2-24	Out of the 78 primers tested, 4 short-listed Got a sex linked marker in one of the hybrid cross but this marker was not universal as it was not reproducible with another hybrid cross.
3	Bartish et al. (1999a)	RAPD	10 population of *H. rhamnoides ssp. rhamnoides*, 1 population of ssp. *mongolica*	15 primers generated a total of 174 markers, out of which 156 (89.7%) were polymorphic and 4 fixed marker. 85% of the variation was within population of ssp. *rhamnoides*
4	Jeppsson et al. (1999)	RAPD	28 genotypes of *H. rhamnoides* ssp. *rhamnoides* and ssp. *mongolica*	85% of the variation was found within population and 15% among population
5	Bartish et al. (2000a)	RAPD	*H. rhamnoides* ssp. *carpatica*, ssp. *caucasica*, ssp. *fluviatilis*, ssp. *mongolica*, ssp. *rhamnoides*, ssp. *sinensis*, ssp. *turkestanica*, ssp. *yunnanensis*, *H. salicifolia*, *H. tibetana*, *H. neurocarpa* ssp. *neurocarpa*, ssp. *stellatopilosa*, *H. gyantsensis*, *H. goniocarpa* ssp. *goniocarpa*, ssp. *litangensis*	Using 9 primers 219 polymorphic markers and 16 fixed markers obtained 37.3% - 59.6% and 48.6% molecular variation among subspecies and among species respectively.
6	Bartish et al. (2000b)	RAPD	55 cultivars of 5 subsp. of *H. rhamnoides* ssp. *rhamnoides*, ssp. *mongolica* ssp. *carpatica*, ssp. *caucasica*, ssp. *fluviatilis*, and intraspecific hybrids between subsp.	Used 8 primers which produced 63 polymorphic markers (86.3%) and 10 monomorphic.

Contd...

Table 19.1. Contd...

7	Ahmad and Kamal (2002)	Protein profiling	4 genotypes of *H. rhamnoides* ssp. *turkestanica*	Significant variability observed among genotypes
8	Sun et al. (2002)	ITS nucleotide sequences of n.r. DNA	*H. rhamnoides* ssp. *carpatica*, ssp. *fluviatilis*, ssp. *mongolica*, ssp. *rhamnoides*, ssp. *sinensis*, ssp. *turkestanica*, ssp. *yunnanensis*, ssp. *wolongensis*, *H. salicifolia*, *H. tibetana*, *H. neurocarpa* ssp. *neurocarpa*, ssp. *stellatopilosa*, *H. gyantsensis*, *H. goniocarpa*, H. *litangensis*	Amount of polymorphism within a taxon was extremely low in most taxa Out of the 690 bp alinged amplicon length 269 sites were variable (39%)
9	Tian et al. (2004a)	ISSR	11 natural population of *H. rhamnoides* in China	Out of the 90 primers tested 8 primers were selected which produced 207 polymorphic loci Genetic variation within and among 11 population was 93.21% and 6.79% respectively
10	Tian et al. (2004b)	ISSR	15 natural population of *H. rhamnoides* ssp. *gyantsensis*, ssp. *sinensis*, ssp. *yunnanensis*	Out of the 90 primers tested 8 primers were selected Genetic structures are different for different subspecies, genetic variation within and among 15 population was 85.5% and 14.5% respectively
11	Ruan and Li (2005)	AFLP	15 Seabuckthorn cultivated varieties from China, Russia, Mongolia and one hybrid	Out 64 combinations of 8 *Eco* RI and 8 *Mse* I primers for screening, 8 primer combination selected A total of 731 AFLP bands were obtained out of which 645 were polymorphic (88%)

Contd...

Table 19.1. Contd...

12	Singh et al. (2006)	RAPD	17 morphotypes of *H. rhamnoides*	Out 20 random primers, 13 were shortlisted High level of genetic distance among morphotypes
13	Sheng et al. (2006)	RAPD	*H. rhamnoides* ssp. *yunnanensis*, ssp. *wolongensis*, ssp. *sinensis*, *H. goniocarpa*, H. *goniocarpa* ssp. *Litangensis*, *H. salicifolia*, *H. tibetana*, *H. Neurocarpa* ssp. *stellatopilosa*, *H. gyantsensis*	Out of 34 primer tested 12 were selected 76.6% polymorphism at species level i.e. 108 loci were polymorphic out of 141 total markers Out of 128 markers 98 loci were polymorphic in population of *H. rhamnoides* ssp. *sinensis*
14	Sun et al. (2006)	RAPD	13 natural population of ssp. *sinensis*	Out of 136 primer tested 15 were selected 88.79% polymorphism

Hybrid plant genome exhibit additivity of parental, therefore molecular marker can be used to identify the origin of hybrid. Analysis of nuclear ribosomal DNA is a useful method for the detection of hybrids. Lian *et al.* (1997) and Bartish *et al.* (2000a, 2002) suggested hybrid nature of *H. goniocarpa* and *H. litangensis*. The ITS sequences of the two species showed high level of polymorphism. In congruent with morphology, geographic distribution, RAPD and chloroplast DNA studies, ITS sequence data strongly indicate that *H. goniocarpa* is a hybrid of *H. neurocarpa* subsp. *neurocarpa* and *H. neurocarpa* subsp. *stellatopilosa* and *H. litangensis* is a hybrid of *H. rhamnoides* subsp. *sinensis* and *H. rhamnoides* subsp. *yunnanensis*. Bartish et al. (2002) analyzed chloroplast DNA and morphological characters of the currently recognized subsp. of *H. rhamnoides* and revealed that the subsp. formed a monophyletic group. In ITS tree, all subsp. of *H. rhamnoides* except subsp. *gyantsensis* form a monophyletic clade, in agreement to Bartish *et al.* (2002) and supports for the monophyly of *H. rhamnoides* proposed by Lian *et al.* (2000) and Rousi (1971) when subsp. *gyantsensis* is excluded. The ITS data points out that three subspecies of *H. rhamnoides* i.e. *sinensis, mongolica* and *fluviatilis* form a closely related group as proposed by Bartish et al. (2002). It also indicates that three subspecies of *H. rhamnoides* i.e. *turkestanica, carpatica* and *rhamnoides* are most closely related

The ITS region within *Hippophae* is remarkably variable, with relatively high sequence divergence between species except the divergence between subspecies of *H. rhamnoides* and *H. neurocarpa* as well as that between hybrids and their putative parents. As indicated by Rousi (1971) on the basis of morphology, *H. tibetana* stood away from the rest of the genus was supported by the greatest sequence divergence between *H. tibetana* and all other groups. In conclusion of the study of Sun *et al.* (2002) on taxonomic classification of *Hippophae* based on ITS region, ITS phylogeny supports the taxonomic treatments proposed by Rousi (1971), Lian (1988), Lian *et al.* (1997, 1998, 2000) and Bartish *et al.* (2002).

19.2 CORRELATION BETWEEN GENETIC DISTANCE AND GEOGRAPHICAL DISTANCE

Some studies have shown that geographical isolation significantly affects the differentiation of population and that genetic distance is correlated with geographical distances (Alpert *et al.* 1993). Bartish *et al.* (1999) found no correlation between genetic and geographical distances in their study of 10 populations of *H. rhamnoides* subsp. *rhamnoides*. Similarly, Sun *et al.* (2006) in their RAPD analysis were unable to detect significant correlation between genetic and geographical distances for 13 subsp. *sinensis* populations. Tian *et al.* (2004a, 2004b) also found no significant correlation between geographic and genetic distance for ssp. *yunnanensis* and ssp. *sinensis* population, suggesting geographic distance does not have any clear effect on genetic differentiation.

Nevertheless, Bartish *et al.* (2000a) observed a significant correlation between geographic and genetic distances when the 12 populations of *H. rhamnoides* were analyzed at a geographical distance ranging from 0 to 700 km. In contrast, there was no significant correlation when the analysis was based on one population from each of the species (*H. rhamnoides, H. salicifolia, H. tibetana, H. neurocarpa, H. gyantsensis, H. goniocarpa*). To be more precise, the association between genetic and geographical distances can change considerably along the range of geographical distances. Similar study conducted by Sheng *et al.* (2006), indicated that genetic distance had a significant correlation with altitude distance and latitude distance among the population within *H. rhamnoides* ssp. *sinensis*. However, Mantel test showed that there is no significant correlation between genetic distance and longitude and latitude distance. Thus, when both the above study by Sheng et al. 2006 is taken into consideration, there is a significant correlation between genetic distance and altitude distance.

19.3 MOLECULAR TAGGING OF SEX DETERMINATION

As seabuckthorn is dioecious shrubby plant, there are separate male and female plants of a cultivar. There is much demand of female plants as they bear berries which are of economic importance. The gender of the seedling can not be determined until flowering that usually takes 3-4 years in field conditions. Therefore plant breeders have to retain the large number of superfluous males for several years. An easily scorable genetic marker would be of great use to plant breeders as it could be used to screen seabuckthorn seedlings at an early stage and discard the males which otherwise costs much of manpower and money.

Gender is the most often genetically determined trait in dioecious plant, attributed either to distinguishable sex chromosome or to alleles at one or several loci on non-distinguishable chromosome (Irish and Nelsion 1989, Durand and Durand 1990). Persson and Nybom (1998) used RAPD markers for gender determination in seabuckthorn. Two F_1 progenies derived from the cross Leikora X Pollmix I and cross between BHi10224 and 2-24 were tested for 78 primers out of which four seemed to be yielding segregation between male and female plants. One marker was found to be present only in male genotypes of the progeny and in father and was absent in all females and mother. However this marker was useful for only one hybrid i.e. Leikora X Pollmix I as there was no specific banding pattern in other hybrid progeny. This finding proved that gender is genetically determined in seabuckthorn. In their study Sheng *et al.* (2006), found high level of genetic identity between male and female plants of same species and subspecies. Though low level of DNA polymorphism was detected between the male and female plants but no special sex linked RAPD marker was obtained. Therefore, further screening of markers is required to get more generalized marker.

19.4 GENE EXPRESSION PROFILING

Despite the high medicinal importance, nutritional quality and its ability to withstand harsh extreme environmental conditions, surprisingly very little efforts have been made to generate any molecular data of this ecologically important plant. Due to the distribution of the species in high altitudes, at places where the annual rainfall is 250-800 mm, the plant is exposed to different abiotic stresses like drought and cold. To cope with such situation the plant either up or down regulates the cold and drought responsive genes and these genes might represent the robust alleles. Due to the up regulation of their expression, their transcript would be in abundance, which can be used to generate molecular data relevant to cold/drought tolerant gene expression.

Techniques for high throughput gene expression analysis can be divided into two broad categories:

a. Closed architecture system which requires pre-existing knowledge of the nucleotide sequence of the gene. e.g. DNA microarrays

b. Open architecture system which does not require any prior knowledge of the gene sequences. e.g. ESTs, SAGE, MPSS.

Microarrays are currently the most popular technique for large scale expression profiling as it allows analysis of thousands of genes simultaneously at a reasonable cost. Thousands of known cDNA sequences or oligonucleotides are imprinted on a microscope slide using an application robot. More than 30,000 sequences can be placed on the surface of the chip. These sequences serve as probes. In this technique mRNA are first reverse transcribed into cDNA. At this stage, DNA of one sample is labeled with a green fluorescent dye, Cyanine 3 (Cy3), whereas DNA from the second sample is labeled with a red fluorescent dye, Cyanine 5 (Cy5). Cy3- and Cy5-labeled cDNAs are mixed together (target) and put on the chip containing single strand DNAs (probes). The array is then scanned with confocal or non-confocal type scanners by using laser emission. Software like ArrayPlot is used to visualize the differential expression of genes. The ratio of signal intensities between control and test cDNA targets reflects the induced/repressed or unchanged expression of mRNA species or genes. One limitation of this technique is that the analysis of genes is dependent on immobilized probes on the chip. However, if probes corresponding to low abundance transcripts are present, this can circumvent one of the main issues with microarrays.

SAGE is a powerful gene expression profiling method, which allows analysis of thousands of transcripts simultaneously. SAGE is based on the principle that a short oligonucleotide sequence tag (10-15 bp) of mRNA is sufficient to identify the whole transcript. A unique sequence is generated by cleaving the transcript with a specific restriction enzyme (anchoring enzyme) at a fixed distance from the poly (A) tail. Although SAGE is a useful technique, it has a shortcoming regarding the size of the SAGE tag. A 15 bp tag is frequently too short to unequivocally identify the gene of origin. To overcome this difficulty of conventional SAGE, Saha et al. (2002) switched over to Type IIS restriction enzyme to obtain 20 bp tags. This SAGE variant is called LongSAGE. To circumvent the problems of SAGE and LongSAGE, Matsumura et al. (2003) developed a method dubbed SuperSAGE which uses Type III restriction enzyme and generates 26 bp tags from transcripts. Therefore, SuperSAGE can provide reliable gene expression profiles based on the 26 bp tags.

A recent advance in tag based gene expression analysis is massively parallel signature sequencing (MPSS). MPSS is based on methods to clone individual cDNA molecules on microbeads and sequence short tags or signatures from these cDNAs, in parallel (Brenner et al., 2000a, 2000b). Using an unconventional but ingenious method of sequencing, large numbers of beads are sequenced in parallel. A series of digestion, ligation, and hybridization reactions are performed in consecutive steps while the beads are immobilized in a flow-cell underneath a high-power microscope so that the reagents flow over and around the beads. The final output of MPSS is a set of abundances for thousands of distinct 17-20bp tags, most of which uniquely identify a particular transcript. The limited length of the tags sometimes hampers the use of MPSS for *de novo* sequencing, but makes it a very powerful tool for expression profiling of organisms with preexisting sequence information. SAGE and MPSS techniques also give quantitative measures of transcript abundance.

The most straightforward and unbiased way to analyze a RNA population is the sequencing of cDNA libraries and quantitative analysis of the resulting expressed sequence tags (ESTs). Exhaustive sequencing of ESTs is a common method for gene expression profiling, although the primary purpose of EST sequencing is usually to generate genic sequence data. EST sequence represents the most extensive available survey of the transcribed portion of the genome. ESTs are single pass, partial sequence reads generated from either the 5′ or 3′ end of a cDNA clone (Quackenbush 2000). ESTs are currently the most widely sequenced nucleotide commodity from the plant genomes in terms of the number of sequences and the total nucleotide count. ESTs provide a robust sequence resource that can be exploited for gene discovery, genome annotation and comparative genomics. ESTs provide the first glimpse at the content and expression characteristics of the genome of any organism. Bioinformatics based sequence analysis tools have extended the scope of EST analysis into the fields of proteomics, marker development and genome annotation. Although EST collections are certainly no substitute for a whole genome scaffold, this "poor man's genome" resource forms the core foundation for various genomewide experiments pertaining to yet unsequenced plant genomes.

Our laboratory took the initiative in generating EST sequence database for seabuckthorn. In the first step, recovering RNA of high quality and quantity is of prime importance to insure the representation of all expressed genes in a cDNA library. However, extracting RNA from seabuckthorn tissue was a difficult task as it is known to be rich in polysaccharides, polyphenols and secondary metabolites. Its leaves contain 11 to 22% crude protein, 3 to 6% of crude fat and some flavonoids (Guoli, 1986). Though these secondary metabolites attribute to the medicinal importance of seabuckthorn plant but on the other hand impede the pure RNA isolation which is a perquisite for cDNA library construction. Several well known RNA isolation protocols (Chomczynski and Sacchi 1987, Salzman *et al.* 1999, Zeng and Yang 2002, Alemzadeh *et al.* 2005, Wang et al 2005) were tried but good quality RNA was not obtained. Our laboratory standardized the RNA isolation protocol with some modifications in the CTAB method. cDNA library was constructed using the SMART cDNA library construction kit (Clontec) and checked for insert size through restriction digestion of the plasmids isolated from transformed colonies. Colonies were picked and glycerol stocks were prepared as the library contained nearly 90% insert percentage. Plasmid was isolated in 96 well format through Perfectprep Plasmid isolation Kit (Eppendorf). The quality of the plasmid was analyzed by gel electrophoresis against lambda *Hind*III DNA marker. Plasmids were sequenced in MegaBACE sequencer which is based on Sanger's DNA sequencing principle. The chromatograms obtained from sequencer are basecalled using Phred Software to generate sequence reads. These sequences are then further subjected to cross_match software which trims off low quality and vector sequences. After quality trimming the sequences are submitted to dbEST in NCBI. Till date we have submitted 2788 sequences in db EST and further sequence inclusion is in process. As the ESTs are representative of the transcripts, therefore, there are high chances of sequence redundancy. Before annotation of the sequences, redundancy was removed by assembling the sequences by CAP3. The contigs and singletons obtained from CAP3 output were than subjected to Blastx. Function was assigned to sequences depending upon the best hit against non redundant protein database (NCBI). As expected, we obtained nearly 50 plus unique sequences which were annotated as genes which might be playing a role in cold tolerance in seabuckthorn.

As an alternative approach to obtain more of cold/drought tolerant genes we are planning to construct subtractive cDNA library. This employs subtractive hybridization technique where the subtraction step excludes the sequences that are common to the two differing population (leaves isolated in different seasons) being compared. This increases the probability of obtaining low abundant differentially expressed genes.

19.5 MICROPROPAGATION

Lummerding (2001) conducted experiments regarding the micropropogation protocol development in two seabuckthorn varieties i.e. *sinensis* and elite Indian Summer (IS). Initially they encountered many difficulties in achieving living elite seabuckthorn Indian Summer (IS) lines in tissue culture. The tissue culture media contained a carbon source, inorganic nutrients and vitamins. Supplementation of plant hormones like cytokinin i.e. Benzylaminopurine (BAP), thidiazaron (TDZ) and zeatinriboside and auxins such as 2,4 diclorophenoxyacetic acid (2,4-D) and indole 3 butyric acid (IBA) are known to be facilitating growth enhancement of plants. As many of their cuttings did not survive for weeks, they attributed this failure to many possible reasons like leaching of anthocyanin and plant fluids into the media from the bottom of the cut stem, excessive ethanol pre-sterilization treatment and absorption of bleach into the cuttings. In response to above problems, auxin was supplemented in media which increased cellular division which in turn closed the phloem vasculature at the stem bottom. Alternatively, bottom of the cutting could be dipped in liquid paraffin to seal off the vascular system before pre-sterilization treatment and cut off the wax after sterilization and insert into media.

Various attempts have been made to optimize media composition like two modified Murashige-Skoog (MS) media, six and four variations in Saskatoon and Chokecherry media, respectively. Variation was also brought about with respect to macronutrients content. Good shoot growth was obtained in 50% Chokecherry media and Paul termed this media as "Seabuckthorn Propogation Medium" (SBM). They could not conclude as whether the cuttings were susceptible to water damage during sterile water rinsing or soil water saturation. It is well known that seabuckthorn don't relish high water levels and frequent watering results in soil saturation. Therefore, Lummerding (2001) proposed to increase the agar concentration to 0.8% or higher instead of normal range of 0.6-0.7% to help bind extra water. Lummerding (2001) also found that soft tissues grew faster, had higher survival rate and was less contaminated upon initiation than hard wood because soft tissue does not have much endogenous bacteria in cuttings and needs lesser sterilization time than harder tissues. The bacteria can remain latent and could grow under favourable conditions such as in tissue culture. Moreover bacteria grow faster than the cutting and can out compete cuttings for nutrients. Therefore Paul Lummerding (2001) used low levels of ampicillin (2 mg/L) in media and suggested the need of higher concentration upto 50+ mg/L for more effective protection against any endogenous bacteria. In shrubs apical meristem on a branch controls the growth of branch's lateral buds until the next growing season. The meristem was removed in the culture to facilitate the immediate dormancy break of any lateral buds, thus waking up the cuttings without dormancy period.

Like shoot, initiation of roots also proved to be quite cumbersome. Lummerding (2001) could not make out as why majority of sinensis plants died during rooting process. In general, *in vitro* shrub rooting requires half of the main nutritional ingredients compared to the propagation medium. Though some roots were observed but neither half nor full strength media produced successful results, even different combinations of MS medium with plant growth hormone didn't gave optimal results. Later on, by using Wood Plant Medium (WPM) some positive result were obtained. Further variation in amount of macronutrient content was also made. These changes resulted in emerging of small root hairs from the top of the callus and carried downwards giving "hairy spider legs" like look and only few produced roots that were longer than one cm.

Thus Paul Lummerding came out with a successful micropropogation procedure for *sinensis* varieties. *Sinensis* plants have been successfully initiated, propogated, rooted and transplanted in green house.

19.6 CONCLUSION AND FUTURE PROSPECTS

Genetic diversity is the measure of amount of genetic variability among individuals of the same species and is an important component of biological diversity. Genetic diversity is the result of long term evolution and represents the evolutionary potential of a species surviving in harsh environment, as a species has to change and accumulate more genetic variation in order to adapt itself to various environmental pressures. Molecular marker study indicated that plants of the genus *Hippophae* possess abundant genetic diversity and distinct variation both among and within species. As suggested by Tian et al. (2004b) study, the genetic diversity of seabuckthorn mainly lies within population therefore it is recommended to collect more individuals in each population to conserve the seabuckthorn resources. The results obtained from molecular marker study can be used as guidelines for improving germplasm collection and breeding. Agronomical as well as morphological traits should also be scored in addition to molecular markers since the core collection must provide a good representation of all traits found in whole germplasm. In the study to obtain a sex linked marker in seabuckthorn, one marker showed promising results but was giving band with only one of the hybrid progeny instead of two tested, therefore there is still a scope of developing a sex linked marker or to tag the sex trait.

Expression profiling gives an insight into the hardiness of the plant to tolerate the hostile environment of nature like abiotic stresses. From our small collection of 2,000 unigenes, more than 50 unigenes were found to be related in acquiring abiotic stress tolerance. Thus by increasing the number of EST sequences to around 15,000-20,000 a fair representation of the seabuckthorn transcriptome may be achieved that may prove to be helpful in further fishing out of abiotic tolerant genes. To reduce the total amount of EST sequencing and getting better picture of abiotic stress tolerance, approaches like construction of cDNA library by substraction and normalization can also be taken into account. High through-put approaches like SAGE and microarrays are also promising in understanding global gene expression profiling.

References

Ahmad S D & Kamal M. 2002. Morpho-molecular characterization of local genotypes of *Hippophae rhamnoides* L. ssp. *turkestanica* a multipurpose plant from northen areas of Pakistan. Online *Journal of Biological Sciences*. 2: 351-354

Alemzadeh A, Fujie M, Usami S & Yamada T. 2005. Isolation of high quality RNA from high phenolic tissues of Eetgrass (*Zostera marina* L.) by keeping temperature low. *Plant Molecular Biology Reporter*. 23: 421a-421h

Alpert P, Lumaret R D & Giusto F. 1993. Population structure inferred from allozyme analysis in the clone herb *Fragaria chiloensis* (Rosaceae). *Amer. J. Bot.* 80: 1002–1006.

Avdeyev V I. 1983. Novaya taksonomiya roda oblepikha: *Hippophae* L. Izv. Akad. Nauk. Tadzh. SSR. *Biol. Nauk*. 4:11–17.

Bartish G I, Jeppsson N, Bartish I V & Nybom H. 2000b. Assessment of genetic diversity using RAPD analysis in a germplasm collection of seabuckthorn. Agr. *Food Sci. Finland*. 9: 279-289.

Bartish I V, Jeppsson N & Nybom H. 1999a. Population genetic structure in the dioecious pioneer plant species *Hippophae rhamnoides* investigated by random amplified polymorphic DNA (RAPD) markers. *Molec. Ecol*. 8: 791- 802.

Bartish I V, Jeppsson N, Bartish G I, Lu R & Nybom H. 2000a. Inter- and intraspecific genetic variation in *Hippophae* (Elaeagnaceae) investgated by RAPD markers. *Plant Syst. Evol*. 225: 85–101.

Bartish I V, Jeppson N, Nybom H & Swenson U. 2002. Phylogeny of *Hippophae* (Elaeagnaceae) inferred from parsimony analysis of choloroplast DNA and morphology. *Syst. Bot*. 27(1): 41–54.

Brenner S, Johnson M, Bridgham J, Golda G, Lloyd D H, Johnson D, Luo S, McCurdy S, Foy M, Ewan M, et al. 2000a. Gene expression analysis by massively parallel signature sequencing (MPSS) on microbead arrays. *Nature Biotechnology*. 18: 630-634

Brenner S, Williams S R, Vermaas E H, Storck T, Moon K, McCollum C, Mao J I, Luo S, Kirchner J J, Eletr S, DuBridge R B, Burcham T & Albrecht G. 2000b. In vitro cloning of complex mixtures of DNA on microbeads: physical separation of differentially expressed cDNAs. *Proc Natl Acad Sci USA*. 97: 1665–1670

Cabrita L F, Aksoy U, Hepaksoy S & Leitao J M. 2001. Suitability of isozyme, RAPD and AFLP markers to assess genetic differences and relatedness among fig (*Ficus carica* L.) clones. *Sci. Hortic*. 87: 261-273.

Cervera M T, Cabezas J A, Sancha S G, Martinez de Toda F & Martiaez-Zapater J M. 1998. Application of AFLPs to the characterization of grapevine *Vitis vinifera* L. genetic resources. A case study with accessions from Rioja (spain). *Theor. Appl. Genet*. 97: 51–59.

Chomczynski P & Sacchi N. 1987. Single-step method of RNA isolation by acid guanidinium thiocyanate-phenol-chlorofrom extraction. *Anal Biochem* 162: 156-159.

Durand R & Durand B. 1990. Sexual determination and sexual differentiation. *Crit. Rev. Plant Sci*. 9:295–316.

Chen X L, Ma R J, Sun K & Lian Y S. 2003. Germplasm resource and habitat types of Seabuckthorn in China. *Acta Bot. Boreal.-Occident. Sin*. 23 (3): 451–455.

Ge S, Oliveira G C X, Schaal B A, Gao L Z & Hong D Y. 1999. RAPD variation within and between natural populations of the wild rice *Oryza rufipogon* from China and Brazil. *Heredity*. 82:638–64.

Guoli W. 1986. Comparative study of Seabuckthorn oil in the Northwest Zone of China., pp: 55-60.

Heum M, Murrhy J P & Phillips T D. 1994. A comparison of RAPD and isozyme analyses for determining the genetic relationships among *Avena sterilis* L. accessions. *Theor. Appl. Genet.* 87:689–696.

Hyvonen J. 1996. On phylogeny of *Hippophae* (Elaeagnaceae). *Nord. J. Bot.* 16: 51–62.

Irish E E and Nelson T. (1989) Sex determination in monoecious and dioecious plants. *Plant Cell.* 1:737–744.

Li F, Xiong Z T, Li F M & Zhu Y G. 1999. Genetic diversity and divergence between populations of *Hemerocallis lilioasphodelus* L. from Henan and Hunan Province. J. *Wuhan Univ.* (Nat. Sci. Ed.) 45:849–851.

Lian Y S. 1988. New discoveries of the genus *Hippophae* (Elaeagnaceae). *Acta. Phytotax. Sin.* 26: 235–237.

Lian Y S & Chen X L. 1992. The ecogeographical distribution of *Hippophae rhamnoides* subsp. *sinensis* and its phytogeographical significance. *Acta Phytotax. Sin.* 30 (2): 394–355.

Lian Y S & Chen X L. 1993. Study on the germplasm resource of the genus *Hippophae* L. International Symposium on sea buckthorn *(Hippophae rhamnoides* L.), August 23-25 1993. Novosibirsk, Russia, pp. 157-161.

Lian Y S & Chen X L. 1996. Systematic classification of the genus *Hippophae* L. *Hippophae.* 9 (1): 15–24.

Lian Y S & Chen X L. 2000. The regular patterns of distribution on the natural components in plants of the genus *Hippophae* L. *Journal of Northwest Normal University* (Natural Science Edition). 36 (1):113–128.

Lian Y S, Chen X L & Sun K. 1995. New discoveries of the genus *Hippophae* L. Proceedings of International Workshop on Seabuckthorn. China Science and Technology Press, Beijing, pp. 60-66.

Lian Y S, Chen X L & Sun K. 2003. A new subspecies of *Hippophae* (Elaeagnaceae) from China. *Novon* 13: 200–202.

Lian Y S, Chen X L & Wang F. 1997. Investigation of intraspecific patterns within *Hippophae rhamnoides* subsp. sinensis. *Journal of Northwest Normal University* (Natural Science Edition). 33 (1): 36–46.

Lian Y S, Chen X L, Ma R J & Li C B. 1998. The primary study of biological and ecological property for the genus *Hippophae* L. *Hippophae.* 11(3): 1–7.

Lian Y S, Chen X L, Yu Z D & Ao F. 1997. Study on the origin of the genus *Hippophae* L. *Hippophae.* 10 (2): 1–7.

Lian Y S, Lu S G, Xue S K & Chen X L. 2000. Biology and Chemistry of the Genus *Hippophae*. Gansu Science and Technology Press, Lanzhou, pp. 1-226 (in Chinese).

Lu R. 1992. Seabuckthorn a multipurpose plant for Fragile Mountains, ICIMOD occasional paper no. 20 Kathmandu, Nepal. pp: 6-7: 18-20.

Lu R. 1997. Eco-geographical distribution of seabukthorn and prospects of International cooperation. In: Lu S., Li M., Hu J., Liu S. (eds.) Worldwide Research & Development of Seabuckthorn. Beijing, China Science & Technology Press, pp. 11-22.

Lummerding P. 2001. Micropropogation protocol development for seabuckthorn (*Hippophae rhamnoides*) selections for commercial orchard production. Agri-Food Innovation fund project# 19980162

Matsumura H, Reich S, Ito A, Saitoh H, Kamoun S, Winter P, Kahl G, Reuter M, Kruger D H & Terauchi R. 2003. Gene expression analysis of plant host-pathogen interactions by SuperSAGE. *Proc Natl Acad Sci USA*. 100(26): 15718-15723.

Morgante M, Hanafey H & Powell W. 2002. Microsatellites are preferentially associated with non repetitive DNA in plant genome. *Nature Genetics* 30:194-200.

Persson H A & Nybom H. 1998. Genetic sex determination and RAPD marker segregation in the dioecious species sea buckthorn (*Hippophae rhamnoides* L.). *Hereditas*. 129: 45-51.

Quackenbush J, Liang F, Holt I, Pertea G & Upton J. 2000. The TIGR Gene Indices: reconstruction and representation of expressed gene sequences. *Nucleic Acid Research*. 28: 141-145

Reddy M P, Sarla N & Siddiq E A. 2002. Inter simple sequence repeat (ISSR) polymorphism and its application in plant breeding. *Euphytica*. 128:9-17.

Rousi A. 1971. The genus Hippophae L. A taxonomic study. *Ann. Bot. Fennici*. 8: 177- 227.

Roy P S, Porwal M C & Sharma L. 2001. Mapping of *Hippophae rhamnoides* Linn. in the adjoining areas of Kaza in Lahuland Spiti using remote sensing and GIS. *Current Science*. 80: 1107-2001

Ruan C & Li D. 2005. AFLP fingerprinting analysis of some cultivated varieties of seabuckthorn (*Hippophae rhamnoides*). *Journal of Genetics*. 84: 311-316

Ruan C, Qin P, Zheng J & He Z. 2004. Genetic relationships among some cultivars of sea buckthorn from China, Russia and Mongolia based on RAPD analysis. *Sci. Hortic.* 101: 417-426.

Saha S, Sparks A B, Rago C, Akmaev V, Wang C J, Vogelstein B, Kinzler K W & Velculescu V E. 2002. Using the transcriptome to annotate the genome. *Nature Biotechnology*. 20: 508-512.

Salzman R A, Fujita T, Salzman K Z, Hasegawa P M & Bressan R A. 1999. An improved RNA isolation method for plant tissues containing high levels of phenolic compounds or carbohydrates. *Plant Molecular Biology Reporter*. 17: 11-17

Sheng H M, An L Z, Chen T, Xu S J, Liu G X, Zheng X L, Pu L L, Liu Y J & Lian Y S. 2006. Analysis of the genetic diversity and relationships among and within species of *Hippophae* (Elaeagnaceae) based on RAPD markers Plant systemic and Evolution. DOI: 10.1007/s00606-006-0413-1

Singh R, Mishra S N, Dwivedi S K & Ahmed Z. 2006. Genetic variation in seabuckthorn (*Hippophae rhamnoides* L.) population of cold arid Ladakh (India) using RAPD markers. *Current Science*. 91: 1321-1322

Sun K, Chen X, Ma R, Li C, Wang Q & Ge S. 2002. Molecular phylogenetics of *Hippophae* L. (Elaeagnaceae) based on the internal transcribed spacer (ITS) sequences of nrDNA. *Plant Syst. Evol.* 235:121–134.

Sun K, Chen W, Ma R, Chen X, Li A & Ge S. 2006. Genetic variation in *Hippophae rhamnoides* ssp. *sinensis* (Elaeagnaceae) revealed by RAPD markers. *Biochemical Genetics*. DOI: 10.1007/s10528-006-9025-2

Tautz D. 1989. Hypervariability of simple sequences as a general source for polymorphic DNA markers. *Nucleic Acid Res.* 17: 6463-6471.

Temnykh S, Park W D, Ayres N, Cartinhour S, Hauck N, Lipovich L, Cho Y G, Ishii T, & McCouch S R. 2000. Mapping and genome organization of microsatellite sequences in rice (*Oryza sativa* L.). *Theor. Appl. Genet.* 100: 697–712.

Thiel T, Michalek W, Varshney R K & Graner A. 2003. Exploiting EST databases for the development and characterization of gene-derived SSR-markers in barley (*Hordeum vulgare* L.). *Theor. Appl. Genet.* 106: 411–422.

Tian C, Lie Y, Shi S, Nan P, Chen J & Zhong Y. 2004a. Genetic diversity of seabuckthorn (*Hippophae rhamnoides*) populations in northeastern and northwestern China as revealed by ISSR markers. *New Forest*. 27: 229-237

Tian C, Nan P, Shi S, Chen J & Zhong Y. 2004b. Molecular genetic variation in Chinese populations of 3 subspecies of *Hippophae rhamnoides*. *Biochemical Genetics* 42: 259-267

Trajkovski V & Jeppsson N. 1999. Domestication of sea buckthorn. *Botanica Lithuanica Suppl.* 2: 37–46.

Vos P, Hogers R, Bleekers M, Reijans M, Lee T V, Homes M, Frijters A, Pot J, Peleman J & Kuiper M. 1995. AFLP: a new technique for DNA fingerprinting. *Nucleic Acids Research*. 23(21): 4407-4414

Wang T, Zhang N & Du L. 2005. Isolation of RNA of high quality and yield from Ginkgo biloba leaves. *Biotechnology Letters*. 27: 629-633

Welsh J & McClelland M. 1990. Fingerprinting genomes using PCR with arbitrary primers. *Nucleic Acids Research*. 18: 7213–7218.

Williams J K, Kubelik A R, Livak K J, Rafalski J A & Tingey S V. 1990. DNA polymorphisms amplified by arbitrary primers are useful as genetic markers. *Nucleic Acids Research.* 18: 6531–6535.

Wu K S & Tanksley S D. 1993. Abundance, polymorphism and genetic mapping of microsatellites in rice. *Mol. Gen. Genet.* 241: 225–235.

Yao Y & Tigerstedt P M A. 1993. Isozyme studies of genetic diversity and evolution in *Hippophae. Genet. Resour. Crop Evol.* 40:153–164.

Yu Z D, Ao F & Lian Y S. 1993. The origin, classfication, populations and resource of the genus *Hippophae* L. in China. *Hippophae.* 6 (1): 19–24.

Zeng Y & Yang T. 2002. RNA isolation from highly viscous samples rich in polyphenols and polysachharides. *Plant Molecular Biology Reporter.* 20: 417a-417e

Zietkiewicz E, Rafalski A & Labuda D. 1994. Genome fingerprinting by simple sequence repeat (SSR)-anchored polymerase chain re-action amplification. *Genomics.* 20: 176–183.

Zhao H, Zhu C, Gao C, Li H, Liu Z & Sun W. 1991. Geographical variation of fruit traits of the Chinese sea buckthorn and selection of provenances for fruit use. *Hippophae.* 4:15–18.

Colour Plates

1. Fruits of H. rhamnoides

2. Fruits of H. rhamnoides bunch

3. Fruits of H. salicifolia

4. Fruits of H. tibetana

5. Seabuckthorn in natural population

6. Female plant at flowering stage

7. Male plant at flowering stage

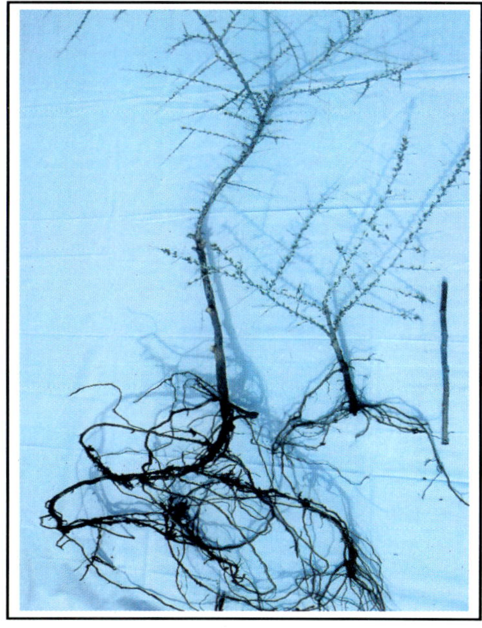

8. Propagation through hardwood cutting

9. Hardwood cuttings in green house

10. Hardwood cuttings in open nursery

11. Propagation through softwood cutting

12. Propagation through suckers

13. Propagation through shoot tip culture

14. Shoot proliferation

15. Orchard layout

16. Genetic diversity in seabuckthorn

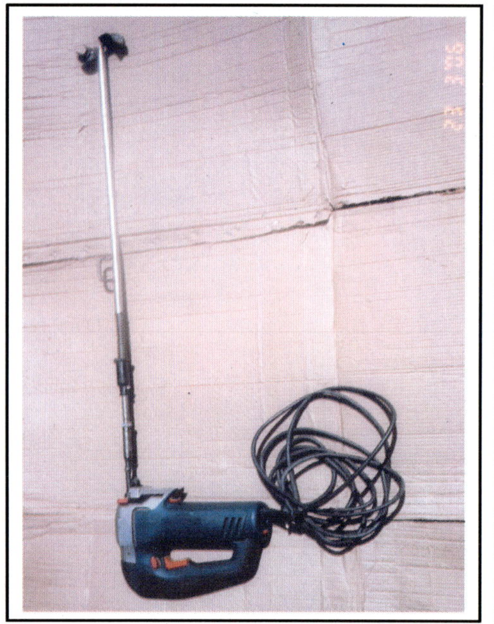

17. Seabuckthorn harvestor

18. Harvesting of seabuckthorn

19. Pulping of seabuckthorn

20. 'Multivit' beverage from seabuckthorn

21. Commercial beverage from seabuckthorn

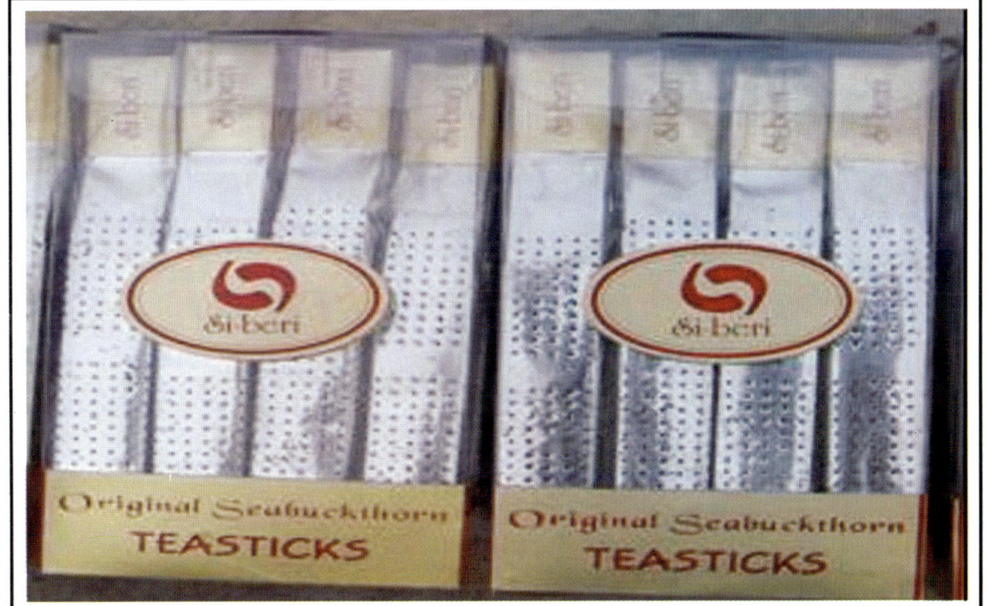

22. Commercial seabuckthron tea

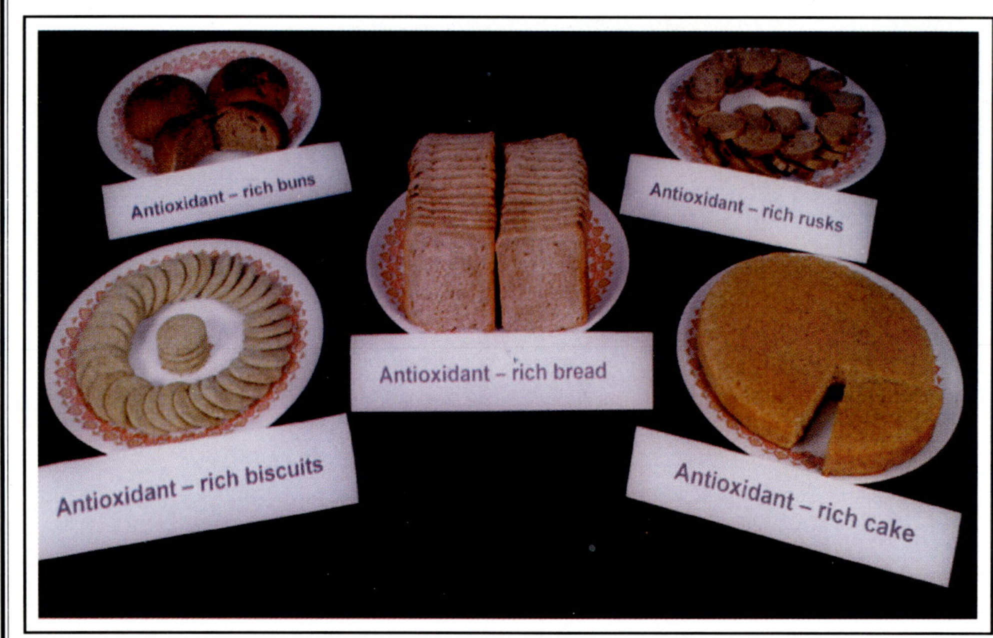

23. Bakery products from seabuckthron

24. Seabuckthorn oil

INDEX

C

D

G

H

I

Q

R

S

T

U

V

W

Y

Z

□□□

About the Editors

Dr. Sanjai Kumar Dwivedi is a Scientist (Horticulture) in Defence Research and Development Organisation (DRDO), Government of India. He has been doing pioneering research on propagation, genetic diversity, cultivation and processing of seabuckthorn since 1996. He has been credited with developing, patenting and commercialization of seabuckthorn beverage in India, which is one of the most popular fruit beverages in the Indian market. He has 05 patents and 50 research publications to his credit besides, 10 chapters in books, 05 monographs/ bulletins and 75 popular articles. Dr. Dwivedi is life member in 06 professional societies, member of the editorial board "Progressive Horticulture", and executive councilor in the Indian Society of Ornamental Horticulture. He is recipient of 4 medals, HSI Golden Jublee Students' Award and two prestigious National Awards namely, Fakharuddin Ali Ahmed Award (ICAR) and DRDO Young Scientist Award.

Dr. Parimelazhagan Thangaraj did his undergraduation from St. Joseph's College, Trichirappalli and received M.Sc and Ph.D degree from Bharathiar University, Coimbatore. He has worked as Scientist in Defence Institute of High Altitude Research, Defence Research & Development Organization (DRDO), Leh and presently he is Reader in Department of Botany, School of Life Sciences, Bharathiar University, Coimbatore. Dr. Parimelazhagan has received National Science Day Medal and DRDO Laboratory Scientist of the Year award 2005. He has also been conferred with Young Scientist award 2008 in the International Seminar on Medicinal and Aromatic Plants held at Sri Venkateswara University, Tirupathi for his contributions in herbal product development. He has two patents and over 20 research publication in peer reviewed journals to his credit. His current research focuses on the Phytochemistry of medicinal plants for value addition.

Dr. Shashi Bala Singh is the Director, Defence Institute of High Altitude Research, Leh. She obtained her Ph.D in Physiology from All India Institute of Medical Sciences, Delhi in 1986. Dr. Shashi Bala Singh joined DRDO at Defence Institute of Physiology & Allied Sciences, Delhi in 1990. She has vast experience in the field of high altitude human physiology. She has made excellent contributions in the field of Neurosciences.

She took up the challenge to find out the solution to the problem of hypophagia and cognitive functions impairment at high altitude. Her work has led to the elucidation of some basic mechanisms involved in hypophagia at high altitude in terms of taste receptor sensitivity changes. The impairment in cognitive functions is mainly due to the oxidative stress leading to neuro-degeneration in brain structure involved in learning and memory. She is the life member of International Brain Research Organization (IBRO), Association of Physiologist and Pharmacologists of India (APPI), Physiological Society of India (PSI), and Indian Academic of Neuroscience. She is the fellow of Indian Association of Biomedical Scientist (FIABMS) and Indian Academy of Neuroscience (IAN). She has over 60 publications to her credit in reputed international and national Journals. Dr. Shashi Bala Singh is the recipient of several awards and honours i.e Dr. J.N. Maitra Memorial Oration award in 2003, Bharat Nirman Talented Ladies award in 1995 for the outstanding contribution made in the field of life sciences, DRDO laboratory scientist of the year award in 2006 and Surg Rear Admiral M S Malhotra Prize in 2007 and Prof Baldev Singh Oration in 2007. Recently she received DRDO Technology Spin off Award -2007 from Prime Minister of India in May 2008.

Dr. Zakwan Ahmed is Director, Defence Agricultural Research Laboratory (DRDO) – Pithoragarh since 01 Sep. 2007. He is serving Defence Research & Development Organisation since last 20 years. Dr. Zakwan Ahmed contributed significantly toward varietal/hybrid development in vegetables including infrastructure development during his tenure in DARL, Pithoragarh. During his tenure as Director FRL, Leh from 2004-2007, Dr. Ahmed has Made significant contribution in lab to land programme which have boosted the supply of fresh vegetables to the army in 'L' sector 14 Corps zone, the most hostile and highest battle field of the world. It is because of his persistent efforts that seabuckthorn herbal beverage and DRDO herbal tea could be inducted in high altitude special ration of army. He has also designed morphological descriptor for seabuckthorn. Dr. Ahmed has 50 research papers in National and International journals, 06 chapters in books, 02 books and 07 technical bulletins to his credit. He has been awarded Laboratory Scientist of the year award 2004, Certificate of Honour of GOC, 14 Corps-2005, DRDO scientist of the year award –2006, Bioved scientist of the year award-2007 and he has also been honoured honorary Fellowship (FRSCS) of the Royal Society of Crop Science-2007. He visited China and presented research paper in the 2nd International Seabuckthorn Association Conference. At present he is actively engaged as a programme Director of DRDO-Army Biodiesel Programme.